the physics of consciousness

the physics of consciousness

Quantum Minds and the Meaning of Life

EVAN HARRIS WALKER, PH.D.

A Member of the Perseus Books Group
New York

A CIP catalog record for this book is available from the Library of Congress.
ISBN-10: 0-7382-0436-6
ISBN-13: 978-0-7382-0436-9

Published by Basic Books, A Member of the Perseus Books Group
Previously published by Perseus Publishing

Text design by Jeff Williams
Set in 10-point Minion by the Perseus Books Group

EBC 08 09 20 19 18 17 16 15 14 13
First paperback printing, December 2000

This book is dedicated to the memories of my parents, Eva Victoria Harris Walker and James William Walker, and to Merilyn Ann Zehnder, without whom there would be nothing.

contents

acknowledgments

I have always felt that a well-developed intellect must embrace both an artistic sense and an understanding of scientific reasoning—and it follows that, fully developed, these sensibilities will, of themselves, ultimately demand their unification. This work is an outgrowth of that vision. Both the meaning of life and the physics of consciousness must, ultimately, become one understanding.

This book looks to both of these sensibilities for its inspiration, and it finds these sources not just in the museums of art and the catalogues of scientific research, but in the people who have influenced and breathed life into this work: Merilyn Ann Zehnder, my parents, Eva Victoria Harris Walker and James William Walker; friends who were a part of this story, among them Tom Whittson and Maitland MacIntyre; family and those who have otherwise added to my life: my brother James (Jimmie) William Walker, Jr., his wife Eileen Newton Walker, and family of James Willam III, Michael Pittman, Lee Newton, Helen Harris and Caroline Irwin, and theirs; Steve Blumenthal, S. Fred Singer, Elsa Pilarinou, Warren Hillstrom, Jo Carroll, Eduardo Palomino, Marian Buchannan, Alice Ferguson Holmes, Catherine Stathers, Gertrude and Ralph Neal; my wife, Helen Marie Moseley Walker; and those who have striven to make this work the book that it has become, Brian Nolan, Eric Edwards, Jay Jackson, Marco Pavia, Amanda Cook, Connie Day, and especially the perseverance of David Hiatt and Jeffrey Robbins.

She dwelt among the untrodden ways
Beside the springs of Dove,
A maid whom there were none to praise
And very few to love:

◈

She lived unknown, and few could know
When Lucy ceased to be,
But she is in her grave, and, oh,
The difference to me!

◈

No motion has she now, no force;
She neither hears nor sees;
Rolled round in earth's diurnal course,
With rocks, and stones, and trees.

—William Wordsworth, "Lucy"

introduction

Beneath the Soil of Jackson

Whom the gods love dies
young.

—Menandes

As I walked down the steps from her apartment, my mind went back over the years. My mind went back to things that have been and that I have done, the things of my life and the things of this day. I will write of this, someday. I will tell what happened here. But time is needed for its meaning to grow clear, to become part of the perspective that gives life meaning. Time is needed for these images to be reflected in the history of my life.

My mind went back to the things I must say here—back to an image, to a terrible image, to a vision of my future and of my purpose. My mind went back there.

She lay there dead. I have spent half a century trying to understand that moment. There had been such wonderful moments. There had been then the times of walks to the park, luscious southern summer days, forever summer in my mind—captured forever in the memories of our play; of games of tennis and walks to town; of my own jaunts to visit, climbing the long hill to Clermont Drive, then up the nearly vertical two flights of stone steps that mounted the steep front lawn to her house. My mind goes back to those times, to the time before she lay there dead.

I had seen and shared with her the full youthful joy of love. She had written for me her simple, direct poems about her love for me, and I had written my poems—awkward, stilted efforts to speak beyond my years or my knowledge, but still speaking of this new emotion. And then this ended. Leukemia came to claim her. There was a last glimpse through the open door of her hospital room as she lay there on her last day, and then death closed her eyes.

My last sight of her was at Rideout's Funeral Home in Homewood. A day later she lay next to her father, beneath the soil of Jackson, Mississippi. She had gone home.

Years have passed, years in which the questions have persisted every day. The pain has given way to the searching, to the quest to find out what we really are and what, if anything, remains when the tissues of the brain and body have ceased their functions. The years have passed, and the questions have persisted. Where is home? Is there any home?

I remember one particular day at Auburn. I sat on an open grassy ridge near the college. Above me spread a blue sky filled with those puffy white clouds that seem never to come anymore. I sat there till the sun set and then till dusk quieted the land to the distant hills. I sat there asking, praying for some sign that she existed. There was absolutely nothing but the sky, the white puffy clouds, the grassy ridge, the distant hills, and the sun that set and ended the day.

The pain faded many years ago, to be replaced by other pains in other days. But the questions have remained. The questions have been there to prod me on in my search for answers, answers that are more than just promises and assurances designed to comfort. And I have found some of the answers. There are answers, answers to the questions like "What are we really?" There are answers to our search for the meaning of our lives and the nature of our universe. There are answers to my quest, to your quest to know the fabric of reality, to know our meaning, our destiny, and to find home—that corner of the world that smiles for us.

We live in a whirlpool of "reality dreams." All the illusions of reality that we face everyday swirl around us, and so we search for a stable resting place—somewhere. So many levels of illusion, so many false realities that we play for real. So many dreams we take to be our lives, till one morning we discover that our life is all over. The illusions fall away in silent screams, a silver-screen Freddy Krueger spooks us, we laugh, and there is silence.

Think of all the levels. Think of the plans we make for our lives, and then we talk to ourselves within our minds about these plans. Our plans feed our daydreams. We stare out windows imagining loves fulfilled, futures blessed with success, riches, fame, recognition, and respect. We hope, and we build ever more elaborate plans to hide our disillusionment as each hope slips away. Then one day, we awaken to find ourselves buying lottery tickets to patch these ludicrous fantasies together as they turn to nothing.

Some, failing to find the reality they wished for, plunge further into the web of illusion, spinning drug dreams with highs and then lows—both empty in the end.

Some dreams leave us changed—knowing, we think, that there is something true out there. There may be some moment when our vision clears, or some moment when our mind's eye sees more than ever before. In some fever-induced delirium, we may see things that change everything we had believed. We glimpse a moment in a past life we once lived: some moment four centuries ago, a TV screen looking in on the past flickering in the mind, a vision of a moment when our own eternal life touched another soul. It feels so true. Is it real? Is it—was it—a part of "objective" reality, or is it only a feeling, some joke of the mind or flaw of the brain that fools us into believing what otherwise would be obvious foolishness? Is it a vision of ultimate truth, a piece of the pattern that makes up reality, or is it a delusion?

Some lose themselves in doing their jobs at work and cutting the grass at home. They close their eyes to a billion years that have passed and a billion parsecs to the edge of the universe. Reality for them is the fantasy of here and the illusion of now.

Others lose themselves in illusions and diversions that are just as far from reality—myths of ages past, games of the moment, things that let us live lives that do not ask too many questions. Things in our heads that let us live in lands on planets roaming among the midnight suns, far away from the realities that control what and who.

But as there are dreams, there are nightmares, nightmare realities that are real. There are psychotic horrors that live in the most ordinary thing. One moment you may look up to see a dull red sun glowing behind overcast skies, nothing more; then suddenly, be filled with a terror that tears out your sanity in one brief moment. It reaches into your being, rolls up everything that holds sanity to flesh, strips you of any place to hold on to, and makes you fear the worst fear of all—that that fear will come back.

We live in a whirlpool of reality dreams, asking all the while, "What is reality?" while all the while it goes unseen before us in the present moment. It is there before you. It is some color you see now, some thought you hold for an instant. Reality is some memory, summoned to mind and savored, of being happy for a moment with someone you once loved. It is a wish you dreamed would come, and the coming true of the wish—your daydream, some future you plan now and look on as if here already. But reality is also a piece of steel with a sharp edge that cuts. Or reality is loneliness.

Reality serves up to us all our meanings, and yet who knows which myth to believe? Who knows the facts or where they will lead us? We cut our piece of reality into boxes, into rooms, into spaces carved by our designs, set by our own selections, and imagined to be whole worlds—entire universes. We put up walls, and because we no longer see the world, we pretend that it is the outside world that is the illusion. The things we create and build are models of illusions about reality. They wall off pieces of the world that we want the world to be. We fence land into plots, into pieces that they are not, trying to make little worlds that we can hold because we are too small to hold the sky and too limited to understand even one small cluster of stars.

We fashion out illusions in many ways. We dress up in a car as if putting on a nice hat. We sit in boxes all day, pretending the walls talk. We weave daydreams, play lotteries, and plunge each evening into the TV screen. It takes us floating into a world of illusory concerns and escapist fantasy. We know it is illusion. But so is the rest. All the things that control the mind—books, magazines, letters, placards, posters, e-mail, faxes, data—all the things that tell us what to think are all a part of our world of fantasy. They are today's religion. Where do we go for salvation?

It was many years ago. The loss that had launched this quest was by then something that defined the background of my life but was no longer felt so immediately. I still wondered every day, "Where is that lost life?" but I had by then taken the first steps toward answers without quite realizing that this road would get me where I wanted to go. I remember as a graduate student arguing with my friend George Hinds about the significance of our chosen field. We were working toward our Ph.D.s, both physicists in the raw. We shared the same convictions and prejudices that physics would enable us to discover the truth about our universe and about the nature of the physical world. Much that we discussed involved philosophical questions, asking ourselves if physics could provide the answers to questions that had long been fruitlessly argued by the sages of philosophy: What is the ultimate nature of reality? Who are we? What is consciousness? Is there a purpose?

I don't remember our arguments from those days, but I do remember the point. Physics is the cornerstone of our scientific knowledge. In the realm of factual knowledge, it provides us with the foundation, with the procedures, with the means for confirmation that we need if we are to search out and find the answers to those age-old questions.

Physics was our world then, and now it provides us the means to find the answers to those philosophical questions. It is the path to understanding what is physically real. It is the path we have to take so that we may discover reality opened up to our vision, naked, like a lovely woman whose beauty and allure are at once mystery and revelation. Physics is the tool we must use to learn about reality. But it has its hazards. If we are not careful, *she* will ensnare us. If we are not careful, we may begin to believe there is nothing else but this *physical* reality.

But there is a new physics, and with that new physics, a new age. There is a physics now that is finally probing those fundamental questions about reality. We have a journey to travel, a journey of discovery. On our journey, we will be going much further than anything in contemporary physics. We will survey that hard-won sense of an objective physical reality. As we travel, we will peer deeply into the atoms of matter, even into the quarks and leptons of which these atoms are made. We will catch a glimpse of the structure of the universe and of its beginnings in the Big Bang birth of everything that exists in the material world. But our trip will take us just as far in yet another direction, a direction that probes the fabric of the mind and the ultimate nature of reality.

As a tool for understanding our world, physics is powerful and majestic. Physics lets us probe the inner pieces of atoms and see into the interior of stars. It gives us the tools to discover how living things function. It lets us understand the fabric of space and time. And as we will see, it will help us discover how brains unfurl their strange visions of reality. It might seem that questions of art, values, morals, philosophies, and religions lie somewhere outside science and physics. Yet even here, we will find that the tools of physics will help us understand and probe further into the nature of the world about us and within us.

There is in this what seems at first to be a contradiction. The power of the current paradigm of physics (and of science generally) is founded on the principle that nothing exists but objective reality, nothing but the physical world, nothing but pieces of matter. By its power to describe accurately and control the world around us, physics has manifestly overridden all other doctrines and beliefs. It would seem ultimately to explain everything in terms of a materialistic philosophy, in terms of material objects in a physical space like so many pebbles in a box, and to sweep away religion, vanquish all spirits, and leave *her* there—no home for the soul—to lie there dead.

But flaws in the materialistic paradigm of science have appeared in recent years. These flaws have grown to a gaping rent, torn across the whole fabric of the materialistic conception of reality. Strained by the conflicts between Einstein and Bohr over the ultimate meaning of quantum mechanics (developing in the Einstein-Podolsky-Rosen paradox), subjected to further stress in Bell's theorem, and finally ripped through in recent tests by Aspect in France, the whole cloth of the materialistic picture of reality must now be rejected.

We will come back to all this later to see how each of these battles altered, little by little, the picture of reality until it became clear that everything must be revamped. The threads must be picked up again and the fabric of reality rewoven. But where science, all of it, has failed most has been in its inability to cope with the question of mind—the question of the nature of consciousness—the doorway to the quantum mind.

We, as scientists, sense a certain inadequacy in our efforts to understand what consciousness is. We recognize that we are reaching for something. We strain to understand it, yet we find that our best efforts to catch that essential substance of life are inadequate. It has eluded science. In *Broca's Brain*,[1] Carl Sagan catches that feeling, that sense of mysteries unsolved—mysteries as elusive now as at the dawn of time.

> And here was Broca's brain floating, in formalin and in fragments, before me. I could make out the limbic region that Broca had studied in others. I could see the convolutions on the neocortex. I could even make out the gray-white left frontal lobe in which Broca's own Broca's area resided, decaying and unnoticed, in a musty corner of a collection that Broca had himself begun. It was difficult to hold Broca's brain without wondering whether in some sense Broca was still in there—his wit, his skeptical mien, his abrupt gesticulations when he talked, his quiet and sentimental moments. Might there be preserved in the configuration of neurons before me a recollection of the triumphant moment when he argued before the combined medical faculties (and his father, overflowing with pride) on the origins of aphasia? A dinner with his friend Victor Hugo? A stroll on a moonlit autumn evening, his wife holding a pretty parasol, along the Quai Voltaire and the Pont Royal? Where do we go when we die? Is Paul Broca still there in his formalin-filled bottle?

And yet with the mystery held tightly in his hand and spoken eloquently on his tongue, Sagan misses the lesson that this mystery has to teach us.

> Might it be possible at some future time, when neurophysiology has advanced substantially, to reconstruct the memories or insights of someone long dead? And would that be a good thing? It would be the ultimate breach of privacy. But it would also be a kind of practical immortality, because, especially for a man like Broca, our minds are clearly a major aspect of who we are.

But Carl Sagan never held Paul Broca's mind. He sensed the existence of something more—more than the brain's gray mass stuffed in formalin, more than the convolution on the neocortex, more than a collection of branching neurons—and perhaps he even realized that Paul Broca was more than his memory traces, more than the abrupt gesticulations by which he might be recognized when he talked. Behind his wit, behind his guise of skepticism, there was something that felt those quiet sentimental moments, something that smiled and was the wit he created somewhere within that gray mass. That had been Paul Broca's mind, and no one knows where that Paul has gone. And yet this is the question we must search out. This is the question we will answer.

What I want to do in this book is to tell you some of the answers that I and others have found in recent years, as directly and as simply as I can. But in order to understand these answers, it is necessary to know what the current and past conceptions of reality are. To advance to a new scientific understanding requires acquaintance with the images, the words and terms, and the errors and strengths of past knowledge. Thus what I offer here is a sequence of pictures showing a progression of steps in comprehending our place not just in the universe, but in the fabric of reality itself. These steps will lead us through a doorway, will lead us to the quantum mind.

These are images that begin with the dawn of mankind and that evolve step by step into the modern age of science. We view primitive gods of rock and wood, a world of person and God, a clockwork universe, quantum atoms, space-time, and then something beyond. Many books in recent years have shown pictures of nature drawn from the pages of modern science—pictures of the new physics and even some speculation about what lies behind the closed doors to the future. We, however, will unlock those doors and peer into the rooms beyond. We will look at the very heart of reality—beyond the edges of Einstein's curved space, beyond the quantum jumps of atoms and molecules, beyond quarks and leptons, and before the beginning of time. We will see what space and matter are made of and what consciousness is, and we will view the power of the mind. We will gaze through an open doorway, looking beyond the lifeless forms that our lives have become—looking beyond, into the very face of God.

1

Where Have the Gods All Gone?

To see a World in a Grain of Sand
And a Heaven in a Wild Flower,
Hold Infinity in the Palm of your Hand
And Eternity in an Hour.
—William Blake, "Auguries of Innocence"

It is easy to imagine fantasy as physical and myth as real. We do it almost every moment. We do this as we dream, as we think, and as we cope with the world about us. But these worlds of fantasy that we form into the solid things around us are the source of our discontent. They inspire our search to find ourselves. In order to put meaning back into our lives, we should recognize illusions for what they are, and we should reach out and touch the fabric of reality.

Although we know that our common-sense understanding of the world is merely fiction, the illusions stay with us. Science has entirely overturned what we know about the structure of the world. But rather than revising our picture of what reality is, we cling to a collage of incongruent shards. We preserve a false assemblage of images, one pasted upon another, so that we can keep unchanged the mental portrait of ourselves and of the world to which we are accustomed. We go about our business despite the fact that the world on which we base our lives is so much in question, so much a mystery.

Even when we have searched out some knowledge, and when we have penetrated into the *jitterbug* world of Mr. Zukav's *Wu Li Masters* or of Carl Sagan's billions and billions of everything, we are left with only so many more unanswered questions about reality. We want to know. We ask. We search for answers, and we are given a box with little pebbles inside. Is that what the world is? Little pebbles, big pebbles, pebbles in a vast box shimmering and shaking about. Have we our answer? Is reality only a box filled with pebbles? Is that it? Is it all just a little box of rocks that holds infinity inside and stretches to the edge forever?

We want to ask, "Is there a God? Does my life have meaning and purpose?" Science, we are told, says that even to ask about God is beyond its scope. But this is not true. Either there is no such thing as God, or science—which embodies our ability to reason—must be able to frame the question and provide us with answers.

We know that science has proved capable of giving us dependable, solid, objective answers. It is the one path that yields answers about the machinery of reality and shows that these answers are valid. When such questions are asked, science must answer.

To many scientists, however, God is only a memory from childhood: a put-off to questions they once asked themselves. "Where do I come from?" was left unanswered with, "From God." Yet perhaps, the great shortcoming of such questions is that the concept of God is so conventional that it too is apt to be as empty as that box scientists give us—that box filled with the universe and yet empty of meaning to what we have asked: "What is reality, really?"

THE OLD GODS

Let us go back to mankind's earliest times. Think of *Homo habilis* looking out into the cosmos, gazing into the blackness of a fearful night with sparkling wonder spread across the vaulting sky. Think of such a man alone in the night's stillness, looking at the stars. He blinks his eyes and wonders. His mind transcends the immediate hazards of the day, and he sees things in the sky that he cannot reach. He sees for the first time the edge of his own being and looks beyond, perhaps forming the first thoughts of some new understanding, the first thoughts of some new knowledge, and then he falls asleep. Somewhere in that early time, in a pattern of stars seen overhead, in the stirrings of an image in the bush, in a lifeless form that did not move from its forest bier, the first troubled, questioning thoughts came to early man and passed into oblivion.

But I can see another, later time, a time when another early man lay more sheltered in a cave sleeping. As the moon rolled in its changing orbit, its full face appeared in the entrance to the cave, its light filling the doorway and jolting the primitive being into a frightened awakening. Such an experience would deepen the mystery of the sky, perhaps forming a memory that would last until the experience recurred months later. Its appearance would spark a need to know what was happening in that subtle other world. Perhaps, wanting somehow to mark what had happened, he picked up a bone and a rock and scratched the first written record.

Ten thousand years later, archaeologists searching the ancient ruins at Gohtzi in the Ukraine found the record he left. There in the ivory of a mammoth's tusk lay etched the incised marks he cut, charting the passing phases and movement of the moon.

I can remember awakening suddenly in the middle of the night to see the shimmering face of the moon on just such a far excursion into the northern latitudes. It peered through the branches and around the corner of my bedroom window as if it had some intent to watch me. Had I not known better, I might have set markers to test the mind of this celestial voyeur. I might have repeated the same observations that my ancient forebears made at Clava, at Kintraw, at Ballachroy, at Avebury, and on the plain at Stonehenge.

Five thousand years ago, our ancestors used small stones and wooden posts stuck in the ground to record the moon's excursions and the constancy of the celestial bodies. These stones and posts, like modern marks on paper, described celestial laws of motion, measured out man's course in the world, and marked his woman's cycles.

These posts, the markers of one age, repeated through a thousand years, have become the tools of ritual and the talismans of the old gods our ancestors worshiped. These were gods in the sky—regular, dependable, knowable. They were worshiped, but they were always out of reach.

The moon has wanderings. More than the sun, it has features, subtleties that suggest the mystical nature of imagination. But having recognized the moon as a goddess of the night, who then could fail to recognize the true carrier of power over life? Who then could fail to see the powerful eye of God?

Think of that earliest time, the time of the *Old Ones*. This was a religion wherein people paid homage to a god who reigned over them and gave them warmth and life, a god they could see, who stood over them, looking at them with his one gleaming bright eye—the sun.

But others in other times created other gods and other pictures of their idea of reality. Others in treacherous forest worlds saw gods frozen in wood awaiting the knife to carve them free. Stones awaited the chisel to liberate their power—a power over the mind, a power to throw terror back into the forest, a power of death. They created images, gods cut from their own imaginings. They made gods of wood and stone. They made images drawn of lines and paint upon the walls of caves. And the lines became words.

What images flashed in the minds of those who painted the Lascaux caves a thousand generations ago? What thoughts beyond mere existence flickered in the minds of those of species *Homo habilis* who left their footprints in the soil at

Olduvai five million years ago? Whatever structured the reality they imagined beyond what they could immediately see, those thoughts were the beginnings of what we are today. Those thoughts were the new trials offered in the struggle for survival.

This clash between early man and nature has woven a pattern of fact and illusion. Evolution and the pressure of survival have endowed us with an ability to understand and to reason, and this has filled us with questions about meanings and values in life. In our search, we have carved stone god answers. Our carved gods have failed us and have been replaced. The worship of Og, Bodb, Llaw; Njord, Woden, Ing, or Sif; and of Horus, Osiris, Amen-Ra, Min, or Thoth has gone. Adad, Ashur, Baal, and Gibil are gone, as are Nintoo, Nusku, Shala, and Sin. Zeuses and Aphrodites have rotted into the soil, and the Jupiters and Venuses are scattered, broken marble busts and torsos that line vacant halls as epitaphs to a world that is now gone. Our fathers struggled against these gods and found their victory.

It was the genius of Abraham, I believe, to have met and triumphed over the superstition of a hundred ages, over the false demands of false gods' priests. Abraham put down his knife. No Baal would take this man's son Isaac. In his act of defiance against religious superstition, he became the father of a new way of faith in a God that spoke more rationally and lovingly to His people. He created a new vision of God: that God, to be worshiped, must have greater love for His subjects than even a father for his son. Somewhere in Abraham's mind or in his heart, some voice did say, "Lay not thine hand upon the lad." Tribes, nations, and peoples have come to follow this Abraham, who was the seed of the three great religions that worship the unseen God—God of the burning bush, God of the Passover, God who parted the sea, God who felled the walls of Jericho—this God who made man in His own image.

Abraham created a new view of the world, a view in which the world of our daily concerns is the creation of a power, a mind, and a spirit that governs our lives, just as God governs the universe. It is a view of reality divided into two parts: God and His creation. It is a view of God who leads His special people, the people of the nation of Israel and ultimately the peoples of all the nations of earth. But what can we believe of this God who would not save His own people from the Holocaust. If there is a God, how distant is He? If there is a God of creation governing the incredible expanse of this universe, what care has He for me? Where is He now for me this moment as I search, hoping to find that whisper of her still-living mind somewhere? Where is this God?

The questions are ancient. The Israelite nation answered with their faith. The Greeks answered with their mind. The Greeks, who with the philosophies of their time could hardly hope to explain the workings even of this world's machinery, certainly could not argue against the existence of some power that would have created the world. But what of it? For the Greeks, there was little to show that whatever God or gods existed had any concern for people and their problems. One might seek after some favor with offerings to some lesser divinity, but to the analytical Greek mind, a supreme God was beyond appellations.

It was into this learned and skeptical Greek world that Christianity appeared as an answer to the question of the relationship between people and God. Christ, the only begotten Son of God, was born into this world to give testimony to His love for each individual person. How incredible is the idea that God could be so infinite in dimension that He could create the universe and yet know the personal needs of an ordinary individual. Jesus came into this world, as an answer to the prophecies of Isaiah, to give wondrous signs, to suffer the death of crucifixion, and then to rise from the dead—showing at once the personal love and infinite sovereignty of God. Paul carried the message to the Greeks and throughout the Roman world: Virgin birth, the lame made to walk, the sightless to see, the dead to rise, and on the third day His own resurrection. This is the faith and the reality that have guided the Christian world for nearly two thousand years.

Now all that has changed. Demons do not cry in the winter wind. Baal does not look down from the sky with one bright eye and take the first-born child. And the walls of our modern Jericho can be brought down by better means than the gods of old ever possessed.

The stone gods did not protect the ancients and were discarded, and the God of the Jews did not protect Jews either. With the advance of science, our knowledge of physics, and an understanding of evolution, we find our explanations elsewhere. The God of Abraham no longer suffices in the secular city.

The story of Jesus is surely inspiring. He surely lived, and he certainly sacrificed His own life for some cause. But what of the rest of His story can we believe in? When we look into the laws of physics, the mechanisms of biology, or the facts of medical practice, where is there any reason to believe that Jesus could make a blind man see or miraculously cure a beggar lame from birth? Do you believe this myth of a virgin birth? Do you believe that some god came down to earth to father anyone? Do you really believe that Lazarus, dead until his body stank from decay, was raised from the dead by anyone? Can anyone who claims to be rational today—when religion no longer serves as an explanation of where we come from or how we got this way—believe that anyone was raised from the dead?

In his book *God and the New Physics*, Paul Davies surveys the necessity of the God hypothesis to explain our existence and the nature of the universe in light of recent advances in physics. He points out that just as evolution theory removed the need to assume God to explain the variety of life forms, physics has recently been able to search back to the very moment of the beginning of time and give us an understanding of even the origin of the universe itself. The successes here have brought science within reach of explaining everything that exists—and the existence of everything—without God. Davies offers one central, telling argument. More than anything else, Davies attacks the idea that God must be assumed to exist to explain the existence of the universe, of matter, of space, or indeed of time. Davies asks rhetorically, "Why this universe, this set of laws, this arrangement of matter and energy? Indeed, why anything at all?" Because physics now has been able to trace the start of the universe back to the moment of the beginning of space and time, matter and energy in a singularity, and map out the course of the future to the far distant heat-death of the universe with no need to invoke God as its creator, where is there any need for God? Davies says, "There is no need to attribute the cosmic order . . . to the activity of a Deity." Darwinism removed the need of God to create the species, and it might seem that modern physics is removing any need to invoke God in order to explain any aspect of the universe. The role of God in the order of things is gone. If that is the answer, then that is the answer. And what Davies describes is a good rendering of modern scientific thought.

◆

I drive past an Episcopal church, much like many whose stained-glass windows look out over the Maryland countryside, and think of the gentler, more certain times its congregations have witnessed. These times were troubled, surely, by crises of life, crises of death, crises of depressions, and crises of wars. Yet they did not suffer the crisis of meaning itself. That church, with its stained-glass windows, always stood there to remind those gentler ones that their struggles had meaning and their questions an answer. And now it is a part of the past.

A few years ago, vandals smashed one of those beautiful early nineteenth-century stained-glass windows. Modern physics has no place for any deity, and the message rings even in the ears of the vandal in the street: "There is no sacrilege—only the moment, only the event."

In *The Seduction of the Spirit*, Harvey Cox paints the change that has come over society. Cox tells us of his days growing up in Malvern, Pennsylvania.[1]

> Whenever I peeked in the half-open doors of St. Patrick's while on my way to Stackhouse's grocery store or the post office, I'd catch a glimpse of a mysterious darkness broken only by an even more mysterious flickering red lamp. Catholic playmates assured me in hushed tones that Jesus Christ Himself was up there on the altar. We didn't even *have* an altar, let alone one with Christ Himself on it. Many times I would like to have ventured into the dim recesses of St. Patrick's, but I was scared. It seemed so foreboding, so dark and awesome.
>
> By high school it was a commonplace among the rest of us that it was just plain useless to argue with Catholics about religion, because no matter what you said, they *knew* they were right, or at least they seemed to know.

> Today, decades later, when I talk honestly to Catholics, I get the feeling that, although they belong to the Catholic Church, they know now how I felt then. For now, even on the inside of their church, that serene assurance is gone. So is that secure conviction that it all goes back directly to God Himself.

Drive through Malvern today. St. Patrick's is aging. It has become an anachronism even to its few parishioners who drag in their children. The world has passed Malvern by and left St. Patrick's in its past. Believers still frequent the place, but the old faith has lost its hold on their souls. And the children leave to search out the secular world's video stores in Philadelphia and beyond. St. Patrick's no longer gives its people quite that same sure faith that they need if they are to believe today.

Today people need proof in order to believe, and they deserve that proof. The degeneration in the values of our society is not due to the waywardness of the people or to the affluence that permits a lax morality. It is not the secular city or drugs or a rebellious youth that has caused society to drift away from God. It is, instead, the message of science borne on the wings of our fast technology. It is the thinking of intellectuals of a century ago that has come down to the streets. The ideas that are today a matter of academic speculation begin tomorrow to move armies and topple empires.[2]

It is the perceptions of our science, the tenets of modern physics so well summarized by Davies, that now instruct our futures—into the streets. But it is all wrong.

◈

I remember her. I remember Merilyn. I remember her so terribly much. I remember her as she looked at me, asking me questions with her eyes. I remember her as she looked quizzically at me, asking with one eyebrow raised, asking. And I answered with my eyes, answering her question of love. I put my arms around her; I kissed her; I felt her body in my hands. I pressed her against me. "I love you," I said. "I love you." And she answered, her words sparkling, "That's funny, I love you, too!"

That had been a year earlier. That had been before she was about to die.

◈

Harvey Cox writes "I have tried to make clear that metaphysical operations cannot be muted by the secular age, but that metaphysical systems will neither again integrate whole societies nor still men's persistent questions as once they did."[3] But Cox is dreadfully wrong. There are answers. The truth does exist, and when the truth is honestly sought, with a mind that is ready to accept the truth, whatever the truth turns out to be, then the answers do come, and the answers change people.

2

It's a Material World

This is a material world,
and I am a material girl.
—Madonna

Gravity catches up with all
of us.
—Marilyn Monroe

What you and I take to be common-sense reality began more than three centuries ago in the mind of one man. Isaac Newton created a conception of reality that is based on the idea of an objective world governed by orderly scientific laws. Newton was not the first scientist to probe, hypothesize, and test speculative ideas against observation, but he created a new way of understanding the universe. His way of seeing the world has changed the way the "man in the street" sees the world. With Newton, an entirely new age dawned.

This world view that Newton originated did not spring into being *ex nihilo*. As he said himself, "If I have seen further, it is by standing on the shoulders of giants." Ptolemy, Copernicus, Kepler, and certainly Galileo all saw in nature regularities in motion that could be described by geometric constructions or mathematical symbols. But Isaac Newton was the first to create an entire philosophy of reality based on quantitative scientific principles verified by observation. From the motions of the planets to the common workings of levers and tools, from the ebb and flow of the tides around the world to the fall of an apple from a tree, Newton saw things no other person had ever seen. He saw universal order in the glint of prismed colors flickering on a darkened bedroom wall. His mind traced these rays of sunlight into the heavens and out among the celestial spheres. Newton moved where no one had ventured before—he moved in the realm of the gods. Copernicus, Kepler, and Galileo could describe what they saw from far off, but Newton opened the door and made it possible for us to enter and dwell there.

Ironically, Newton's new perspective on the nature of the universe turned into something alien to his personal philosophy. Newton had hoped that he would show more clearly that God governed on earth and that every event on earth was, in some small way, affected by even the most remote heavenly body. Instead, his brilliant reasoning showed that the laws of nature formulated and proved here on earth are not limited to our world but govern even the stars. It is a world of planets that move about their stars, of billiard balls that collide and roll away on their given paths, of human bodies that walk and talk as the mechanics of nature dictates. It is a world of rocks and pebbles—but no gods.

The scope of Newton's achievements is breathtaking. Consider for a moment just one of the fascinating exhibits in Newton's gallery of achievements, a formula called Newton's drag law. It is a formula that we use today to calculate the aerodynamic drag on the surface of a hypersonic aircraft. The mathematical expression, which is not exactly a trivial formula, describes the drag force acting on each piece of an aircraft moving at very high speed. By means of the calculus, which Newton also invented, the individual forces acting over the surface of the plane can be integrated to get the resulting aerodynamic drag force on the entire aircraft. Take this drag force, along with Newton's laws of motion, and you can describe how that airplane will fly—everything about how it will fly. His equations are the basis for describing the motion of the aircraft you fly in today, and his drag law will be used to design the shape of the aircraft that will carry you at hypersonic speeds into the twenty-first century. Newton created a vision of reality that has spanned the centuries from his day to the present—and into your future.

◈

The vision that Isaac Newton created began long ago, in a small town in Lincolnshire, England.

It was the summer of 1972. I was on my way to Scotland. I was traveling from London to Edinburgh, driving pell-mell down the wrong side of the roadway, racing north along the A1. This was my first time in England. I had rented Avis's tiniest Italian vehicle for the trip north. I wanted to drive so that I would be able to see something of the country, but for all I was seeing, I might as well have been on the New Jersey Turnpike as in "merry ole England." I had to get closer to the country, closer to its history, closer to that England rolling away from me in all directions. I could see another, quieter road, following along the same way north, meandering in and out of little towns and villages as it weaved over the nearby countryside. It was a road that showed more of what I had traveled here to see, so at the next junction I left the A1 for this picturesque route north.

Almost at once the road stretched off in a new direction and in a few miles carried me to the main street of the small city of Grantham. I was tired, so I pulled over to the curb in front of the city park. I had fortuitously stopped near a bronze statue of some man standing on a block of marble (altogether it was some 15 or 20 feet high). The figure caught my eye. The man looked familiar, like someone I had known from a textbook a decade or so earlier. I got out of my car and walked over to check the name. It was indeed he, Isaac Newton. I was surprised. As a physicist, I was certainly familiar with Newton, but city governments seldom take note of physicists, even great physicists.

I asked a man walking past about the statue. "That's Isaac Newton," he said, which I had already confirmed by looking at the bold bronze letters fixed to the marble base, letters that said simply, "NEWTON."

"Why is there a statue of Newton here?" I asked. "This is where he lived," came the answer. "He went to school here. He was born just a few miles down the road from here in *Chomp-Chomp-Chomp*." My ear was not yet finely tuned to pick up the subtleties of the British manner of swallowing their own place names as though they were sacred syllables to be hidden from the rest of the world. Only later, when I checked my map, was I able to discern that the great Sir Isaac Newton had been born just eight miles south of Grantham, just about where I had left the main highway.

"He attended the King's Grammar just down that road there." The man continued telling me about Isaac Newton's connection to the town while my mind was busy trying to catch the name of the town he had just swallowed. It was getting late. I had to have my supper somewhere, and a place to spend the night, so I thanked the Grantham gentleman with a smile and a comment to show that I, too, knew and appreciated his fellow townsman. I then excused myself to look for supper and lodging.

At the far end of the business district, which was only about three or four blocks long, I found a nice restaurant and inn. I dined in the restaurant, but the inn seemed a bit expensive. I felt I could do better; besides, there was the eleventh-century Gate House Inn at the other end of Grantham that I thought would be an interesting place to stay. As it turned out, however, its prices were even higher. I finally decided to forget the trappings of history and simply stay at the George Hotel. It looked like a nice hotel. It reminded me of a hotel in upstate New York I had once stayed in, a building constructed in the 1920s or 1930s.

Before ending the day, I walked down the road that the gentleman in the park had pointed out. There was a pub open a couple of blocks away. Inside I asked the bartender about the school Newton attended more than 300 years earlier.

"Oh yes, The King's Grammar," he said. "It's just down the street. My daughter just graduated there this spring, head of her class. It's the same school that Isaac Newton went to." Strange how well time had held onto this place, I thought. But it had held on far better than I even imagined.

My hotel room was large, sparse, but comfortable. I lay down on the wrought-iron bedstead and lost consciousness. I slept with dreams of looking for the traces of Isaac Newton the next day until the sun opened my eyes in time for the English breakfast that came with my room. In the hall, as I waited for a table, I picked up a card that mentioned something about the hotel. It was named the George because it was built in the Georgian style and because it was built during the reign of George III. Notables who had slept there included Dickens and, yes, Isaac Newton, who as a student had stayed in a room above the even older apothecary shop. The apothecary's facade is still there with its trade shield on top of the building, though the druggist's shop itself is long gone. Now Pearls was holding a sale on women's dresses there—two flights below the room in which Isaac had lived. Somehow, Isaac Newton's world was beginning to infiltrate around my mission to Edinburgh, as though, I imagined, there might be some hidden link between the thoughts of that great figure in history and the message I had to carry north with me.

After breakfast, I walked again through the park and down the street, past the pub I had visited the night before. A little farther, I found the school. There I saw a pink rose in bloom beside an ancient yellow building. Windows with clear leaded glass were set in brown stone casings, each graced on top at either side by crests of the realm. Between the large windows were solid buttresses, more than a match for the centuries they had endured. I looked in through the diamond-shaped pieces of glass, but I could tell only that this was the library. Further on, I found a somewhat less ancient structure. It was the main school building, built in a similar style with stone and leaded windows. At the end of the building I found an entrance that led down to a cloistered hallway. Desks piled in storage lined the passage. Descending the steps, I entered the dark interior of the school.

At first there seemed to be no one around. Then I heard someone typing. There was a woman alone working in the school office. She was the vice principal. She invited me in. When I mentioned my interest in Newton, she told me about his days at the school. She mentioned that the part of the school we were in was the "new" addition and that the school in Newton's day consisted only of the building I had first seen. It had been turned into the school library. She asked whether I would like to see it. I was delighted. There were no crowds, no tourists. I was so easily reaching back in time, was so graciously being shown this treasure of England.

Inside the library, I found that the building was now one tall story high. The second-story floor had been removed. Looking up to the ceiling, I saw the same type of oak timbers, black with age, that hold up the roof at Westminster Abbey. The windows were high, having once lighted both stories. This was where Newton had been schooled. The walls everywhere were covered with the names of the hundreds of students who had also been there—before and since. I asked my guide if perhaps Newton had carved his initials into the building somewhere. "Yes," she said. "You can see it, if you want to climb up on top of one of the book cases." It was a bit precarious, but I wanted to see where the boy Isaac had carved his name in the woodwork of his grammar school. A few steps up, and there among many other initials I saw "I Newton" carved into the window sill.

After lunch, I drove the eight miles south to Woolsthorpe Manor, where Isaac Newton, no more than a pint in size, had been born on Christmas day in 1642. The three-story stone building is still occupied. With its high sidewalk around the compact building, it looks as if it had been plucked from the streets of London to be placed in this tiny, rural setting. Just as you might guess, there is an apple orchard in front of the house; it seems to wait for Isaac to return. The parish church stands just a couple of blocks away. When I had finally mustered the courage to knock, the parish priest kindly invited me in. I told him I was a physicist. I did not have to say any more for him to know why I was there. As we sat in his study, I mentioned that the church looked as though it was there when Isaac lived in the manor. I made some comment about it being difficult for me to tell, because I had learned how easily one can be off by a few centuries in England. He told me that not only had it stood there then but it was much older than that.

There was a stack of leather-bound books, ledgers, sitting on the window seat behind him. The priest rose and stepped to the window. He pulled out one of the books, about the fourth from the bottom, and opened up its yellow parchment pages. There he pointed out the entry for Isaac Newton. It was fabulous to see how easily he could reach back 300 years to touch the living history of this great man. He then turned back one or two pages in the book. There was a dark red splotch that covered several entries on that page. "Ox blood," he said. "They used to use ox blood to blot out entries that someone wanted hidden." He paused a moment for me to look closely. "The ox blood is beginning to fade. You can almost read the entry now." There, wearing through the centuries that had hidden the scandal, was the record of the birth of an illegitimate child and the names of its parents. He smiled slightly and said, "In time, the truth comes out."

We walked next door to the church. He showed me the sundial that Isaac made when he was a boy. It has now been set into the wall of the church. Inside, there are columns on both sides of the sanctuary. The styles of these columns and the arches between them testify to the age of the church. Construction and repairs have introduced Gothic and more modern arches to the earlier, Roman work. The priest showed me a Celtic Cross that had been found walled up in a circular structure behind the pulpit. He said that it was discovered there when workmen had to open the wall for repairs. "The previous rector had to go to London just after it was found, so he left it to be removed by the workmen while he was away. They broke it," he said. "The old man was never quite normal after that. Did something to him. Changed him, to the day he died."

The priest told me that the church stood on ground that had been holy for thousands of years. Before Christianity came to Britain, Druids had buried their dead there. He showed me some excavations beneath the floor of the church begun earlier that year to install a heating system. Skeletons of these people had been found there. The bones lay stacked in several wooden boxes. "Come on down and take a look," he said.

Late in the afternoon, just before I left, he stood at the altar gazing off into the distance, as though the back wall of the church were not there. He said, "Sometimes when I come here alone, it seems like I can see the people who have been here all through the past centuries—see them moving about, dressed as they were dressed during all the past ages." I felt as though he had added the phrase "it seems" because I might not understand were he to tell me that these had been visions of spirits still there. Perhaps he thought that I, a physicist, might not understand that there are other realities, things that are still there to be seen floating through the sanctuary, if only one would sit quietly and listen to the past.

Perhaps there are things just beyond the physical reality we physicists know—things other than the material fibers Newton wove into his laws of motion.

I left Lincolnshire feeling I had almost been able to reach Newton himself. Perhaps his spirit still haunts the old parish church or sits beneath one of the apple trees across from his manor house.

Reading Isaac Newton's writings at times can give one a jolt. I have been stunned to discover that he had figured out things that he simply should not have known when he lived. *Science*, as we understand the term, simply did not exist then, and yet Newton could observe that the particles of violet light were smaller than those of red light. (We speak now of the wavelengths of light, and the wavelength of violet light is about half that of red light—0.0004 millimeter for the shortest violet light versus 0.0007 millimeter for the longest red wavelengths.) But it goes further than this: Newton not only recognized the corpuscular nature of light but also proposed that corpuscles of light travel as though guided by a wave or field. This is a very modern idea; indeed, such ideas did not develop until the twentieth century. Newton made this suggestion to explain the reflection and transmission properties of light corpuscles as they pass through layers of glass. These guide waves he called the "fits of easy reflection or easy transmission," indicating that they occur in cycles, depending on the thickness of the glass. Later generations would reject his conception of light in favor of a wave theory of light, and it would not be until the turn of this century that the modern theory of quantum mechanics would reveal that light *does* consist of particles—photons—that move as though guided by waves. But we are getting ahead of our story.

REALITY AS LAW

Newton's conception of reality is our starting point. His conception of reality is the first truly scientific portrait of existence. It has become our modern world picture of what reality is. This is why it is important for us to understand the elements of Newton's physical reality. Understanding his view of the world makes it possible to understand the changes that physics has undergone in the twentieth century. Only in this way can we grasp the strange reality of the new physics. Newton, then, will provide us with the foundation we need for our work.

Newton stated that there are three laws of motion. It is not important for us to study these laws, but let us at least state them. Newton's first law of motion states that if an object is left alone, it will continue undisturbed, either remaining at rest or continuing to move along in a straight line with uniform motion. In the absence of any force acting to cause a change in the motion of an object, the object maintains its state of motion. This principle, known as inertia, is now recognized as holding everywhere and for all things. By means of this simple statement, Newton dismissed mankind's accumulation of nymphs, spirits, demons, ghosts, goblins, and gods.

Newton's second law tells how a force alters the motion of a body. A force applied to a body will cause a change in the momentum of the body. Here Newton made it clear that momentum is something quite specific. It is the product of two numbers: the mass of the object and its velocity. The greater the mass, the greater the momentum. Obviously, a freight train going 10 miles per hour will not be stopped if I sprint headlong into the thing; its greater mass gives it greater momentum. Similarly, a baseball bat swung full tilt will produce a greater effect than when it is used to bunt; its greater velocity gives it greater momentum. Both the mass and the speed determine the momentum of an object, then, but Newton's second law of motion is more specific. It tells us precisely how the force will change an object's motion. It says that the effect of a force acting on a body is exactly given by the rate at which the momentum of the body increases with time. The precision of this law has led to the precision that has made physics such an exact philosophy. The precision of Newton's statement lets us determine the nature of matter in ever-increasing detail. Newton's second law is perhaps the most powerful statement ever uttered, for it has changed man's earth more and it has moved us closer to a precise knowledge of physical reality than any other single statement.

The third law of motion says, "To every action there is always opposed an equal reaction; or, the mutual actions of two bodies upon each other are always equal, and directed to contrary parts." That is, any time two bodies collide, each exerts a force on the other, the force exerted by one of these bodies on the second has exactly the same magnitude at every instant as the force exerted by the second body on the first—except of course, these forces act in exactly the opposite directions. Moreover, this equal and opposite force principle also holds true any time that two bodies act on each other, whether in a collision, through friction, via gravitational attraction, or by any other means.

These then are Newton's three laws of motion. To these, Newton added his law of universal gravitation, which specifies that every body in the universe exerts a force on every other body. This force depends proportionately on the masses of the bodies and falls off as the square of the distance separating them. The reason for selecting such a hypothesis for the way the force of gravitation works is simple. Obviously, the larger the object, the stronger its force of gravitation. The earth surely has a greater pull on me than an apple has. The gravitational pull must depend on how much mass is there. If it didn't, the force on a *heavy* body would not be proportionately greater than the force on a *light* body, and Galileo's famous experiment at the leaning tower of Pisa, where he dropped objects of differing densities to show that they fall at the same speed, just would not have turned out the way it did. So the gravitational force of the earth on me has to depend on how much material—how much mass—there is in my body, and it also has to depend on the mass of the earth. I pull on the earth with exactly the same force with which the earth pulls on me. The relationship is symmetrical. The force of the earth on me and my body's gravitational pull on the earth are identical. This is just Newton's third law telling us again that objects exert equal and opposite forces on each other. Newton discerned that the gravitational force falls off with the square of the distance by noting that this is how the apparent size of an object falls off with distance. He astutely observed that this may be how distance dilutes the force of gravitation of such very massive but very distant objects as the sun, the planets, and the stars.

But our purpose is not to discuss Newton's mechanics in such detail. What I want to point out are the main features of his picture of reality, which will serve as our point of departure. Newton's understanding either has been the basis of, or has strongly colored, all modern scientific, philosophical, and even theological efforts to represent reality. This "classical" picture of reality that he gave us had to be there in Newton's mind before he stated any of his laws of motion. Indeed, his laws of motion are the product of his very clear understanding of the nature of reality. Newton's picture of reality presumes the following three basic ideas.

First of all, it is clear that Newton's laws presume that the condition of an object can be given by specifying the object's location, its speed and direction of motion, the forces acting on it, and certain characteristics of the object, such as its mass. To put this differently, Newton's laws give us a picture consisting of objects, each of which has a specific *state*. The object is not possessed by any spirits. Its nature does not change depending on the will of some capricious god. Each object has one innate position, one speed, and one constant mass.

Second, Newton describes a deterministic universe, a universe in which every effect has a specific cause and each distinct cause produces one distinct effect. Objects that are left alone stay at rest if initially at rest, or stay in steady motion in an undisturbed straight line if initially in motion. Only when a force acts on the body does the body change its motion. And its response to that force is immediate and absolutely specific.

Third, Newton's philosophy describes the world in entirely objective terms. The world is real and consists of discrete objects. There is no place for mind, consciousness, or will. There is no place for a soul to survive—no place for God.

This last statement would have rankled Isaac. He was a devout believer in God. But the philosophy Newton created has led us elsewhere. It is a picture of reality populated only by pieces of matter that differ simply in shape, mass, location, and motion. The universe is made of nonsentient objects that change according to three exact laws of motion.

Light, too, in Newton's conception consists of tiny pieces of matter. The rainbow of colors that enliven the world around us simply reflects the differences in the sizes of the different grains of light and the way in which those corpuscles bounce off of, or are absorbed by, the porous surfaces of objects. Newton's picture of the world contains forces, but he removed the distinctions of the *vis viva* and *vis morter*, the animate and inanimate forces, and banished as superstition the supernatural forces lurking behind shadows that jump in the night. His mechanics showed that all forces, whatever their origin, have one effect: to accelerate the body they act on; and one cause, one origin: the presence of other objects that are otherwise inert except for their mechanical properties.

Yet for Newton and for most of the scientists of his time, reality also had another dimension. They believed there was, in addition to and removed from material objects, a hidden agency—an omniscience, an omnipotence—a personality that existed as creator and first cause. For Newton, human souls existed as real things. But Newton's success was greater than even he would ever have imagined. His philosophy pared out the possibility that a deity or a human soul could play any role in the world. After Newton, natural philosophy passed by the questions of God and of the human soul. Thus even though Newton saw things differently, and even though this was never his own intention, the elements of deity and soul must be omitted from the scientific picture Newton produced. The scientists of Newton's time did not presume that their paradigm was adequate to deal with all aspects of reality, nor to fill the robes of God. The conditional stipulation of a supernatural order was still an essential proviso of philosophic thinking at that time. For Newton and the scientists of his time, God had set up the universe and set it in motion. Newton's laws simply governed the running of the universe. That is how Newton saw the workings of God and the workings of God's universe.

Newton's belief in a personal God has been abrogated by the very success of his mechanics. His laws of motion have worked so well that this philosophy has pushed back the unknown frontiers where once Saint Paul's "Unknown God" could hide. By the end of the nineteenth century, scientists had constructed a picture of the universe that could be thought of as very nearly complete and self-consistent. The scientific picture of reality fleshed out during the two centuries following Newton's *Principia* added thermodynamics and Boltzmann's statistical mechanics. With these tools based on Newton's mechanics, physics could reach down into the molecular and atomic world. Newton's laws could explain the behavior of matter at work in steam and combustion engines and could account for fire, heat, and the thermal properties of matter.

By the end of the nineteenth century, physics had added Clerk Maxwell's electromagnetic theory, embracing the diverse phenomena and forces of electricity and magnetism. Now physics could explain the detailed nature of light and reach far beyond Newton's own achievements to measure the distances to the stars. With a basic knowledge of chemistry and the synthesis of a few organic compounds, science could conceive of a chemical basis of life. With Darwin's handy adaptation of Wallace's theory of evolution, even the origins of life could be tentatively traced back into the dim beginnings of time.

By the end of the nineteenth century, science had put into place a comprehensive materialistic philosophy of reality.

Toward the end of this "classical" period of physics, the whole world seemed to fit so well into these few simple laws that physicists could boast that the future of physics lay only in the next decimal place. There seemed to be nothing fundamentally new left to be discovered. There existed only the task of further detailing this perfect philosophy. In its proudest moment, classical physics could look upon nearly everything one could see with an imagined total comprehension. In this view, the universe consists of matter made of elements with orderly properties. In combinations, these elements form all the materials of every object in the universe, including our own bodies. The heavens are filled with stars like our own sun; these and the planets and their moons move, as does the earth beneath us, under the force of universal gravitation according to Newton's laws. Light is understood to be a wave of electromagnetic energy; its speed through space can be calculated from theoretical principles.

The knowledge of the workings of the world coming out of Newton's philosophy spawned the machines of the new technologies: electric motors, telephones, electric lights, electric generators, telegraph, and radio. Electric forces acting between atoms accounted for chemical reactions that released the heat to drive gasoline engines or produced the steam for the great locomotives and ships of war. The age of the secular, technological city, the world we live in today, was created by this philosophy. It shapes what we think and believe to be reality. All of this wonderful, terrible creation began in the mind of Isaac Newton.

Newton, with his prism and silent face,
The marble index of a mind for ever
Voyaging through strange seas of
Thought, alone.
William Wordsworth, "The Prelude"

THE LATHE OF GOD

The significance of science lies not only in its philosophy and technology, but also in its simplicity and beauty. Understanding what the physicist means by this is difficult without going into the intricacies of the mathematics that has given rise to scientific achievements. But let's focus on one particular aspect that can convey a sense of the rigor and power that mathematics brings to natural philosophy.

Is there anyone who has admired the beautiful case of a Renaissance chamber clock, or a Tompion or Terry tall clock, who would not wish to look inside to see the design and craftsmanship of the inner works? One needs to see that beauty and workmanship, or much of the reason for the antique's appeal, will never be appreciated. It is the same with classical physics. Just a look at the works can reveal things that otherwise would be overlooked.

The momentum of an object—what Newton called the "quantity of motion"—is actually obtained by multiplying an object's mass by the velocity. For convenience, we call the momentum P. Newton's second law tells us that it is this momentum that is changed by the action of a force on an object. Newton expressed the idea of *the rate of change* by placing a dot over the quantity that would be changing with time as a result of the action of a force. The force that causes the change in momentum can be any kind of force: gravity, friction, or a man pushing with his hand. Whatever the source of the force, the force can be represented by the symbol F, which also represents the number that tells us the magnitude of the force.

In these symbols, Newton's second law looks like this:

$$F = \dot{P}$$

This equation just says that the force F causes the momentum P to increase as time passes. This is the primary gear in the clockwork of Newton's universe. The extraordinary thing, however, is not what it says but that it is so simple and so incredibly precise as to have given us most of our understanding of the world. Of course, it takes many mathematical manipulations to turn this simple statement into the final equations that describe, say, a planet's orbit, but it all begins with this simple expression.

For most problems in astronomy, the gravitational force is the significant force acting between astronomical bodies. Newton's law of universal gravitation says that this force varies proportionately with the masses of the bodies pulling on each other, and inversely with the square of the distance between them. We write down the masses, we put in a symbol r for the distance between, and we have an expression for the force that moves the worlds. Just a few steps, and we have the equation for any planetary orbit problem. After that, a little geometry, some calculus, and we have the answer—the world described by so many numbers. True, the math is hard. But the physics is really just that simple. And this simplicity is the reason why physicists speak of the beauty of these ideas.

There is something else here, too, or perhaps I should say that something is absent; something has been removed from all mankind's previous realities. The things we see in the world are moved about not by spirits or gods but by inanimate forces. The force F that causes movement is nothing but the "equal and opposite" physical forces of one object pushing on, and being equally pushed back by, another object. It is just a billiard ball world. Even when something more at first seems to appear, any such hope proves to be an empty expectation. Where we might at first see in Newton's universal gravitational force acting through the vast reaches of space the hand of God holding His creations in their celestial orbits, we are forced to conclude only how limited that hand would be. Bound by Newton's law, it has only the strength that the masses allow; it weakens in a precise way with distance, and it can act in only one direction—a simple pulling of objects together. It is a simple mechanics that everything obeys—even the gods.

It is all so beautiful and easy, it is all so simple, that one might think it must be true—it must be *the answer*. It is like holding a beautiful work of art and thinking that it is lovely but forgetting where the thought began and where the thought must end. An object carries only the stamp of a thought from one mind to another. The beauty lies in the mind. If the world is beautiful and the laws of the universe sublime, must there not then be somewhere someone else's mind? Must there not be somewhere something beautiful that still remains behind, lingering, waiting to be found?

Somewhere, just beyond my desire, something else beyond the billiard balls that scramble our lives, something besides the rocks and pebbles, causes and effects—something else, perhaps, is moving the world; something else amid the stars.

Somewhere, perhaps, *her* voice remains still. I have been trying to remember when I first met her—what she did, what I said—trying to remember just where we were, but it is gone. Too many years. She was full of life and full of beauty. That much I still remember. But the words she spoke, they are almost all gone—only a few comments jotted down remain as they were then, things written in the margins of dusty school books saved from the boxes of my father's house when he died.

There is an old high school textbook with foolish notes scribbled on every possible free surface. "ΔΑΔ Θ*K* ΦΔ ΦΔΠ Θ*K* ΔΑΔ" is penciled around the edge and across the back cover. Exuberant Greek letters shout allegiances to high school fraternities and sororities. Tucked away on page 143, brown and fading, a School Days picture: 1949–1950 Love, Ina. It must have meant something to Ina. Perhaps I meant something to Ina, but I was looking elsewhere. Ina did not mean much to me then. I look at it with a moment of regret—she was pretty—and put Ina back between pages 142 and 143 to age some more.

Another book, and another page: "SIXTEENEEEEE!" It is not my handwriting, and the "spelling" is better than mine. I think it is hers—that memory. Along the margin, there is one of those sketches she liked to draw, a drawing of a woman's legs in shear silk stockings and high heels, very high heels. "CAHABA!!" That was the name of the old high school, Shades Cahaba, the name before it was changed. Even then, the name was something that had already passed into the past, and for all I know, the name has changed again. Phillips, Ramsay, Shades Valley, these were the names of an infamous high school heraldry marked in indelible pencil lines across the book's yellowed pages.

More drawings, now showing full-breasted Marilyn Monroe figures in evening gowns and more detached curvaceous legs. My name begins to appear in her handwriting written in the margins of *Consumer Economic Problems*, and someone else, Skipper, whoever he was, begins to appear there less often.

> Page 99: "I hate people who lie."
> Page 112: "I love someone besides Skipper because he is hateful!"
> Page 116: "Cheer the Mounties on to Victory tonight. Buy a shaker! 15 cents."

I had not realized how long ago it was. "Get hep! Buy a shaker 15 cents tonight at the game." I probably did buy a shaker, whatever a shaker was. I probably would have bought anything. I am sure I was at that game. She already had me by this time—shaker, 15 cents, game with the Mounties, and all. This was the beginning. These were the memories, just the smallest fragment of what I was looking back to find.

More legs—up to the waist—crossed legs—high heels. My memories begin to come back. I begin to hear her voice from long ago speaking to me across an abyss of time, speaking my name. "Well, are you coming tonight . . ." a pause for me to get enough composure to answer, "to the game tonight?" "I'll be in the top row in the middle," she told me. I took my shaker, whatever it was, and with it some wonderful new excitement grabbed me by the sleeve and carried me off into a new life. "*Les Amies*," there it is. That was the name of her sorority, *Les Amies*, on page 143. *Les Amies* was written over and over as if to remind me all these years later. On the next page, there is a newspaper clipping—just a short notice from the newspaper in a small black-lined box:

> 180,000 Reds killed and wounded since May 25. U.S. Eighth Army Headquarters, Korea. Sunday, Sept. 30—(AP)—The United Nations field commander in Korea said Sunday his forces have killed or wounded more than 180,000 Red troops since May 25. The figure equaled 18 Communist divisions or six Red Army corps.

That is all that is said. She had clipped this item out of the newspaper to show me. As best I remember, it had not made an impression on me. Perhaps she was more perceptive than I. Perhaps in this she left her mark on me and on my life. This little clipping has left a brown stain—caused by the residual sulfuric acid that newspaper stock contained then—a brown stain on the pages of the book where it has been for 50 years.

3

Into Eternity

It is a riddle wrapped in a mystery inside an enigma.
—Winston Churchill

In its proudest moment, classical physics could look upon almost everything one can see with what was believed to be total comprehension. The universe was envisioned to be an infinite space containing objects, stars, planets, moons, and the things on them, all made of matter, which in turn consisted of the fundamental elements of nature. All the matter of the universe consists of the elements, their compounds, and mixtures of these compounds, whose properties result from the various combinations of the orderly elements. These things are the constituents of everything seen in nature: the hills in the distance, the stones underfoot, the dust of the air, and even our own organic bodies. Chemists have ordered the forms of matter; biologists have classified the species on the land, in the air, and under the seas. Wallace and Darwin have laid down the principles that we use to understand our own origins and form. The heavens are filled with stars like our own sun, planets, and moons of planets that move like our earth under the force of gravity according to the laws of motion. The motions of wheels and gears, the flight of birds, the sounds of songs, of whistles—all are understood in mechanics, acoustics, and thermodynamics governing the cycles of work and the balances of energy expended and conserved.

During the nineteenth century, Maxwell formulated the equations of electric and magnetic forces. As a consequence, light was revealed to be waves of electromagnetic radiation oscillating and rippling a fluid called the *ether*, like waves on a pond that fills all space. Light consists of the stuff of the force fields that surround magnets and of the static electricity that makes garments cling, but set into waves of motion that alternate from a magnetic field in one direction to an electric field perpendicular to the magnetic field, both cycling back and forth as the wave moves away in yet a third direction. These electromagnetic forces, as light from the sun, as the glue between atoms holding molecules together, fueling chemical reactions, creating the warmth of our bodies, powering machines, and making the blaze seen in fire—these electromagnetic forces and the force of gravity on earth and in space provided an explanation of how things move about us. Classical science had found its picture of reality. Our bodies, the objects around us, the smallest things we see, and the farthest distant objects in the sky—all are the matter of that classical reality. Their structure, their changes and motions are explained by the concepts of Newton's laws, Maxwell's equations, and Darwin's evolution. At its pinnacle, classical science seemed to have within its reach a complete picture of reality built upon the ideas of material objects set on an infinite stage.

By the end of the classical period of physics, it seemed that the whole world could be surveyed with full understanding issuing from the principles governing the motions of bodies and the radiation of light. There was seemingly a complete body of knowledge explaining how everything worked and embracing every aspect of scientific thought.

But then, in the span of a few years at the turn of the century, Jean Perrin showed that matter is not infinitely divisible but rather is made of atoms. Rutherford showed these atoms to be made up almost totally of empty space. The heavens were pushed back to the remoteness of galaxies, islands of billions upon billions of stars, themselves but specks of light in the vastness of space. Einstein showed the relativity of matter, of space, of time—and even stranger, he, along with Max Planck, helped to show us the beginnings of quantum theory, which has danced the nineteenth century's sense of reality, as in a measured, ballroom waltz, through a looking glass into the twentieth century's jitterbug of quantum jumps. It all happened so quickly that the mind has not yet quite grasped the new reality that science has found.

The change in our picture of reality is nowhere so great as in our understanding of the atomic nature of matter. Just before the turn of the twentieth century, scientists such as Wilhelm Ostwald, Ernst Mach, Pierre Duhem, Mercelin Berthelot, and even Max Planck, whose later discoveries would be the turning point for modern physics, had in one way or another taken a stand against what they believed to be a philosophically preposterous hypothesis: the idea that matter could be divided into some smallest chemical constituents—divided into molecules and atoms. Many believed that the innermost structure of matter could be built up out of infinitely divisible electromagnetic fields. Others simply believed that any hypothesis about atoms, something that could not possibly be seen, must be foolishness. The conflict was sharpened by the rise, particularly in Vienna, of the philosophical doctrine called *logical positivism*, which sought to purge philosophy and science of concepts not supported by direct sensory experience. Strong passions animated these thinkers. The opposition to the atomistic theory was not merely fierce, it was vicious, and it chose as its victim the man ultimately most responsible for the development of our understanding of the real nature of heat phenomena and thermodynamics in terms of the motions of atoms and molecules, Ludwig Boltzmann. In a moment of despondency on September 5, 1906, when all others seemed to have rejected and ridiculed his life's work, he threw himself to his death from a tower in Vienna. Such were the passions.

But the opponents of the atomic picture of matter were to lose. What had begun with Dalton's premise of different kinds of particles of matter for each element, and of their combination in multiple proportions to produce new compound chemicals to explain the chemical properties of matter, had been extended by Boltzmann to explain physical and thermodynamic properties of gases, liquids, and solids. By the turn of the century, it had all come down to one final contest. The last argument against the atomic hypothesis was simply that these atoms cannot be seen, can never be seen, and that therefore they are, and will always be, merely hypothesis.

In 1905 Einstein published a theoretical description of Brownian motion. Brownian motion is a jerky motion of microscopic particles, such as the black carbon particles in India ink, that can be seen under a microscope to be in a perpetual dance, almost as though they were themselves alive. Einstein showed that this random motion could be understood and mathematically described as caused by the collisions of individual atoms and molecules with these tiny particles seen in the microscope. Hence these particles enable us to *see* atoms.

Between 1908 and 1911, Jean Perrin, working at the Sorbonne, bombarded the scientific community with brilliant experiments that verified Einstein's formula describing Brownian motion. Perrin marshaled a splendid array of evidence proving the existence of atoms, and, in the words of Max Planck, "Its opponents gradually died out."[1] Time passed them by and left them to sleep in history. What a short time the world would wait before these unseen things of hypothesis could shatter a world. No one says atoms do not exist anymore! A hidden power was unleashed from inside those hypothetical unseen things, and now, no one sleeps so well anymore.

A CONNECT-THE-DOT WORLD

But what are these atoms? Incredibly, we now have atomic force microscopes and scanning tunneling microscopes that let us see and even manipulate individual atoms. But at the end of the nineteenth century, little was known beyond the fact that atoms must consist of pieces of charged matter. To explain that different materials, when heated to high temperatures, glow with different-colored spectra, scientists proposed that atoms must have some kind of structure consisting of electrically charged parts. The structure was thought to look something like a marshmallow of positive electric charge with raisins of negative electrons stuck about the surface. In turn, these marshmallow atoms stuck to each other by the trillions upon trillions to form the objects of the world. But at the hands of Rutherford, even these ultimate pieces of matter shrank into a void. Substantial objects such as tables and chairs became mostly empty space. Everything about us, all the seemingly rigid, solid objects that make up our common-sense world—we now know to be something quite different. Suddenly, scientists could look into matter and see it as it actually is, tiny specks in a vast void—you and I, tables and chairs—all specks lost in a void of space held together by electric fields. We and all the things we see about us are like a connect-the-dot picture spread thin through space.

As Rutherford peered with his beams of atomic particles into the atomic structure of matter, he found that the marshmallow atoms with raisin-like electrons stuck about did not exist. Instead, his beam of particles zipped through matter as though nearly all of it were empty space. Unexpectedly, he discovered an intricate structure within the atom. He found that the mass of the atom resides in a nucleus of positive charge so small that if the atom were the size of a bedroom, the nucleus at the center would still be invisible to the eye except as a glint in a beam of sunlight.

This more powerful probing of matter's structure revealed a new scientific material reality behind the illusions of solid matter. The probing revealed an atomic structure like a tiny solar system in each minute fleck of matter. And yet there was a flaw. These whizzing electrons, like tiny electric currents, should, in a flash of light, surrender all their energy and collapse everything to a millionth of its size. The electromagnetic theory of light, a crowning achievement of nineteenth-century physics, demanded that these electrons radiate light and fall to the center of the atom, still and cold. But if they did so, the earth itself would have collapsed long ago. There was a flaw somewhere, and as scientists at the turn of the century scrambled to fix it, other flaws in the great classical picture of reality began appearing. By the turn of the century, these flaws seemed to be everywhere. They signaled the great revolution in physics that has since altered our understanding of space, time, and matter.

PUZZLING LIGHT

At the turn of the century, two separate conceptual systems existed: one based on Newtonian mechanics, with its ponderable matter, and the other based on

Maxwell's equations. The question was natural. Might not matter actually consist entirely of electric fields—obeying and explained by Maxwell's equations—with Newtonian mechanics nothing but a summary of the electromagnetic laws of physics? In this design, it would be possible to view everything as formed out of electromagnetic fields that act as stresses in this medium, this fluid that has to be there, filling all space for light to exist, just as ripples on a pond need water for their waves to exist. How close could Spinoza have dreamed we would come to realizing his idea of God than to have discovered that the substance of which everything consists is at once matter and, at the same time, the stuff that fills all space—the *ether*. Perhaps Albert Abraham Michelson, who designed the experiment to detect the ether as the earth orbited through it, was too practical ever to have entertained such heady thoughts. Had I been doing that experiment, I could hardly have helped but think these things—think them and then, certainly, say nothing of them.

According to Maxwell's equations, light is a wave undulating a fluid that fills all space and is called the ether. As the ether wiggles in its wave motion, the stresses that keep the undulation going give rise to the electric and magnetic fields we spoke of earlier. The earth, in moving through space in its orbit, must move relative to this ether fluid. For example, when a light is turned on, the energy of the light begins to move through the ether as a wave that shakes the ether and travels with a speed characteristic of the "stiffness" of this ether. So, if we measure the speed of light as it moves through the ether, in the same direction as that in which the earth moves with the ether or counter to the motion of the ether, and then again measure the speed of light crossing the ether perpendicular to the motion of the earth, we should be able to see a difference in its speed measured in these two directions. By measuring differences in the speed of the waves of light sloshing through the ether, we should be able to determine the true motion of the earth in space (that is, we should be able to see which way the earth and sun are moving off toward their endless destination in the night of space).[2]

But Michelson and his assistant, Edward Morley, found nothing! No matter which way they set up their apparatus to detect the smallest differences in the speed of light over different paths, they could not detect light undergoing any drag upon its progress that would give evidence of an ether fluid flowing with the earth or across the path of the earth in any direction.

Maybe the earth is stationary in the ether? No, the earth moves in an orbit around the sun at 30 kilometers per second. At some point along the earth's orbit, it might be moving with the ether, but everywhere else it would have to be moving upstream, downstream, or across the ether stream. Still Michelson and Morley found no motion. Perhaps the earth drags along the ether immediately around it, like a wet sponge dragged through water. But this would result in easily observable consequences: Starlight would bend as it passed through from the outside of this ether to the ether being dragged along with the earth. We would see the stars shift in the sky at different times of the year. These effects are not observed, so there must be another answer to this mystery of the ether.

In 1892 an Irish physicist named George Francis FitzGerald offered an outrageous explanation for the absence of any discernible motion through the ether. He reasoned that because all objects, including the Michelson–Morley apparatus, consist of matter held together by electric and magnetic fields (fields that are actually stresses in this ether), motion through the ether should distort these same fields just as much as this motion would affect the propagation of the electromagnetic waves. If so, then the ether's motion would also cause the atoms of the apparatus to bunch together. The light waves measured with those yardsticks would seem to travel farther in a given time than they actually did go, and exactly compensate for the slower speed of light measured by Michelson and Morley. Shortly after FitzGerald's proposal, the Dutch physicist Hendrik Antoon Lorentz independently reached the same conclusion and proposed equations to explain the effect. This contraction hypothesis became known as the Lorentz-FitzGerald contraction. The equations are known as the Lorentz transformations. The difficulty, however, was that this solution seemed just a bit too pat.

In 1905 Einstein had a better idea, an answer that has revolutionized our conception of space, time, matter, and energy. It is a picture in which space and time, matter and energy are relative—they depend on the *observer* in a very curious way. The curious thing about the way the picture of reality has to be altered is that the space and time in which objects exist—the very basis of the idea of objective reality—must be made conditional upon the vantage point of the *observer* who looks upon this reality. For the first time in physics, we discover that these basic constructs of objective reality are contingent upon the conditions under which we observe space, time, matter, and energy. As a result, the theory of relativity gives us the first indication that the picture of an absolute objective reality is false and must be changed.

Remember now the purpose: the physicists at that time were looking for a way in which Maxwell's electromagnetic theory could explain all matter. For other physicists at that time, this suggested an underlying, all-pervasive ether that would account for the properties of the electric and magnetic fields. From this idea, perhaps, a theory of matter as frozen chunks of electromagnetic stress in the ether—chunks of ether—might explain all the properties of matter. But for Einstein, Maxwell's equations were fundamental. Why try to explain them in terms of something else? Why concoct an ether? For Einstein the question was not "How does motion through the ether affect the propagation of light?" but "How should Maxwell's equations look to an observer traveling near the speed of light?" As Einstein thought about this problem, he hit upon a very simple, unexpected answer: Perhaps Maxwell's equations are exactly the same for a moving observer as for a stationary observer. That is the way things work for ordinary Newtonian mechanics. In fact, it is an old principle in mechanics known as Galileo's principle of relativity. All the laws of mechanics that are valid in one frame of reference, such as fixed to the ground, will be found to work exactly the same way when tested in a uniformly moving, constant-velocity frame of reference, such as in a smoothly traveling airliner. If you drop your pen while flying in a plane, it falls to the floor just the same way it would fall if you were sitting at your office desk. Think about it a moment. That is quite remarkable. The fact that you are zooming through the air at 500 miles per hour has no effect on the way the laws of mechanics look to you.

Now, as Einstein saw the issue, if Maxwell's laws are to be so fundamental that they might replace classical mechanics, then surely Galileo's principle of relativity must apply to electromagnetic phenomena as well.

There was just one problem with this idea. In Maxwell's theory, the velocity of light is a constant—always the same speed. It does not depend on anything. It does not depend on space or on how we move through space, and why should it? Space is nothing! If space is nothing and Maxwell's equations are absolutely correct, then the velocity of light must be constant under all conditions of uniform motion through space. Even when we move relative to the source of light with any velocity, the light always goes at the same speed. Even if we travel in the same direction as the light, at a speed almost as great as that of light itself, we must still find that it races ahead of us with the same speed! How could it be possible for light to travel at the same speed no matter how fast we travel? Whatever the answer to that, the Michelson-Morley experiment showed without a doubt that the velocity of light is a constant. Every observer, no matter what his or her speed, will find that light travels at the same speed relative to him or her. Ptolemy, Copernicus, the Greek sages, all the medieval theologians, and even modern astromancers had it wrong; the earth is not the center of the universe. You are. I am.

But if we are going to get ridiculous results like that—if the speed of light must stay the same no matter how fast we move—then something else that comes into play when we measure the speed of light must be relative and not absolute. Thus, Einstein made the brilliant deduction that measurements of space and time (the things we measure in order to determine the speed of light) must be relative concepts, whereas the speed of light itself must be a constant, an absolute.

When Einstein calculated the changes in space and time that would be necessary so that all observers would measure the same speed for light, he found that all objects traveling through space must undergo a contraction. When we use our measuring rods to measure the lengths of objects moving very rapidly past us, we find that those objects are all shorter. But they are shorter not because of any ether causing them to contract but because space itself contracts. Also, moving clocks run slower than our own clocks, not because of any ether affecting their works but because time itself runs slower for objects moving relative to us.

The well-known Twin Paradox illustrates just how strange this is. One twin on his twentieth birthday sets out from earth aboard a rocket that will carry him at near the speed of light, say 90% of the speed of light, or 165,000 miles each second. The other twin stays on earth. Watching his twin in the rocket, the earthbound brother will see his twin moving and aging more slowly than he himself ages on earth. After 40 years have passed, the 60-year-old man on earth will greet his 37-year-old twin brother returning from his great voyage in space. Even stranger, had the rocket-bound twin watched his brother back on earth while his rocket coasted along at a constant speed, he also would have seen his earthbound brother moving and aging more slowly than he! That's why it's called the Twin Paradox: Each twin sees the other aging more slowly than he, yet when they are reunited, the earthbound twin is the older. The reason is that the twin in the rocket must change his speed in order to turn the rocket around for the return trip home. Despite the queer appearance of the Twin Paradox, this effect has been confirmed in many laboratory experiments. Indeed, J. C. Hafele and R. E. Keating demonstrated the effect in 1972 by flying clocks around the world.[3] They said, "There seems to be little basis for further arguments about whether clocks will indicate the same time after a round trip, for we find that they do not."

Appearances take on a reality we might not have expected. We could understand that our yardstick might look shorter to an observer whizzing past us and that this traveler's yardstick might look shorter to us as well. We could see that clocks might seem to run slower. It is easy to acknowledge that appearances might change. But Einstein has shown us that reality itself is contingent on our vantage point. We see that this local, this personal meaning of time has a real impact on what happens in the "objective" world. Twin A is older than twin B. They look at each other and see an isolation greater than what mere distance had written across their faces. This is an isolation each of us has from everything else. We are separate—utterly separate. We do not even share the same space or time with others.

But there is much more of a puzzle here. The most basic property that objects have in common is their mass, the stuff of their substance. We measure the mass of an object by taking some reference force, like the force of gravity, that can be used to move it. We measure how long it takes the mass to increase its speed under the action of this force. That is, we measure its acceleration. If we keep the force the same and use different objects, we can use this measurement of acceleration to determine the mass of any object.

Thus the result we get when we measure an object's mass depends on how we measure space and time. Consequently, Einstein's discovery that space and time have meaning that is relative to the observer's measurement means that mass is also contingent on these measurements.

If we continue to apply this force, the object moves faster and faster. Assume now that we have taken an object and measured its mass by giving it a push. Now let us apply a force to accelerate the object to a very high speed. At this new speed, we make a new measurement of its mass. We observe what happens as the object flies past, and we figure out what its mass is. But because space and time are distorted for objects that are moving relative to us—distorted in such a way that the speed of light is always constant—the mass we obtain by this measurement is also *distorted*. The mass is greater now that the object has been accelerated to a higher velocity. Einstein showed that merely increasing the velocity of an object increases its *substance*. The increase in mass that Einstein calculated is directly proportional to the increase in the energy of the mass due to its higher velocity. Subsequent events, Hiroshima and Nagasaki among them, have amply proved the equivalence of mass and energy and have confirmed the validity of Einstein's conception of space and time as relative to the observer.

This equivalence of mass and energy has become such a cliché that it is easy to lose sight of its bizarre meaning. In Einstein's equation relating mass and energy, we find that the mass, the very substance of an object, is actually energy. What is energy? It is simply a measure of a substance's positions at the different times when we observe it. Note that through all these explanations, there is an undercurrent of implications all leading to the same place: *The way things appear to us has become something of the substance of what they are.*

This process implies an unexpected connection between energy and mass. Objects are the very potential to move. Or, to take further liberties, objects are their motion—motion in a space that is itself nothingness. The very answer begs the question "What is reality?"

Somehow this enormous triumph of probing the nature of matter has left us with more of a puzzle, more of an enigma, than we ever had before. And yet isn't this what we should have expected from the beginning? When we seek objective answers about the world in which our basic facts reduce to statements about the location of objects at various times, is it so surprising that the philosophic meaning we uncover has no more depth than that objects *are* these motions? Is it surprising to find that reality is in fact our observation of that reality? But why should objective reality have an existence that is at all relative to our observation of it?

This relativity of mass, space, and time is the first scientific indication that physical reality depends in some way on the observer. In relativity, the vantage point from which we make an observation—a measurement—plays a role in the description of physical reality for the very reason that space, as such, has no substantive reality.

It may seem that I am exaggerating the significance of measurement making and the role that observation plays. It might seem more appropriate simply to describe the way space and time are distorted under various conditions of measurement and let it go at that. But Einstein, in his original paper on relativity theory titled "Zur Elektrodynamik bewegter Körper" ("On the Electrodynamics of Moving Bodies"), shows us that measurement is fundamental. Any limitation that affects our ability to measure a physical quantity, and that persists no matter how we try to devise equipment for that measurement, not only represents a limit on us but also shows a limitation that must exist when any two physical objects interact. This limitation on me as the observer is also a stricture on nature. If I cannot get at the physical quantity no matter what physical tool I employ, then perhaps that *physical* quantity never takes part in any physical interaction. And if that is so—if it plays no role in physical processes—it is not a part of physical reality. The conventional wisdom in physics has been to concentrate only on this latter fact, only on the implications as a stricture on nature. But our purpose is to understand the basic fabric of reality. As we will see later, the observer will play a much more important role in understanding reality. By emphasizing the importance of the observer that already is to be found in the theory of relativity, I am laying the basis for us to discover a much broader picture of reality that will unfold as we go along.

I remember mostly the rain. I spent most of the day driving across Alabama—through Andalusia, Greenville, Butler. The purple lace of wisteria draped through the bristly pines painted long, secluded avenues beneath the backdrop of the dark sky. Deeper in the woods, I could see white dottings of dogwood blossoms heralding the end of the short southern winter. Here and there, where the road cut through a rise in the landscape, the gallery of an orange clay bank would fill my view. Then suddenly, the silver skeleton of a steel bridge reaching over the Tombigbee loomed and passed. I sped west—into the rain—into the long, dark afternoon. I was going where she had finally gone. I was going to Jackson.

The place lies in that ill-defined realm so many cities harbor that is neither town nor suburb but something between, with a mix of fast foods, gas stations, vacant lots, boarded buildings, homes not quite poor, a church—then, an iron fence edging maybe a hundred acres that slope away down a gentle hill.

They take care of the place in a fashion. I am not sure what I had expected, only what I had hoped. Something more than a gravel road, something more than the tags identifying block numbers. Something more than ragged grass. Something more than just a place to leave the dead.

They lie there now side by side, there on that hillside cemetery, waiting the ages. There is a long marble slab on which two stone "pillows" mark their names: MAURICE L. ZEHNDER on the first, MERILYN ANN ZEHNDER on the second. Beneath these the dates: 1899–1947 under his name, 1936–1952 under hers. She lies there beside the father I so often heard her say had died too young, the father she had loved so much. She lies there now in the place she had told me she would some day lie, beside her father, beneath the soil of Jackson.

It was five in the afternoon when I got there. The rain had slackened to occasional showers. I found the grave, eventually. It was odd. There was no sense of the time. So many, many years, and yet, it could have been, oh, just a few days, a few months. I am not sure what I had expected. On the way, I wondered what I might find. Leaving, I guess, one always wonders what was found. It does not matter. We find only one thing, emptiness. Emptiness—that cloth that covers over the ages of disappointment, that cloth that is never expected despite the fact that it was always the only option. Just earth, marble slabs here and there, marble slabs scattered over the grounds almost in some order, and the cedar bushes standing by twos about the hillside. It is a cemetery. It is a place where the dead do nothing. It is a place where the living come to learn that lesson once again.

I stood there, believing I was fulfilling some obligation. The rain did not care. I stood there in wonder at all the time that had gone by—feeling wet, looking at the gray sky, and . . . , and what? Wishing that there might be something more. Wishing that this time, for this visit, there might be some bit of a miracle, some whisper, a whisper in my head—something. But she was not there. If she had any existence at all, she was not there.

I left. I drove east into the night, into a pouring, weeping rain, into the black night.

So much we will never know. So much more we have known that is gone. Fragments of life, of consciousness, the pieces of our having been here at all, of having lived. Those pieces of her life, as lovely and wonderful as she was, have all left only such fragments as these, and the rest of her has slipped away into eternity.

4

The Light Fantastic

Trembling with tenderness
Lips that would kiss
From prayers to broken stone
—T. S. Eliot, "The Hollow Men"

As the twentieth century got under way, sobering discoveries in relativity and atomic physics electrified the new age. These were wonderful discoveries that excited imaginations and reordered our appraisal of our place in the universe. Here was the discovery of the atomic nature of living matter, consisting of parts—atoms that determine everything that you and I are—and discoveries of the diminutive reality of that matter. Here were the discoveries that showed the insignificance of mankind and the earth as but a mote in the vastness of space. These were discoveries that displaced and diminished man's self-image. These discoveries followed hard on the heels of the Wallace-Darwin theory of evolution, and they left man—already suffering from the remoteness of God—with no meaningful place for God in the emptiness of space, no place for Him to be, no role for Him to play.

All these discoveries left little absolute permanence in the picture of reality. There existed only the elements themselves, vast empty space, and time. Even these had no permanence. Matter itself, not just its shape and form, could be entirely converted into nothing but the energy of motion; space and time were made elastic, dependent on one's particular point of view. Certainties became relative; man became animal; soul became id. But for all that had been lost, much had been gained. Gone were myths, and with their passing, new freedom gained, and some small found part of the permanent truth.

Michelson and Morley's observing changes in the speed of light and Einstein's postulating that the speed of light is a constant were signal events in understanding the nature of space, time, and matter. But the study of light was to reveal even more startling facts about its nature, and the nature of reality itself.

QUANTA

At the turn of the twentieth century, the study of light raised other problems besides the issue of an ether. A number of physicists had been trying to figure out how light is radiated from hot objects. The way light and heat are radiated from an object, of course, depends to a great extent on the condition of the object's surface. But every object that is left alone to come to a constant equilibrium temperature reaches a point where it radiates as much energy in the form of heat, and, if it is very hot, light, as it absorbs from its environment. A white object reflects most of the light that falls on it and usually radiates only a part of the heat that reaches its surface from its interior. But a black object has rather ideal properties. A perfectly black object absorbs all the light that falls on it and radiates all the heat that comes up to its surface. This makes it a perfect radiator. A black object heated in a very hot furnace radiates heat in a fashion that depends on nothing but the temperature to which it has been heated. More than this, anyone who has ever been around furnaces knows that as the temperature goes up, the color of glowing hot black coals goes from a dull red, to a cherry red, to orange, yellow, and white and then as it gets still hotter, the furnace interior takes on a bluish hue. The question that physicists wanted to answer was "What is it in the nature of this process of radiation of heat and light that determines just how much infrared heat and how much red, yellow, and blue light is radiated by a perfectly black body when it is glowing white hot?" Now if this seems like a perfectly esoteric and academic problem, that is just what the physicists thought as well. It was a perfect problem to demonstrate their mathematical acumen. The problem should be straightforward. Indeed, there seemed to be many different ways to work out the details of this blackbody radiation problem.

There was just one difficulty. Every time the problem was worked out, the result proved to be preposterous. The attempts predicted that more and more radiation would be emitted at shorter (bluer) and shorter wavelengths of light. The theory said that the total power (kilowatts of radiated energy) would be infinite. Such results are clearly wrong, but all efforts that conformed to the classical laws of physics kept yielding this same result. In 1900, Max Planck finally supplied the correct answer. After more than a dozen tries that led to the same incorrect answers everyone else had obtained, he figured out a way to finesse the correct answer. To do so, he had to make an assumption that even he did not believe. But it worked. He assumed that light is radiated—not continuously, as had long been assumed, but in jumps and spurts that release packets of energy called quanta.

In a way, this hypothesis was not so strange. It was already known that when gaseous material is heated, it emits its radiation at specific wavelengths. Of course, each atom emits its little contribution of energy at some specific energy-dependent frequency. What if this always happens? What if each atom emits a discrete amount of radiation so that it rapidly cools down as it releases its excess energy, until it again absorbs more energy? Planck assumed that the radiation emitted by the walls and objects in the box had a discrete amount of energy that was proportional to the frequency of the radiation. This simple hypothesis of light as quanta turned out to describe perfectly the radiation streaming from a white-hot blackbody.

Despite appearances, Planck's theory of light was not a reinvention of Newton's corpuscular theory. It had been assumed that light waves were produced when electric charges were moved back and forth, much as waves are produced when a stick is moved back and forth in water. In Planck's new description, the radiation of light was more like the effect of tossing a stone in the water. The energy went out in one lump. Planck thought these quanta would radiate out and behave just as Maxwell's theory of electromagnetic radiation predicted. It was simply necessary to assume that when the energy of light or heat is emitted, it comes out in little chunks of energy.

Five years later, Einstein brought this description of the quantum nature of light full circle. He showed that experiments carried out by Lenard at Heidelberg proved that light is also *absorbed* in chunks of energy.[1] In these experiments on the photoelectric effect, ultraviolet light of different frequencies was found to be knocking electrons out of metal surfaces. What was of interest was the fact that the electrons did not behave as though they took time to build up enough energy from the light wave to break free but, rather, would pop off the metal surface as soon as exposed to the light. Moreover, the energy of the electrons coming off the surface did not depend at all on how strong a light was shining on the metal surface; it depended only on the frequency of the light. The bluer, higher-frequency light kicked the electrons out harder. The lower the frequency of the light, the less energy the electrons had when kicked off the surface of the metal. This showed that the energy of these electrons depends on the light's frequency, just as Planck had assumed in his solution to the blackbody radiation problem. Thus, both when emitted and when absorbed, light behaves as though it consists of *quanta*—chunks of energy called photons.

And yet light must still obey Maxwell's equations. Light is still a wave. It may be radiated as a chunk, and it may still be a discrete chunk when it is absorbed, but it is also a wave. Light acts both like particles and like waves. This prickly pear would cause no end of mischief. How can electromagnetic energy spread out like a wave to the ends of the universe and yet still be deposited all in one neat package when the light is absorbed? Strangely, in 1905 Einstein had elevated Maxwell's equations to the height of an absolute principle in his theory of relativity and yet cut away that principle's underpinnings with his theory of the photoelectric effect. Together, Planck and Einstein had created the greatest paradox ever to exist in science.

If you aren't a physicist familiar with the intricacies of the wave-particle duality problem, the fact that one experiment gave wave-like results and another gave particle-like results for light might not even seem odd. But let us look a bit closer to see what kind of fantastic quandary this wave-particle duality of light leads us into.

Consider the double-slit interference experiment that Thomas Young performed. In that famous experiment, light from one source illuminates a screen with two narrow and closely spaced slits to let a small bit of light through to a second screen. These two slits now serve as two light sources that, because they are so close, permit the waves of light from one to interfere with the waves from the other. As the light passing through the two slits falls on the screen, the place where the crest of the wave coming from the right slit meets the wave crest from the left slit produces a bright band of light. Where trough meets crest, however, the two cancel out and the screen is dark. Thus the presence of the second slit produces a pattern of bright and dark bands on the screen. This interference pattern of light and dark bands on the second screen serves as a signature for the presence of a wave and proves the wave nature of the light.

But what happens when we do this experiment with single particles of light? What happens when the light intensity is so low that only one photon at a time goes through the apparatus? Now we have a situation in which a single photon must go through either the right or the left slit in the apparatus. With such a dim light source, we cannot see the interference pattern directly but must use a photographic plate to find out where the photons are going. After enough photons have gone through the slits to build up an image on the plate, we discover the same interference pattern as before. If we close either slit in the screen and repeat the process, however, we see no such interference pattern. We are forced to conclude that every single individual particle of light that goes through one slit in the screen interferes with itself as it goes through the other slit on its way to the photographic plate. This single photon—this single piece of light—is at once a particle that causes a single grain in the photographic film to darken at one single point and a wave that passes through both holes in the screen and fills all space in all directions at the speed of light.

Think of one photon being emitted. It is a brief train of electromagnetic waves that speeds away at a speed that Einstein's relativity has shown is the fastest anything can go. The wave passes by object after object—perhaps going into space, perhaps passing planets and stars as its spherical waves move out—passing by things without disturbing them as though the wave had not passed that way at all. Suddenly it encounters some single atom somewhere, and all the energy of the entire wave is instantly deposited in this one spot. The only way we can imagine that the entire energy can be instantly gathered into one place is if it was always in one packet of energy in the first place. That leads to several problems, however. First, waves spread; no matter how tight the packet of waves, they would rapidly spread enough to allow us to split off some of the energy. But this doesn't happen. The photons act like discrete pieces of energy. Second, what would that idea do to the wave behavior of light? How could we explain the interference of light in the double-slit experiment? What complicates this picture even more is that if, in the double-slit experiment, we attempt to find which hole the photon actually passes through, we either fail or we destroy the interference pattern of the double-slit experiment. Efforts to pin down whether we have a particle or a wave end in defeat. In certain types of experiments—wave phenomena experiments like the double-slit experiment—the photon seems to be everywhere at once, whereas in other experiments, the photon is a particle to be found in one tiny spot only.

But things are even stranger. In 1967, two physicists at the University of Rochester in New York, R. L. Pfleegor and L. Mandel,[2] carried out a variation on Thomas Young's experiment in which *two* light sources, lasers—one for each slit—were used as the source of the photons rather than a single light source shining through two slits. They found that a single photon created in and coming from one laser actually interferes with *itself* as it is created in and comes out of the other laser!

What makes all this so much more confusing is what happens in such experiments when they are carried out using ordinary pieces of matter instead of light. Light is a form of energy. But as Einstein has shown us, energy and matter are just different forms of the same thing. If light waves of energy behave like particles, is it possible that ordinary particles of matter behave like waves of energy? Does it make any sense to speak of, say, a football wave?

What is it that was going on so many years ago? What do those happenings tell me now about the nature of reality? The answer must lie in these memories somewhere. But what happened then?

Somehow, pondering everything I have learned about atoms and wave-particle duality has jogged other memories of long ago. What has become of all the moments we spent together? What has become of that love I dimly know was once the thing that filled every day? What did we do and say? These are the questions that decades later leave me searching so many blank pages of my old memories. They are questions that echo about at times half-noticed until a word cues a memory of something important, something that makes me say, "How could I have forgotten? How could I have forgotten the diary?" I kept a diary that year—only that year. Every day I made an entry and then returned the diary to the back of the top right-hand drawer of my dresser. I left that dresser years ago when I moved away to college, returning only for visits. I moved many times over the years, as did my parents, but eventually, with their deaths, all their things were parceled out, and that dresser came back to me. I set it aside for storage space. But what had I done with the diary? Where was it? Probably in some box among a hundred others in the basement, where water had damaged some beyond recovery . . . or might it still be in that dresser drawer? Climbing back into my storage, I searched out that drawer. And there was the diary, after almost five decades still sitting in the same spot at the back of the top right-hand drawer, just as though I had put it there the day before. I opened it: "Tuesday, January 1, 1952." There it was, my life. There she was once again, living in my mind, once again:

> Wound up an evening of a show (TV: My Favorite Spy) and a stay at Merilyn Zehnder's house at 1:40 A.M. Awoke at 10:45 and made a date for a tennis game at 2:00; doubles. M.A.Z. wore shorts, red shorts [the temperature reached 80° F that day]. After tennis we went to her house; it was a lovely day. Ceil came to her house and took a picture of me holding M.A.Z. in the air by her ankles. I left about 5:00. Tommy Whitson [best friend, cohort in crime] called and we went to the show Distant Drums and a leadout [fraternity promenade dance]. While I was out a girl called for me.

Red shorts, that was meaningful; that was the time of short-shorts. I had forgotten about that picture. I did not get a copy of it, so it is now just a dimly refreshed memory of the three of us in Merilyn's front yard. I was just barely able to clear her 106 pounds from the ground. Somewhere in that faceless memory, I catch images of the pines and of laughing.

> Wednesday, January 2: Telephoned Merilyn. At about one o'clock my mother and I went to Elvina Loman's house for lunch. [I am sure that was not what I would have freely chosen to do. In Elvina's house, one touched antiquity. Eighteen-ninety was palpable. Elvina was slim, tall, pleasant, and old; Elvina's world was something made out of darkness, somber colors, oppressive decor with black crinkled lacquered Empire pieces that then were still too recent to be anything but old furniture.] I went to town, bought a record—O Soave Fanciulla and *E Il Sol Dell Anima*; fifty cents. [I didn't like opera, but the price was right.] I called Maitland twice at about seven, no answer. She called me at about eight and asked me to come over. I did.

Maitland had been my girlfriend for about a year. We were entirely unsuited for each other in every conceivable way, and I think in some ways we thought we were equally unsuited for anyone else. That kept us together, longer than we should have been. She was a year ahead of me in school. Her father was rather well-to-do. She belonged to the right sorority and drove—with total abandon—her father's brand new, ivory-colored Chrysler convertible with tan leather seats.

She claimed she would have preferred three Dodges for the same money. I can hear her to this day: "Three times as many to wreck!"

She was from the social set in which, when she turned 16, her mother's friends would ask when she would start smoking. I did not like her smoking. I liked even less its use as a social litmus of priggery. But we had our clandestine *affaire*, such as it was. We would go to the Mountain Brook Country Club for lunch on Saturdays and play tennis. She was athletic. She was coached by the club pro. In tennis, she beat me. She always beat me. And then, we would go out onto the golf course, cross into the woods that lay between the club and her house, and there in the grass spend our time together.

At other times she would fill the convertible with her girlfriends and me, and we would fly around the hills of southern suburban Birmingham as though we were trying to kill her father's Chrysler in its first year of life. But before long, mother wanted daughter to see more of "other boys" and less of me. We still saw each other. She would phone me to come over at night, and I would walk the back-way streets to reach her house.

One time I had come over while her parents were out. They pulled in without us being aware they were there. The first we knew, we heard the key in the front door. I bounded for the stairs to the basement garage. Maitland quickly hid the flush in her face, and whatever else. I hit the top of the steps just as her father rounded the entrance to the den, only 5 feet away.

I was trapped there. I could not go down the steps. The steps would have made too much noise. They had come into the house so fast that I had not even had a chance to close the door to the basement. I was standing there against it. He paused. His wife spoke to him, and this gave me a moment to act. Slowly, quickly, ever so silently, I pulled the door almost closed as her father, a ponderable man, entered the room.

He entered. I had not quite shut the door. The latch would have made too much noise. I was not held in high regard by her mother. I was hated by her father. I shudder even now to think what her father would have done had he found me there. I held my breath. *Breathing* would have made too much noise!

"Maitland!" he yelled. "You leave this door open?" His heavy body twisted the floorboards into sounds of incrimination as he stepped three more steps toward me. He stopped. He put his hand on the door, paused as though Hitchcock were directing the scene, and then closed and snapped the latch.

I stood there for an hour. The voices eventually quieted into the distant rooms. I took those steps one by one, each step cut into a hundred partial movements. I took those steps listening, sounding the house out with each muscle flexed, taking a full hour for the eighteen steps. And finally, finally, out the garage door, through their yard, I dashed—hoping no one would glance out a window—running through their backyard, their neighbor's backyard, everybody's backyard, and out into the next street and safety. It is, I now judge, attributable to that fair Maitland that today I must take Tenormen daily for my blood pressure and monitor the progress of my heart's arrhythmia.

But all of that was before I met Merilyn . . .

Before the turn of the twentieth century, when even the reality of atoms was still being contested, the raisins-on-a-marshmallow picture of atomic structure was replaced by the Rutherford picture of the atom. This is the concept of the atom as some Lilliputian solar system, with a positively charged, sun-like nucleus surrounded by a swarm of infinitesimal planetary electrons scurrying about in the empty space around the nucleus. But how do these electrons keep from falling into the center of the atom? The electrons should be moving rapidly around that nucleus on orbits, the way the planets move around the sun. If they did that, this moving around and around should stir up light waves in the ether. But that would cause the electrons to lose energy and fall into their "sun." If this were the way the world worked, everything would disappear in an instantaneous flash.

The discovery that light is emitted only in packets of energy, however, made it possible to explain, in part, how atoms could radiate only under special conditions. Atoms, it seems, radiate only when there are packets of energy available, enough to create a whole photon. With this idea, Niels Bohr found a way to explain the structure of simple atoms. He proposed rules that determined special stable orbits for the electrons to circle in without radiating away their energy.

The scheme Bohr invented worked, after a fashion, but Bohr could not explain why these special rules should work. An adequate answer had to await the development of quantum mechanics. That development greatly depended on the insight of the French prince Louis de Broglie, who was studying for his doctorate at the Sorbonne. He had begun by studying medieval history but had changed to physics when told of the wonders of relativity theory by his older brother Maurice, then in the vanguard of new investigations in physics.

Basing his ideas on Einstein's equivalence of mass and energy and the interpretation of light as both wave-like and particle-like, de Broglie proposed that *everything* must have both a wave and a particulate nature. "I had a sudden inspiration," he said. "Einstein's wave-particle dualism was an absolutely general phenomenon extending to all physical nature, and, that being the case, the motion of all particles, photons, electrons, protons, or any other, must be associated with the propagation of a wave." De Broglie presented his thesis to Paul Langevin, at the time France's leading physicist—and also, at the time, having an affair with one Marie Curie. Despite such distractions, Langevin found time to send a copy of de Broglie's thesis to Einstein—for the humor of it. But Einstein quickly realized the great significance of de Broglie's theory. He wrote back to Langevin that de Broglie had "lifted a corner of the great veil." And so he had. (Some years later, Einstein had the opportunity to lift another veil—alas, that of this same Marie Curie, as he and Marie vacationed together in the Swiss Alps. But then, that is another story.)

In his momentous thesis, de Broglie showed not only that light is a wave that acts like particles of matter but also that electrons, protons, and indeed all types of matter act like waves of energy. The reason why electrons inhabit these special Bohr orbits in atoms is that the electron waves can resonate in these orbits just as sound in an organ pipe can resonate at certain frequencies and not at others. The de Broglie wave of the electron fits exactly an integral number of wavelengths into the electron orbit.

But de Broglie's very success turns out to confirm facts that simply do not make sense. What does it mean for a particle to behave like a wave? How can a quantum of light, a photon, also behave just like the electromagnetic waves that Maxwell had described. As more and more experiments have ferreted out this wave-particle behavior of photons, electrons, and even the atoms that we are made of, it has become clear that reality is nothing like what it at first had seemed to be.

Stitch by stitch, the whole fabric of reality, so carefully woven together by classical physicists, was unraveled into a plethora of confusing facts. Matter was not matter but waves and energy, the stuff of no more than motion. Waves of light were as much particulate as waves of energy. And neither space nor time had any more than a relative meaning. There ceased to be any cohesive picture of reality. Physics had run out of pictures to explain its tapestry of colors.

One by one, the discoveries of physics undermined every vestige of the classical picture of an objective physical reality made of immutable matter existing in the infinite box of space and running to the tempo of a divine clock. In Einstein's hands, matter had become mutable to the energy of its own motion. Matter had become somehow a wave, the whole world but ripples on a great pond, an undulation of itself in a nonexistent ether. Light had oscillated from Huygens's waves, to Newton's corpuscles, back to Young and Maxwell's waves, and then to the packets of Planck and Einstein. Paradoxically, it had then finally become both at the hands of the little French prince. The picture was dizzying, stupefying—a Picasso collage pasted together from a collection of shattered ideas. Much would have to be done before waves and particles, matter and energy could all be put back together again in one coherent equation. But as quickly as this old quantum theory had come, it too passed away, inadequate to explain any but the simplest atoms. Bohr's picture of the atom's structure, only a model of what might have been reality, became merely a step along the way, a transition to quantum mechanics.

Thursday, January 3, 1952: Awoke to a cloudy day; called Merilyn. She had a party for 2:00 P.M.–4:00 P.M. [probably for her sorority sisters]. I went to town, bought this diary, and again looked at records, but found nothing. Came home and called the Imeru [M.A.Z.] and set a date to come over at seven. I walked over. We talked and read poetry and did other things—yes. [The word "trouble" is inserted here in small letters together with some sort of scale indicating "2." I don't know what the scale meant, but I believe the top of the scale would have been about "4."] Watched television spasmodically. Left my book of poems there.

Friday, January 4: A gray day again today. I called Merilyn. We decided to play tennis. I put out the Christmas tree [my chore] and left for the tennis courts. I beat her six to two, and then we went to her house. I picked up my book of poems, and a poem she had written which said in effect that she thinks she loves me.

Saturday, January 5: Went to Maitland's. She was packing to leave for Stephen's. She cried because she did not want to leave. I felt sorry. She was listening to the "Hangout." Strange how music can put you in such a mood. I left about 2:25. Called Merilyn from Hill's in Mountain Brook to tell her I would be there in 30 minutes. We played doubles and won. Then at her house, we played chess. I walked home and called Maitland to say good-bye. She called back at seven. I went to Homewood to the movie *The Golden Horde*. I won't see Maitland again till June. I feel sorry, but I love Merilyn.

What a store of memories, and of emotions. I had forgotten so much. I had forgotten I used to call Merilyn my *Imeru*. It's a Babylonian (Akkadian) word that means "donkey"! It used to make her mad, so I teased her with it. Funny, all these memories. Strange how deeply they reside and how one memory pulls out the next like links in a chain, until at times you find that even the subtlest thoughts you had decades ago are still hidden somewhere inside of you. Even the way the words were written on the page is tied to thoughts that went through my mind as I wrote down other things—and now even those subtle thoughts come back.

Yesterday, when I finished writing these pages, I had to stop suddenly. As I rushed out of my office, I felt for a moment, as I turned, as though I were back there still in that past memory rushing out to play tennis with Merilyn, rushing to meet her again—just for a moment.

5

Jitterbug World, Jitterbug Reality

My God, it's full of stars.
—Arthur C. Clarke, *2010*

The music of the spheres, the motions of planets silently sweeping through space, the forces of celestial bodies touching the remote depths of the universe and moving forever—these are the beautiful images of classical science.

By the time of the roaring '20s, physics had new pictures of reality: a world of Lilliputian solar systems with positive and negative bits of matter for suns and planets, darting and jumping specks by the trillions of trillions forming the structures of everything everywhere; atoms structured into crystal or disordered into glass, but always atoms in motion—in constant, quaking, nervous motion. And where the atom dance is frenzied enough, bonds yield and break, and atoms lose their pretense of structure, flowing off into liquid pools or gaseous atmospheres.

Scientific study of this maelstrom of atomic activity has shown that matter has a wave-like nature; but the search for the source of this wave-like nature, however, has found nothing that could serve to underlie these waves—no pond on which there could be rippling waves that could be the particles themselves. All efforts to penetrate to an ultimate reality have yielded two mutually contradictory answers that must both be true: (1) the ultimate reality is that matter consists of particles, not waves of energy; (2) the ultimate reality is that matter consists of waves of energy, not particles!

Usually when science—or philosophy or religion, for that matter—makes statements that are internally inconsistent, reason demands that the claims be judged false. But in this case, the experimental evidence has been overwhelming. Matter presents both a wave-like and a particle-like reality. Despite the clear inconsistency of such an interpretation of the experimental evidence, we are forced to recognize the fundamental truth of this duality, a truth that calls for an entirely new understanding of reality.

THE MATRIX OF REALITY

Resolution of the wave-particle duality problem, when it came, was less a response to Bohr's philosophic conundrum than a response to the need for a more accurate method for handling the problem of atomic structure.

Like the puzzle of wave-particle duality they were to solve, two geniuses, Werner Heisenberg and Erwin Schrödinger, heralded the revolution in scientific thinking that brought quantum mechanics into being with two entirely different solutions that seemed, at first, as impossibly contradictory as the original wave-particle paradox itself. In keeping with the ever-unfolding mystery of reality, their solution to the problem of wave-particle duality has created a new and deeper mystery that has come to be called, cryptically, the measurement problem.

Heisenberg was born on December 5, 1901, in Würzburg, Germany. An interest in Bohr's planetary atom led him to study physics at the University of Göttingen and then, in 1924, drew him to Copenhagen to study with Bohr at the Universitets Institut for Teoretisk Fysik, housed in a weird, rambling building that still suggests the ghosts of its former glory may lurk somewhere within its odd shadowy halls and narrow passageways. In 1925, while recovering from an attack of hay fever on Helgoland, an island in the North Sea, Heisenberg discovered a new and an exact way to find the discrete energy states of atomic systems. He discarded the method that had guided physicists to triumph after triumph for 200 years. He gave up entirely the idea of attempting to explain atomic objects in terms of models—mechanical pictures of objects and their interactions. For Heisenberg, there were only *observations*. What we know are only the events—the outcomes of our experiments that probe that other world, the microworld of the atom, and so our representation of that world should be a mechanics of observations.

Heisenberg noticed that the discoveries of Planck, Einstein, and de Broglie implied a peculiar constraint on our efforts to probe the world of the atom. To predict a particle's path, one must measure both its position now and its velocity now. Of course, when we measure either the position or the speed of an object, we do this by shining a light on the object or by sending some other beam of particles to illuminate its position so that we can get the numbers we need to calculate its future path. When we do this, however, we disturb the object a little. If the object is small—like an atom, an electron, or a molecule—such disturbances may be significant. This fact had long been known to classical physicists, but it was always assumed that a probe that would produce only a negligible disturbance would be used.

Because, as Planck and Einstein had shown, light consists of discrete quanta, the best we can do in making the light dim is to use one photon to determine the position of an object. But even using only one photon, we cannot make measurements more accurate than about 1 wavelength. Remember that light is both particle and wave. To get accuracy in a measurement on an atom's position requires photons with shorter wavelengths (photons with a higher frequency). But Planck and Einstein had shown that higher-frequency photons have higher energies, and a higher energy means that the measurement will disturb the object's motion that much more. The more we try to find just where the particle is by using higher-energy photons, the more we disturb the particle being observed. What's more, de Broglie had shown that what was true for light was also true for matter. Particles also have a wave-like character that limits any efforts to use electrons or other particles to find exact positions and exact velocities. There is an ultimate limit to the accuracy of the data we can hope to get by any means in order to determine the path of an object or to see the inner workings of the tiny components of our world.

There are two lessons here. First, Heisenberg showed that if we take the uncertainty remaining in the most nearly perfect measurement of a particle's position, which we call δx (δ for variation or uncertainty, x for the location of the object), and multiply that by the uncertainty remaining in the momentum of the object, δp (where again δ means variation in the value p, the object's momentum), we obtain a number that is about equal to Planck's constant. More simply,

$$\delta x \delta p \approx \mathrm{h}$$

where $\approx$ means "approximately equal to" and h is Planck's constant. This is Heisenberg's famous uncertainty relation. Nature sets rock-hard limits to our knowledge about the location and motion of things. That is the first lesson.

Heisenberg's uncertainty relation is as widely known as it is misunderstood, however. That measurement disturbs things measured—that the observation of particles, people, and societies disturbs what we observe—was already known, even if perhaps it was never of much concern before Heisenberg. But it is the second lesson, Heisenberg's second discovery about measurement, that is the real bombshell. Heisenberg pointed out that if there is a limit to the precision with which we can know an object's path, then to that extent, *path has no meaning*. We may picture in our mind that a photon goes this way or that, or an electron passes this slit or that, but if our imagination exceeds Heisenberg's limit, we are imagining a fantasy. Objects do not take a path in going from the event at point A to the event at point B, because path itself does not exist.

Of course, when Heisenberg formulated this principle, it was a conjecture calculated and formalized by the equation we gave previously. But it was a conjecture born of the insight of Heisenberg's genius, and it has yielded a wonderful bounty of precise physical theory confirmed by experiment.

What Heisenberg discovered was that the limit to our ability to observe the universe determines the boundaries of reality. Physical reality and observability are tied together. If you and I cannot observe it, it does not exist . . . or is it perhaps, if it exists, it is because you and I observe it?

EVERYWHERE AT ONCE

Heisenberg showed that, contrary to our common-sense conception of physical reality, the idea of path is neither fundamental to nature nor ultimately even a valid concept. It would seem that any effort to analyze the real world out there would compel us to use the idea of path as a basic ingredient of that analysis. When we talk about matter, when we speak of physical reality, the very concepts seem to cry out that one has to go from one place to another; an object must move position by position and thus describe some path. Surely it cannot be everywhere at once, here now and over there in the next moment, completely disjointed from where it was before, or nowhere one moment and back again the next. How can one hope to construct—it might seem, completely from scratch—the whole of physics all over again and achieve a new understanding of the motions of material bodies by throwing out this essential part of our conception of reality?

But Heisenberg did it. In 1925, Heisenberg wrote, "In this paper it will be attempted to secure foundations for a quantum theoretical mechanics which is exclusively based on relations between quantities which in principle are observable." That paper gave us the first formulation of quantum mechanics. In this new formulation, Heisenberg did not work with numbers that represented where an object was at a given time, which in the old classical mechanics would mean assuming that the object was located at a specific place. Instead, he worked with collections of numbers. At any time, the object could be in any place, represented by an infinite collection of numbers that tell us its possible locations. Conversely, if we were to look at any particular place, we might see the object go past at any time. To represent this reality calls for more than just a number. The complete idea of the location of the object calls for a whole array—a *matrix* of numbers. The same goes for an object's velocity or momentum. The object has an array of momenta, and it has an array of locations. In the new quantum mechanics, these matrices replace the old ideas of position and momentum.

But if we forget about "modeling" the atom, which is something that exists only in our imagination anyway, and look instead at the facts of observation, we see not electrons with position and momentum in given orbits but photons that signal the relationship between orbits. What we know of the electron's position is in the jump it makes between orbits. Position actually is a relationship between one orbit and another. We do not know the coordinate x of the electron, but rather its transition: from orbit 5 to orbit 1, from 9 to 3, from 14 to 2. These pairs of numbers represent positions in a matrix. The coordinate *position*, if we are to look only at what we know from our observation of the radiated photon coming from the atom, must be a relationship—a relationship between the set of orbits from which the electron jumped and the set of orbits into which the electron jumped. This gives us an array of numbers; this array is the matrix, and this matrix is the position of the electron in Heisenberg's new mechanics. All of this is true also for the idea of the *velocity* of the electron or the idea of its *momentum*. Momentum now is not an absolute quantity. It has meaning only with respect to a measurement, and the only way to see it is as it is expressed in the emission of a photon. Thus momentum is also an array of numbers that represents possible orbital changes when a photon is emitted.

Now let us see how this Heisenberg idea that there is no path actually works for atoms. To do this, we need to revisit classical physics for a moment. In classical physics, Newton's equations boil down to a prescription that tells us the path of an object. If we ask where an object goes, Newton's equations tell us where, when, and at what speed and what momentum anywhere along the path. That location and that momentum go together to describe a point on the path. If the location is given, Newton's equations enable us to determine the path—and from that the momentum. And if we know the momentum, our knowledge of the path lets us find where the object is.

But in the world of the atom, Heisenberg's uncertainty principle tells us that path does not exist. Given the position, there will be uncertainty about the momentum; given the momentum, there will be uncertainty about the position. Heisenberg therefore proposed that there would be a difference between multiplying x times p and multiplying p times x. We must remember that x and p are no longer just simple numbers for the position and momentum of an object, as Newton had meant them to be, but that they are now arrays of numbers that tie the positions of any beginning orbit of an electron to a final position on another orbit and that tie the beginning momentum to the final momentum.

Heisenberg proposed that the difference between $2\pi x \times p$ and $p \times 2\pi x$ should be Planck's constant, h, the constant connecting the frequency of a photon to its energy.[1] This is the same constant de Broglie used to determine the wavelength of matter-waves, and it is the same number that Heisenberg used to express his uncertainty relation.

One more thing should be mentioned about this new mechanics that Heisenberg proposed. He also said that this difference between $2\pi x \times p$ and $p \times 2\pi x$

would be imaginary. He set the difference equal to *ih*, where *i* is the square root of −1, rather than to Planck's constant as such. The fact that there is an imaginary number in Heisenberg's equation is very important. The significance of its imaginary character may seem obscure at the moment, but we will come back to this later. For now, note that it does not mean the number does not exist, but rather, in a sense, that it exists as a *potentiality* to affect the things we see.

Using this way of understanding the meaning of position, of coordinate locations of objects in the atomic world, and of describing their speeds and motions, Heisenberg was able to reexpress the classical equations of motion and to give them a new meaning in terms of arrays or matrices of possible locations and possible motions. His formulation does not follow the particle from one location to another but, rather, creates an array of values representing all the possible measurement results that might be found when a measurement is carried out on the object. It is as though the way in which an object gets from one place to another, from one energy to another, or from one orientation to another matters much less than the idea that eventually the object will be observed, and that when it is observed, it may be found in any one of a whole collection of possible states. Physics no longer describes where, but the potentiality of where; not how energetic, but the possibilities for an object's energy. In Heisenberg's case, the conception of reality had changed to a picture in which the things that actually are real are the measurement events. Things between observations become a bit fuzzier than they had ever been in physics. In a way, things not seen *now* become "imaginary" potentialities to affect what we will see.

Within months of Heisenberg's discovery, Erwin Schrödinger came up with a totally different solution to the problem of reformulating all of physics so that it would account for the world of the atom. Schrödinger was born and reared in that cauldron of twentieth-century thinking, Vienna, Austria, in 1887. He began studying physics at the University of Vienna in 1906 and, after receiving his doctorate, remained until 1921, when he moved to Zurich. There, at the age of 39, he produced his great new formulation of physics. To note that his discovery would earn him a Nobel Prize risks trivializing the enormity of his achievement. Yet his new formulation was a strange child of his mind; it was something about which he would later say, "I don't like it, and I'm sorry I ever had anything to do with it!" We will see why shortly.

Where Heisenberg used matrices—whole arrays of numbers to represent the positions and motions of an atomic particle—Schrödinger took de Broglie's conception of matter as waves and developed an equation that would describe matter exactly in terms of these waves. Where Heisenberg gave us matrix mechanics, Schrödinger gave us wave mechanics. The paradox of wave-particle duality gave rise to two paradoxical formulations of physics: one in which the discreteness of particles extended even into the description of space and motion as arrays of discrete numbers, and one in which everything was seen as waves. Incredibly, both gave the same accurate answers.

If both Heisenberg's and Schrödinger's formulations give the same correct answers, then somehow, underneath it all, they must merely be different mathematical attire dressing up the same reality that each formulation only partly reveals. A year after Schrödinger developed his equation for wave mechanics, he showed that his theory and Heisenberg's matrix mechanics are basically the same. They are two special ways of talking about a much more general and a much more abstract picture of reality that is described by something called *Hilbert space*, a world that is really an infinity of imaginary worlds!

SCHRÖDINGER'S PICTURE

Schrödinger's equation also describes the diminutive world of the atom and the things of that remote microcosm. But it describes that world in a fashion that is totally different from either that of classical physics or that of Heisenberg. Whereas Heisenberg dispensed with the classical concepts that lay beyond what could be measured, the Schrödinger equation does the opposite. It introduces something entirely new. Schrödinger's equation introduces something that is a part of everything in this world, but something we have never seen or even suspected might exist. Schrödinger was looking for a wave motion to describe matter in a way that would be similar to the de Broglie matter waves. To do this, he introduced the idea of the height Ψ of this wave, thinking that this is what matter is really made out of—de Broglie waves. What he had really found was something quite different.

Schrödinger found the wave equation describing the amplitude of matter waves, and he showed that this new equation agrees with Heisenberg's equations and with experiment. But if Schrödinger can be likened to a Columbus sailing a sea of mathematical physics, it was someone else, Max Born, who figured out where Schrödinger had gone and what he had discovered. Born showed that these matter waves are actually waves of *probability*. Schrödinger's wave equation is a formula that enables one to find the amplitude of these probability waves. It does not give probabilities directly. This, as we will see later, is of the most fundamental significance in our effort to discover the nature of reality.

THE IMAGINARY WORLD OF THE QUANTUM

Actually, wave equations in physics have always turned out to be formulas that are expressed in terms of the amplitude of the waves involved. For example, the equation for waves on the surface of a pond is expressed in terms of the height of the wave at any point above the average level of the pond surface. Moreover, if one squares the amplitude and multiplies by some appropriate juggle factors, one obtains a number that gives the quantity of energy carried by the wave at that point. Adding this up for every point on the surface (that is, integrating the function over the surface) yields the total energy in the wave.

Maxwell's equations—the ones Einstein judged to be so fundamental that he changed space and time to keep the speed of light constant—can also be expressed as wave equations, one for the electric field and one for the magnetic field. These are the equations for light waves, radio waves, microwaves, and so on. In these wave equations, the electric and magnetic fields appear as wave amplitudes. Again, if at any point we square the electric field and add it to the square of the magnetic field at that point, and multiply by the appropriate juggle factors, then we will obtain the energy density at that point in space.

But if this electromagnetic wave actually contained only one photon, then the energy for the whole space would be the energy carried by this one photon. If we were to divide the answer by this energy, we would obtain an answer of unity, i.e., just the chance that the photon is somewhere in all space. Thus the square of the electric and magnetic fields at any specific point simply gives us the probability of finding the photon at that particular point. As a result, the electric and magnetic fields for a photon are a kind of probability amplitude, just like the ones we use in quantum mechanics.

That is the meaning of this new quantity that occurs in quantum mechanics, this matter-wave that physicists call Ψ. But there are further things to discover about this strange new *creature*. In the case of the amplitude of a wave on the surface of a pond, one can actually measure the height of the wave. That is to say, one can directly measure the strength of the electric and magnetic fields in an electromagnetic wave (at least for low-frequency waves containing large numbers of photons). These are real quantities as we mean the idea mathematically. But the Schrödinger wave equation contains an imaginary factor, *i*, which is the imaginary number $\sqrt{-1}$. This *i* is a number that does not exist but that, multiplied by itself, gives –1! Thus Ψ will always turn out to contain terms that are imaginary—that is, terms that are *complex* functions (or quantities). Now the fact that Ψ contains imaginary terms does not really mean that it does not exist. We must be careful on this point. The complex character of the wave function in Schrödinger's wave equation means that what is there is in a sense hidden from us. It means that what is there is a kind of potentiality.

Let us say I have $1 million in the bank and I wish to build a store that costs $100 per square foot. For simplicity, let us talk about stores that must be as deep—to store inventory, say—as they are broad across the front. Obviously, then, my $1 million is enough to buy a store with 100 feet of frontage. Now let us say that I had a bad week at the stock exchange and lost $250,000. That would still leave enough money for a store 87 feet wide. But what about the lost money? What is the length of the store front that I lost? If I had that $250,000, I could use it to build a store with a 50-foot front. But because this is a money deficit, I should talk about the imaginary 50-foot store front—that is a 50*i*-foot store front. I will never see a 50*i*-foot store, of course, but that lack of money will affect the look and size of the store that I do finally build. Tongue in cheek, a mathematician might say that what I have left after losing the quarter million is a building 100 + 50*i*-feet-wide.

In the same way, incredibly, matter can have an imaginary existence. It can exist as though money were short at the bank. There is an imaginary aspect to these quantum mechanical waves that we must remember, but we must also be careful when we speak of these imaginary numbers as not real. Both mathematicians and physicists are inclined to view such an understanding as naive. The mathematician's "imaginary" is not the lay person's imaginary. Still, the observable quantities—those things that we can measure or see and that are described by wave functions—are always obtained from that function by a procedure that turns these imaginary numbers into real numbers. We can talk about the 100+50*i*-foot-wide building, but when we build, we will be putting up a 7500-square-foot building. It is that actual floor space that we finally build.

◈

It is easy to look at this emptiness, this wisp of motion in a relative space, and understand how memories vanish with the years. There is so little of us even now. We are such small creations, and there is so much emptiness ahead of us, so much emptiness and so many years that it is surprising that I hear now any echo at all of her voice, and it is stranger still that something so slight as feeling should have endured.

She had pride in her name, pride that her name was special. I remember her showing me how to spell her name, with an "e"—Merilyn, not Marilyn. She was a special creation, unique; her lips were full and red, the bright red of the '50s. She did not like seamless stockings. They lacked that touch of excitement. She wore stockings with dark seams that had to be kept straight up the back of her legs, like racing stripes painted on a fast car piping out its sexiest curves. She was alive, vital, vibrant. There was something that the rest of creation, whether stars or atoms, could not be, something that once lost could never be made again.

She loved her mother and respected her stepfather. But she kept her real father's name. She remained Merilyn Ann Zehnder, honoring and cherishing the memory of her father. She kept his name to keep his memory alive, just as I try to keep something of her alive through my memory of her. Yet where is the reality of any of this?

Ralph, her stepfather, carries a photograph of her in his wallet with him to this day. Ralph is old now. He has been retired for many years. But Merilyn is still 16. She still has a gardenia pinned in her brunette hair; the white petals of that flower complement her own petal-like skin. The gardenia is so delicate that if one touches a petal, the petal turns brown. I remember her skin with brown blotches as she lay dying in the hospital. Merilyn and the gardenia, two southern flowers, delicate and beautiful—memories of long ago.

What resemblance does this description of physical reality bear to the woman I knew then? My feelings are strained by the thought that these strewn speckles of atomic matter—this dust of particles fixed in a void—that this empty structure could ever be all that there was to the verve of her glance. What relationship can there be between this story of atoms giggling at 10^{15} hertz, in the etherless void of twisted time and space, and her languid smile, her breath breathed near me, her lips touching, slowly touching my own? If these are two different stories—the one of atoms and particles, the other of this image of her—one must be a fable.

And yet, there is another side to the story, one that is filled with the excitement of ideas themselves. Ideas that let us piece together what our reality is through layer after layer of mechanism. It is this mechanism that shows the workings of atoms that, when put together, make up bits of organic macromolecules, in turn combined into the materials that form flesh, bone, muscle, nerve, and skin. It is this mechanism that brings movement to bodies enlivened by the neural networks and the brain. All these facts of science let us understand how these things work.

As much as it seems that these two worlds cannot be bridged, this story of science works too well at each level for it not to be true. Each mechanism at each level of the story folds so well into the next. These are the stories of atoms formed into molecules, of molecular reactions and conformational changes, of chemical kinetics forming, shaping, and moving the structures of the cells. These are the stories of how photons, encountering the scarce electrons in atomic structures, reradiate light that enters my eye and cascades into a neural response that travels down the optic fibers into my brain, sending impulses that represent the shapes and patterns of her body. And then these impulses become a memory. All of this works so well, yet why is it that I feel that I had to bring the existence of those atoms into some contact with my own brain's activity for me to be able to justify that those atoms and Merilyn could have ever been part of the same story? The lesson is that the atoms themselves never form tissue, never form the woman. The atoms are atoms. They are not parts in themselves. Only because I see the woman and feel her, and only because I know of the atoms, are they the parts of something that I build in my mind and can imagine is she.

I remember one day, long ago. I rather nervously rang the bell. I always rather nervously rang the bell. I never knew whether her mother or stepfather might answer the door, and I always felt more as though I had come to steal something than simply to visit their daughter. Of course, I had. I had come to steal her away, and there was no way to hide it. I was sure they must feel I was the thief in the night, as far as their daughter was concerned. Her stepfather was a tall, strong steel worker. Her mother was small, like Merilyn, and except for their difference in age might have been her twin. Both her parents were always pleasant to me, but I knew it was a veneer, a thin, brittle, transparent sheet of mica.

I rang the bell and Merilyn opened the door. Quietly I slipped in. "Where's your mother?"

"You want to see my mother?" she asked. "No . . . just wondering." I always started by checking the logistics; I always hoped we would be alone, of course, but that was unlikely. I just wanted to know through which door one of them might suddenly appear. Whichever direction it was, I would always listen intently for the first sound of footsteps, any unexpected squeak of the flooring.

"She's in the kitchen making us some cookies. You want some? We can have some later, if you want."

This concession to southern hospitality did not alter the terms of my visit. At best, I could feel only about as welcome as the Japanese fleet off Oahu.

I was shown into the living room as usual. I remember the thick carpet covering the whole floor. My parents' house was carpeted in the living room too, but this carpet seemed to quiet the whole world. It was heavy and soft. The living room was perhaps 12 feet wide and 20 feet long across the right front of the house. Merilyn usually showed me to an upholstered chair just inside the living room, a chair reserved for the man of the house. There was a picture window to my left, a fireplace that was never used on my right opposite the window, and, always beside me, Merilyn, who would sit on the arm of the chair, on the floor in front of me, or with me in the chair. When I knew her better, there was always a kiss for me when I came in, and perhaps a second. I would usually go over to that man-of-the-house chair, sit down, and motion for her to join me. She would sit on the arm of the chair, lean across in front of me, as if about to kiss me, pause, and look at me.

"Did you miss me?" I would ask.

"Did you miss me?" she would demand.

"I asked you first. Did you miss me?"

"Yes." She was soft, warm toward me. For her, the games quickly ended.

"I missed you, too," I would say.

"How much?"

Once I held up my hand, I remember, thumb and forefinger showing a gap of about a quarter of an inch. "That much," I said.

"That's not much," she answered.

I grabbed her shoulders, pulled her close, and said, "There is an infinity of points between, and space enough to slip the sun through."

We were, after all, foolishly young. This was the best my youthful mind, adrift in its newfound land of love, and poetry, and science could do—a quote from something I had just read in *Mathematics for the Millions*!

"Do you like Ceil?" she asked.

"She's OK, I guess. I don't know. Why?"

"I think she's got a crush on you." Ceil was her friend, but the statement was couched as if she were an expendable commodity.

"I don't like Ceil. She doesn't mean anything to me."

"She's just been coming by a lot more since you've been around. I think she's got a crush on you." She smiled, pleased as a woman will be when she knows she has someone another woman would like to have.

"I don't care. She doesn't mean anything to me." I didn't say anything more. I pulled her an inch closer and kissed her.

ORTHOGONALITY AND OTHER THINGS THAT GO BUMP IN THE NIGHT

Let us see how this quantity Ψ that comes out of Schrödinger's theory is used to obtain actual probabilities. To obtain the probability that an electron in an atom can be found in a particular volume located, for example, close to the nucleus of the atom, we first solve the Schrödinger equation for the problem. That is, we describe mathematically the collection of particles and forces that interact in the atom on the electron. This is put into the Schrödinger equation. Then we carry out the math needed to obtain Ψ free and clear (that is, to get the results in the form Ψ = whatever). If we specify a particular location in space and put the numbers for that location into the formula for Schrödinger's wave amplitude, Ψ, then it becomes simply a number, the value for the wave function at that location. But that number is a complex number made up of a real part, R, and an imaginary part, iS (where again I represents the imaginary factor). Thus the value of Ψ is $R + iS$ at the specified location.

Because Ψ is a complex number, it is awkward and difficult to understand. It is like that store we built (in the preceding section) that had $250,000 of lost frontage. Just as there, where it made more sense when we talked about the actual square feet of floor area in the store, here too Ψ has more meaning in terms of numbers and in terms of what it represents if we talk about the quantity when it is squared. Squaring Ψ, if it is done right, gets rid of the imaginary part of the number. Mathematicians and physicists show that they know how to square the thing by writing $|\Psi|^2$ for this square.[2] When we square the magnitude of Ψ, we get something useful. It is called the probability density, and it enables us to find the probability point by point in the region around the atom. Thus the probability, p, of finding the electron in a small volume near the nucleus of the atom will be the square of Ψ multiplied by the volume of space we are talking about.

To make it clear that the volume has to be kept very small—or the value of Ψ will change from one side of the volume to the other—we write dv rather than just v for the volume. The expression for the probability of finding the electron in the volume dv is then just $|\Psi|^2 dv$. If we do all the algebra,[3] we get $(R^2–S^2)dv$ for the probability of finding the particle in the volume dv. If we want to get the probability for finding the electron in a larger volume, then we have to add up all the little pieces for all the little dv volumes throughout the larger volume we are interested in.

Now let us see what we have. Here R^2 is like the number of square feet that the $1 million could buy, while the $–S^2$ is the $250,000 in store space that was lost. When we get down to any real business about the electron, we have to talk about this quantity $(R^2–S^2)dv$, which says that the probability that we will find the electron in the little volume of space, dv, is obtained from the squares of the real and the imaginary parts of the wave function.[4] It is as though we had a new kind of electromagnetic field in which the magnetic part competes with the electric contribution rather than adding to it

But this probability wave function Ψ is stranger even than merely being made out of imaginary parts. It is not only a probability wave, but it is, at the same time, the complete representation of any object or physical system for which it has been obtained as the solution to the Schrödinger equation. This latter fact was originally postulated by Bohr, and it has been exhaustively verified. Ψ turns out to describe not only the probability for something happening but it describes also exactly what can happen. Not only does it describe the electron probability distribution in the atom, it provides all the information for all the possible things that can happen—all the possible configurations, all the possible orbits, all the information needed to find the energy of the electron, or the momentum, or how it will interact with a photon that strikes the atom.

How is this all contained in one function? It happens that the Schrödinger equation yields solutions of a very special type. These solutions can be expressed as a series of functions—that is, a series of pictures of the atom that can be written in this form:

$$\Psi = \psi_1 + \psi_2 + \psi_3 + \dots$$

Each of these quantities (ψ_1, ψ_2, ψ_3, etc.) is like a photograph that pictures each possible state.[5] Each is a mathematical picture of the electron in its particular "orbit" or special configuration in the atom, each with its own particular energy for that configuration. And like a photograph, which can be under- or overexposed, each of these quantities can show the probability for its occurrence by means of a multiplier, a coefficient, that is smaller or larger depending on the probability for that picture becoming the reality we find when we look.[6]

All of this can be found by solving Schrödinger's equation. Thus the whole procedure for finding the permitted configurations that the atom or the system can be in, as well as the chance for finding it that way, is given in a straightforward way by the Schrödinger equation. The wave function *is* the description of all the possibilities—all the allowed configurations of that atom simultaneously. This wave function of Schrödinger's equation, containing these simultaneous pictures of all possible configurations, gives us what is called a *linear superposition of states*. This means, first of all, that these individual states, or pictures, of the atom are simply added together and, second, that this superposition still manages to keep the separate parts separate. This is the property called *orthogonality*. It means that at the same time the ψ_1 picture exists as a true picture of the atom, the ψ_2 picture also exists along with it, as does the ψ_3 picture, and as do all the other equally valid pictures of the possible configurations of the atom.

To understand this better, imagine that we have a transparent cube, and on each of the three sides of one corner we paste a photograph of each of these three ψ_1, ψ_2, and ψ_3 pictures. By turning the cube, we can see front-on one, but only one, of the three pictures fully without distortion. That is what the term *orthogonality* means. If we calculate the extent to which the atom can be found to be in both the first configuration and, at the same time, in the second configuration, we get zero. These two states are "perpendicular," or orthogonal, to one another. The object can never actually be seen in both states, but it is in both as a potentiality. If on measurement, however, it is found to be in the first, then the second will become just a bump in the night that never was!

Two more points and we will be done. In the example of the cube with pictures pasted on, we can do this with only three pictures. That is because we live in a world limited to three dimensions. Using mathematics, it is possible to do this with more pictures and even with an infinite number of pictures all at once. Also, we have to note that when we turn the cube slightly, we can see two or even all three of the pictures fairly well. But if instead the cube were under the surface of some murky water, then turning the cube could bring only one of the pictures up even with the water's surface. In that case, we would not see the picture until it was even with the surface, and the other pictures would not be visible at all. That is how it really works. That murky water world is like the imaginary quantity that appears in the mathematical expression. Rotating the cube is the same as rotating the mathematical function out of the imaginary space. It lets the function become real so that we can see an image. This is something like the process of measurement, or of observation—something we will be particularly concerned with later on.

To sum this all up, the function Ψ that we got from the Schrödinger equation packs in a whole lot of information—information that amounts to an infinite number of pictures of what the atom really *is* all at once.

THE HOP-POP-JUMP QUANTUM WORLD

Matter is not really both particle and wave, but rather discrete packages of energy that dart from place to place in a frenzy of quantum jumps, that ebb and flow in waves of chance. It is a world in which nothing stays long where it should be but only stays where it could be. The atoms dance and the electrons hop, zipping about from the wildly random thermal commotion and from the blurry speed with which probability waves speckle the tiny electrons around their nuclei, painting fleeting mazes into solid things like you and me—and her.

And yet none of this is what is there. All of this motion is frozen. All this darting and all these quantum jumps exist only as potentialities. The jumps and darts happen when measurements take place, when we observe, and when things interact with other things. We cannot view the probability waves as arising because real discrete particles of matter in precise locations with exact speeds dart about so fast or so discontinuously that we cannot follow their true paths. They have no path. They have no exact place and no exact momentum. They exist as potentialities at all these places at once: a frozen static frenzy, the silent excitement of nature. Particles do exist, but their states are represented by waves of probabilities. That is the resolution of the wave-particle duality paradox. That is the solution that cloaks still more mysteries than we ever thought might hide there.

6

Hunt for the Tin Man's Heart

Phaedrus was a master with this knife . . . . With a single stroke of analytic thought he split the whole world into parts of his own choosing, split the parts and split the fragments of the parts, finer and finer until he had reduced it to what he wanted it to be.

—Robert M. Pirsig,
Zen and the Art of Motorcycle Maintenance

I look for reality and try to tell you what I see there. The words and pictures come out of the past. My mind goes back as I drive past the creek where as a child of 5 or 6, I was the expendable "test pilot" to try a raft the older children had made from scrap lumber and empty oil cans. I drive over that culvert built at the edge of town in the '20s, past a part of my life—in half a second—a place that was once a whole world surrounded by thick briars and stubby green trees with gravel beneath the spring waters where crayfish and red, white, and blue minnows darted and hid. There was a world of mystery and wonder down that square concrete corridor beneath the street—for those who dared. In that half-second, I feel like many people, remembering other lives, strolling down corridors, looking through doorways into other worlds than my world here and now. And I hear a voice from the past calling my name: "Bookie, come on in now," my mother who died a decade ago says to me. "Come on in. It's time for your supper." I turn again, and I am back here with you searching for reality. Searching and thinking about the things that physics has found and what they mean. What they tell us about what reality is.

The Heisenberg and Schrödinger equations have led us to a wondrous understanding of reality, far more penetrating than had been revealed by classical physics. It is safe to say now that nothing that we see in this world lies outside the breadth and scope of quantum mechanics. With this analytical tool, we can examine the detailed workings of the atoms, determining how they are put together, how they radiate light, and how they combine into molecules. We can put these atoms and molecules together so that they make the materials of our world and, in doing this, understand what makes their strength and their color; know why metals feel cool, why the stars shine, and why glass is brittle, the oceans wet, and the sky blue. We use this knowledge of quantum mechanics to make computer parts that automate our work, to create lasers and optical fibers that carry calls from door to door, and to develop all the wonderful superconductor technology. Quantum mechanics explains all these things *and* all the things classical physics had explained before.

Quantum mechanics is being used to design new drugs to heal and medicate our bodies. It is being used to understand the delicate machinery of life, of reproduction, of evolution. Its offspring fills our newspapers with tales of nuclear power, worries of nuclear war, designs for weapons for future wars, and defenses for future peace. Its products, such as magnetic resonance imaging (MRI) and positron emission tomography (PET) scans, let us peek into the structural and chemical make-up of the living tissues of the body, and with the aid of the scanning tunneling microscope (STM), we can see individual atoms. The consequences of quantum mechanics—the fruits, the benefits, and the liabilities—permeate our lives. The results are all around us. Because quantum mechanics not only describes the atomic world but also embraces the equations of classical physics, Newton's equations, it describes everything in our world. It seems it has all the answers. We need only enumerate the basic pieces of matter, write down the forces acting on them, and turn the crank. Out will come the answer to every question about the nature of our reality. All we have to do now is find out what the basic pieces of matter are, and the equations of quantum mechanics will describe all the patterns in nature. That is the way it looks, and that is the strategy of modern physics.

VIOLETS BY THE YELLOW BRICK ROAD

Quantum mechanics works in a strange way. The way it should work is not the way it does work. It is not set up to describe how things actually are but to work in terms of the way we think about these things. To solve the problem of the hydrogen atom, we pretend that the hydrogen atom really looks the way Bohr described it. We say that we have a nucleus consisting of a single positively charged proton and a point-like, negatively charged electron at some position near the proton moving in some direction that is specified by a set of coordinates. We add to this an expression representing the force acting between the charges on the proton and the electron (actually a term that represents the potential energy in the electric field that surrounds the proton), and then we say that this result tells us the energy of the whole thing as though we were describing the motion of a planet moving around the sun. When this very classical picture of the parts and pieces of the atom is complete, we make some prescription-like replacements. The momentum is replaced by a term that spells out how that probability amplitude changes—its rate of change as we move from place to place; the potential energy term representing the forces has to be multiplied by that probability wave height; and the energy is replaced by a quantity that says how rapidly the wave varies in time. Then we turn a big mathematical crank, and this strangely distorted classical picture begins to spit out a series of quantum pictures of the hydrogen atom.

Each picture has its own pattern that describes not where the electron is, but where it might be. The simplest of these arrangements is just a spherically symmetric puffball, a dandelion pappus of a pattern; other patterns look like very symmetric flowers with petals that reach out in every direction. There is a whole series of these electron probability patterns. With each successive picture in this series of probability patterns, the electron is extended into more and more probability petals that reach farther away from its nucleus so that the atom has a higher energy. When an atom is in one of these higher-energy levels, we say that it is in an excited state.

That, then, is how quantum mechanics describes the hydrogen atom, and the description of all other atoms turns out to follow nearly the same procedure, but with a rather peculiar and quite significant exception known as the Pauli exclusion principle.

This principle, named for Wolfgang Pauli, turns out to have a very basic role in determining the way the world works and looks. It causes the atoms of different elements to take on the various flower-like forms that the Schrödinger equation describes as possibilities, rather than all of the atoms looking the same, as would otherwise be the case. In the hydrogen atom in its ground state—the way it ordinarily exists—the one electron sits in the lowest-energy *shell*, a spherical blob of uncertainty surrounding the atom's nucleus. In helium, the next element up Mendeleev's atomic ladder, there are two positively charged protons in its nucleus, along with two uncharged particles (called neutrons) and two negatively charged electrons surrounding the nucleus. Although there is a repulsive force between the two electrons because of their electric charge, they still both squat down into the same ground-state configuration of a ball-like fuzziness around the helium nucleus. Because this nucleus with its two protons has more positive charge on it than the hydrogen atom with its single proton, that fuzzy helium ball of electronic uncertainty is even more tightly held and smaller than the hydrogen atom.

Lithium has three protons in the nucleus tugging on three electrons. Something new goes on there. The first two electrons are in an even tighter, hydrogen-like fuzzy ball around the nucleus, but the third electron cannot drop down near to the nucleus with the first two. The first two electrons are in the lowest atomic shell, but the third electron occupies an electronic shell outside of these. It occupies a fuzzy double-lobed blob of an orbit. With each step up the chart of the elements, we add one proton to the nucleus and one electron to one of the possible shells. As we do this, we find that the electrons fill up each of these possible fuzzy orbits *by pairs*. Once a pair of electrons has entered a particular level, that level is sealed off to any more electrons. This is due to the Pauli exclusion principle. The reason the electrons can seemingly go into the same state in pairs is that electrons have a property called spin, which is something like the spin of a top, but as with everything else in this topsy-turvy quantum world, not quite.

When we spin a top, we are free to choose any direction we wish for its axis to point. Electrons with their spin are rather different little beasts. The electron has a spin, it always has a spin, and it always has only a single "speed" for that spin. It can spin neither faster nor slower than this one speed. Moreover, when we set up an apparatus for measuring that spin along any arbitrarily chosen direction, we find that only two possible values can occur: The spin can point "up" along the instrument's axis or "down" along that axis! It spins with only that one possible speed, pointing only up or down along a measurement axis that *we* have chosen arbitrarily. Why does the electron's spin depend on the way we choose to measure it? That is another one of those strange quantum mechanical puzzles—and another clue to what reality actually is.

Because the electrons have two possible spin states, two electrons (one with spin-up, one with spin-down) can reside in each possible level in the atom. Thus the shells build up two by two to form larger atoms.

Much later on, we will have more to say about the strange nature of electronic spin, and we will show how its odd nature is much like the very odd nature of something one might not expect to be any puzzle at all—the way light is filtered by polaroid lenses. It is something that is very commonplace and yet it will tear away the picture of reality we have always accepted. We will come to that much later. Right now we must look at another principle of quantum mechanics that is a kind of first cousin to Pauli's principle. This is the principle of *indistinguishability*. Granted, it sounds like nomenclature for service decorations presented to soldiers on the losing side, but it accurately describes the physics of some atoms.

Originally, an Indian physicist by the name of Satyendra Nath Bose advanced the idea of indistinguishability in order to understand certain things about the capacity of objects to hold heat. Einstein extended Bose's idea to establish what is now known as Bose-Einstein statistics; Enrico Fermi and Paul Adrien Maurice Dirac extended Bose's idea and Pauli's exclusion principle to establish what is now known as Fermi-Dirac statistics. Among other things, Bose-Einstein statistics is the first step in explaining how superconductors allow an electric current to flow forever with no loss. Fermi-Dirac statistics explains why metals feel cold. Both of these ways of understanding how matter behaves depend on the idea of indistinguishability. This principle says that two identical particles (for example, two electrons or two helium atoms) have no separate identity. There is no way in which one can place a sign on one of these particles in order to distinguish it from another of the same type. One cannot put a sign on an electron that says, "Richard" or "Joe."

To see what this means, let us say that John Smith and his wife Jane, who own a house in the city, decide to buy a house away from town by the sea. Now there are four ways in which John and Jane can occupy their two houses. John and Jane can both stay in their city house; they can both stay in their country house; John can stay in the city house while Jane is at the country house; or Jane can go to the city while John is off by the sea. Four ways. Now if instead these houses were owned by Jane and her identical twin sister Jill, there would still be four arrangements, but we might be hard pressed to distinguish more than three of them. We would have trouble telling for sure whether Jane was in the city and Jill in the country, or if Jill was in town with Jane by the sea. But the reality would remain that there would be four possible arrangements anyway.

Where helium atoms go and what they do when put in a bottle—and where electrons go inside a piece of metal as it conducts electricity—is a little like this business of Jane and Jill. These things can occupy different places, and as the heat is turned up on them, they switch around faster and faster. The fewer the number of places for the particles to go, the faster the temperature rises and the sooner we notice the effects of the heat on the object. Quantum mechanics tells us just how materials will behave when heated, provided we take this fact of indistinguishability into account. Because any two particles of the same type are indistinguishable to us, switching two of them around does not even exist! Matter behaves as though the fact that we cannot tell these two arrangements apart means that they cannot make such a switch. As the object is heated, it goes right past such arrangements in the rapid switching around of the particle positions in the object. Oddly, the fact that we cannot distinguish these states apart means that such things do not even exist.

Why should the observability of such differences matter so much? This is another of those mysteries about quantum mechanics and about the nature of reality. It has become exceedingly clear that observability of differences matters and that it describes something vital about the nature of reality, yet it seems somehow not to fit our understanding of what this world is all about. It paints a picture of a world that is less objectively real than we usually believe it to be. But quantum mechanics describes reality—in the heart of an atom and in our own macroworld. If it does not make sense, then that is because we do not understand our world quite so clearly as we had thought. Why should it matter to nature whether we can distinguish two particles apart? Why should observability matter so much? The questions hint at the ultimate answer to this riddle.

There is another oddity in physics that we have to discuss while we are on the subject of the idiosyncrasies of quantum mechanics: something known as quantum mechanical tunneling. Let us talk about this phenomenon of tunneling in terms of baseball, even though "we all know" that quantum mechanics plays its role only in the world of the atom. If a batter hits the baseball hard enough and high enough, it will go over the outfield fence some 400 feet away, and, well, that's a home run. If the batter doesn't hit the ball quite so hard, or quite so high, then the ball may go out and hit the fence, but it won't go out of the park and it won't be an automatic home run.

But in the world of quantum mechanics, sometimes a ball that was not hit hard enough to go over the fence just rolls up to the fence . . . and keeps on rolling. Sometimes it just keeps on rolling beyond the fence and out of the ball park. In this quantum world, the ball does not jump over the fence. It does not knock a hole in the fence. It does not squirt between the boards or go under the fence, and it never goes inside the fence. It simply rolls up to the fence and then rolls on outside the fence.

It is important to understand that this now-you-see-it-now-you-don't quirk of atomic reality is not a magic trick but a part of the way nature is. This phenomenon of tunneling is so often met in quantum mechanics that it is in essence a hallmark of the quantum domain. Among the places where one commonly encounters tunneling effects is in natural radioactive decay in atoms—of uranium, for example. There the nucleus contains almost a sea of frothing particles held inside by powerful nuclear forces. None of these particles has the energy to overcome this nuclear force in order to escape, and yet occasionally an alpha particle (the nucleus of a helium atom) comes barreling out anyway. It has tunneled through the outfield fence for a home run. There is another place we will meet this business of tunneling. We will come back to this in a later chapter when we talk about the way the brain works and about what mind actually is. There we show how our brains use this phenomenology of tunneling effects in synapses to carry out the vastly complex activities that create our thoughts.

We have talked about the accomplishments of quantum mechanics, and we have talked about some of the puzzles with which quantum mechanics leaves us. These puzzles define something of the task we face in our search for reality, but we should not misunderstand what these puzzles mean. The existence of these puzzles does not diminish one whit the vast power of quantum theory to explain our world. Quantum mechanics is a firm platform for us to stand on as we search for the nature of reality. Its puzzles are but clues to discoveries that are waiting for us.

Throughout so much of life, nothing really ever happens. Looking through the pages of my old diary, I see the slow pace of that southern upbringing. Days were whiled away with boring and pointless activities, or so it seemed. Looking back, knowing the beginning and knowing the end, I can see that in the meaningless details of life a larger fabric was being woven.

"Monday–another week." My diary lifts a corner on another long lost memory. "No homeroom teacher today–American History, Mrs. Borders, debating class"—nothing happened. Then a small adventure: Going for my motor vehicle operator's license, the rite of passage.

There was one problem: My parents had not exactly been apprised of my intentions. I was rushing off to carry out a mission demanded by the circumstances of the times and of youthfulness. And all the better, I had an accomplice—Merilyn. As soon as school was out, Merilyn and I took off for the Medical Center on Eighth Avenue South, where my mother worked and, more importantly, where the family automobile waited.

Absconding with the family vehicle was easy. No one stood guard. No one noticed the deed. Quickly, and with utmost skill, the vehicle was started and moved from its place of parking. Expertly we powered the 4,000-pound, four-door Pontiac Grand something-or-other onto the public streets and made our way—quite illegally—to Sixth Avenue North, where I let Merilyn out to wait safely in the Birmingham Public Library. She was, after all, a woman. I was out to do man's business, to test my steel before the civil authorities. On Seventh Avenue, I parallel-parked for the first time directly across the street from the Motor Vehicle Registration office. From my vantage point, I could see through the large storefront-style windows into the interior of the Driver's Testing Division.

The men in blue uniforms looked bigger than they had when I scouted out the place. One of them looked up. He looked directly at me. Yes, I was sure of it. Panic began to set in. I thought, °°The line, it's too long. Mother will come out before I can finish. She'll find the car missing—stolen! She might call the police. They're all looking at *me*.°°

I sped off. Merilyn was waiting. "The line was too long," I blurted. "I'll have to try tomorrow." We drove long terrible miles back to Eighth Avenue South. Merilyn must have sensed the mortification I felt. She saw me in this moment of ignominy, in this my *Farewell to Arms*, my manhood not won. She knew I had chickened out. I pressed harder on the gas pedal.

And then we were there. As I turned into the parking lot, again, terror! Someone had taken my mother's parking space. Panic raced through me. My heart pumped audibly. Hairs on my head trembled. Fifteen minutes to five. My mother would be coming out at five. It started raining. We waited.

Then, as if it were a sign from God, the owner of the car emerged and vacated the space. I parallel-parked for the second time in my life. Perfection. Inwardly I smiled. I had recovered that measure of lost esteem. There was now a new feeling that I felt, a feeling of self-assuredness, of, yes, of maturity. Merilyn and I looked at each other and giggled.

WHEELS WITHIN WHEELS

By 1912, physicists had finally proved the existence of the atom, which had been suspected and speculated about since the days of Democritus. By 1920, physicists had figured out the probable structure not only of the atom but also of its nucleus. The simplest atom, hydrogen, has a single particle, the proton, for its nucleus. Other atoms consist of a nucleus made up of a collection of protons that are equal in number to the electrons in "orbits" around the nucleus, so that the negative electric charge of these electrons exactly balances the positive charge of the protons. But except for hydrogen, this picture leaves about half the mass of the atom unaccounted for. Physicists reasoned there had to be another kind of particle, a neutral particle, in the nucleus along with the protons. Rutherford, who had proved that the atom consists of a massive, positively charged core surrounded by electrons, searched in vain for this neutral particle. Finally, in 1932, James Chadwick, who had worked with Rutherford in the Cavendish Laboratory in Cambridge, England, proved the existence of these "neutrons." He was able to show that the sought-for neutrons had been produced 2 years earlier in a series of puzzling experiments conducted by Bothe at the Kaiser Wilhelm Institute in Berlin. The new particles had just the right mass to explain the structure of the atomic nucleus.

The result of this was what finally seemed to be a complete picture of matter as made up entirely of three elementary particles—protons, neutrons, and electrons—from which all ponderable objects are formed. These three particles made possible an explanation of all the 92 elements that occur naturally in nature. But there was still a missing ingredient: "What holds the nucleus itself together?" There must be another force that binds all the protons and neutrons into that very tight nuclear ball. Otherwise, what is to prevent all those nuclei, with all their mutually repelling protons, from flying apart in an instant? There must be a nuclear force.

The search was then on to find the secret of the strong force within the atomic nucleus. This new kind of force would have to be totally different from either the gravitational or electromagnetic force. This new force would also have to be far stronger than either of these other well-understood forces, and yet, unlike these other forces, the nuclear force would be something totally undetectable outside the nucleus. In other words, it has an incredibly short range. In many ways, the strong force is more like a glue that sticks things together than a force. It acts powerfully up close, but falls off exponentially as the particles are separated. The force falls off so fast that it is not possible to do conventional experiments to find it. We cannot very well hold two neutrons a trillionth of an inch apart and measure the force between them. Instead, perhaps the thing to do is to look for a radiation caused by this force acting between two particles as they collide—in other words, look for the "photons" of this strong nuclear force. What was needed was some idea as to what to look for. This is where Hideki Yukawa entered with an idea that was to entirely change the way physicists conceive of the basic forces of nature.

Yukawa, while working toward his doctorate at Osaka University in 1935, proposed that the strong nuclear force was carried by a particle having a ponderable mass—that is, that the particle produced by disturbing the nuclear force field would be more like ordinary matter, having a rest mass as opposed to the photon that only has a mass that is equivalent to the energy carried by the electromagnetic radiation. For the photon, there is no such thing as a rest mass. The photon always travels, obviously, with the speed of light. But Yukawa's carrier of force would have a rest mass and could travel at any speed—any speed, that is, that Einstein's relativity would permit. Yukawa was able to say just how much mass this new particle would have to have if it were to explain the strong force within the atomic nucleus. The mass would have to be intermediate between that of the proton or neutron and the mass of the electron, something like 300 times the mass of the electron. For this reason, the proposed particle was called a *mesotron*, later simplified to *meson*. Two years later, particles fitting this description appeared as faint tracks in cloud chamber experiments. It looked like things were complete.

At last, the whole universe seemed to be explained in terms of these few particles and these few forces. Everything from galaxies to station wagons, uranium to barking dogs—everything. The cosmic onion had been peeled away to show all the pieces that made up the basic building blocks of reality. All the answers. The proton, the neutron, and the electron, together with Newton's universal gravitation, Maxwell's electromagnetic force field, and Yukawa's strong nuclear force. It looked pretty good.

But the looks were deceiving. Before the ink had dried, everything fell apart. The Yukawa meson found in the cosmic ray tracks in the cloud chamber experiments turned out to be the wrong meson. It simply did not interact with matter strongly enough to provide a glue that would hold a nucleus together. The search was on to find the right particle to make Yukawa's picture of the strong force work. The right particle (or rather the right particles) were found, but in the process the search turned up a whole series of new particles and strange discoveries about these elementary particles that made them appear less and less the elementary building blocks originally envisioned.

While all this was going on, the mathematical machinery that physicists use to understand how these particles and fields of force interact also developed rapidly, giving us more certainty and accuracy about the way we describe nature. P. A. M. Dirac, whom we last met extending Bose's indistinguishability principle and Pauli's exclusion principle, succeeded in fusing Schrödinger's wave mechanics and Einstein's special theory of relativity into a single powerful tool that has stood to this day as the most precise achievement in all physics. To show just how successful this theory has proved to be, let me quote from Richard Feynman's book *QED, The Strange Theory of Light and Matter*.[1]

> Just to give you an idea of how the theory has been put through the wringer, I'll give you some recent numbers: experiments have Dirac's number at 1.00115965221 (with an uncertainty of about 4 in the last digit); the theory puts it at 1.00115965246 (with an uncertainty of about five times as much). To give you a feeling for the accuracy of these numbers, it comes out something like this: If you were to measure the distance from Los Angeles to New York to this accuracy, it would be exact to the thickness of a human hair. That's how delicately quantum electrodynamics has, in the past fifty years, been checked.

One of the significant achievements involving Dirac's equation was his use of it to predict a positively charged electron. Dirac came to this conclusion in 1931. A year later, Carl Anderson in the United States found this positive electron, or positron, as it is now called. Dirac's equation gives us further insight into the mysteries of the wave-particle duality. Dirac's quantum electrodynamics, or QED, is a quantum theory of force fields. Just as in the wave-particle duality every particle also behaves like a wave, in the quantum theory of fields every force field also has its own kind of particle that carries that force. The electromagnetic field that produces a repulsive force between two electrons, or that causes two magnets to repel each other when like poles are brought together, is caused by photons, virtual photons emitted and absorbed between the two objects. It is somewhat like two swimmers tossing a beach ball back and forth between them. The repeated impulses caused by throwing and catching the ball will slowly cause the two to drift apart as if repelled by some beach ball field. If instead of throwing this beach ball through the air, the swimmers were to try to "throw" the ball between them under water (the only way one can do this is by rapidly paddling the water out of the ball's way), they would be drawn toward each other, just as the attractive electric force between a proton and an electron pulls the two together in forming an atom of hydrogen. The photons emitted and absorbed by the electrons can, of course, be real photons, just as we have been picturing these pieces of electromagnetic energy. But even when light is not being radiated in the conventional sense, the electric force between two electrons is there and is caused by the virtual photons being exchanged back and forth. These photons are called virtual because, even though they exist to produce the force, there is not sufficient energy in the system of particles for them to exist as free particles, as photons.

This change in our perspective about what a force field is makes it all the more difficult to talk about elementary particles. Particles are continually being created and annihilated in order for the fields to exist. The electron itself must be thought of as surrounded by a cloud of photons that in their turn are continually creating and destroying electron–positron pairs. As the electron moves through the vacuum of space, this cloud of photons, electrons, and positrons is forming, being polarized by the passage of the electron, and then disappearing. And all of these particles being created and annihilated are the cycles and epicycles of quantum vibrations in these fields.

The situation is even more complicated than that. The meson that was originally thought to be the particle of the Yukawa force turns out to be a kind of heavy electron. The particle has nothing to do with the force that binds the nucleus together. It is just a fat electron. It has been renamed simply the *muon*. It can decay into an electron and two neutral particles called neutrinos. These particles are also a part of the cloud of virtual particles surrounding the electron. Protons and neutrons, the particles that make up the nuclei of atoms, also exist as clouds of particles, but their structure is more complicated and involves a cluster of new particles that we will address a bit later.

In addition to all of these particles, there are the "real" mesons. These particles turn out to be the particles predicted by Yukawa that produce the strong force within the nucleus. But the story turned out to be more complicated than had been envisioned. Rather than there being one meson that would have explained the strong nuclear force inside the nucleus, it turned out that there were many mesons. The search turned up particles called pi-mesons (pions). The search found a positive π, a negative π, and a π with no charge at all. Physicists discovered the ρ- meson. They also found the K mesons; they found the η, the ϕ, and the ω. They found the ε, the υ, the ξ, and the ψ. There were so many that they switched from Greek and named them by Roman letters and used asterisks S* and K* and primes like f' and subscripts 1, 2, and 3. They found A_1, A_2, and A_3. They found B, f, and g; they found M, Q^1, Q^2, S, and U. They found mesons that were "strange" and mesons they called "charmed." They found these particles, and they found their antiparticles, all these to decorate the meson particle charts and each sporting various characteristics to distinguish each in this long list of mesons. Inside the nucleus, all these mesons seethe in a maelstrom of creation and annihilation, cycles and epicycles.

In addition to this long list of meson particles, we must add the list of the "heavier" particles; particles which are more or less similar to the proton and neutron originally envisioned as the sole inhabitants of the atom's nucleus. These particles are known collectively as *baryons* and include, as species, the Λ, Σ, Ξ, Ω, and Δ particles. The Λ particles alone come with masses of 2180, 2605, 2750, 2975, 3130, 3270, 3305, 3520, 3550, 3580, 3640, 3935, 3955, 4110, 4130, 4550, 4600, and 5060 as measured in terms of the electron mass, and the list of Σ particles is even longer. Each comes in an assortment of types, charges, and antiparticle forms.

The list of subatomic particles once thought to be the elementary particles of the universe has now grown to include more than 400. The cosmic onion has been peeled back to reveal . . . more onion. But if the perplexity revealed by unfolding this superfluity of particles has seemed bewildering, steady and sometimes brilliant progress has been made in divining the secrets deep within nature's hidden machinery. Physicists using machines that are miles across and cost billions of dollars have probed still deeper into the interior of even these subatomic particles to find the pieces of which they are made.

QUARK, QUARK, QUARK!

Looking deeper into the structure of matter, physicists have found a new set of particles that like to cluster into groups of twos and threes. These are the particles that are called quarks. This fanciful name comes from a line in James Joyce's *Finnegan's Wake*: "Three quarks for Muster Mark," which Murray Gell-Mann quoted when he first presented the theory of their existence. To see how these quarks simplify our understanding of matter, let us list in groups all of the particles that we have talked about. There are four groups to keep in mind at this point. This number will change as soon as we see how two of these groups can be explained in terms of quarks, but for now there are four: (1) quanta, (2) leptons, (3) mesons, and (4) baryons.

The quanta are the particles that explain the forces like electromagnetism and gravitation. We have already talked at some length about electromagnetism and gravitation. Electromagnetism involves the photons, the particles of light. Physicists believe gravitation is caused by particles called gravitons. We will talk later about the quanta associated with the other two kinds of forces, the strong nuclear force and the weak force.

The second group of particles is called the leptons. The name comes from the Greek for "small," in particular, a small coin (presently equal to a hundredth of a drachma), and refers to the fact that these particles are very light. The ones we have already mentioned are the electron and the muon. There also exists the neutrino, a neutral particle that is emitted along with the electron in the radioactive decay of some atoms. The muon acts in every way as though it were just a heavy electron. It even has its own neutrino associated with it, called simply the muon neutrino. A third pair of these particles are known as the tau particle and its tau neutrino. Within the limits of the present experimental evidence, neutrinos have no mass, or if they have any mass, it must be extremely small.

The third group consists of the meson particles, which we discussed as perhaps playing a role in keeping the nucleus together. The operative word here is *perhaps*. The way things actually work turned out to be somewhat different, and somewhat simpler than Yukawa had thought.

The baryons, which constitute the fourth group, are rather heavy particles that include protons and neutrons and interact with other particles as though they were protons or neutrons. These last two groups of particles, the mesons and baryons, are so numerous that they overstuff the atomic cornucopia. Furthermore, these last two groups can now be understood in terms of more fundamental constituents, revealing a basic simplicity in the structure of matter. It turns out that all the mesons consist of a pair of the particles called quarks. Baryons, including protons and neutrons, consist of three quarks.

No free quark has been found. So far, quarks have always been found in groups of at least two. Because of this, the types of quarks must be inferred from the various possible interactions that can occur among the collections of quarks—that is, among the various mesons and baryons. It is now fairly well established that there are six kinds of quarks and no more. It is believed that all six kinds of these quarks have been found in particle collision experiments. The quarks are named *up, down, strange, charm, bottom,* and *top*. Originally, the names *beauty* and *truth* were given to these last two. This led to too many lovely quips, such as, "There is no *truth* to be found in quarks." The uses to which the names *beauty* and *bottom* could be put, had they both continued to be used, might have resulted in some ultimate decline in the productivity of theoretical physicists, had not steps been taken in time. As for me, I still prefer the names beauty and truth. A particle containing a single quark of a given type, such as beauty, is said to show *naked beauty*. Better a naked beauty than a *naked bottom*. If one is to use absolutely preposterous names that have nothing to do with the internal structure or dynamics of the particles, it would be nice if one could at least use a modicum of literary taste! Lately, physicists have taken to using simply the letters *u, d, s, c, b,* and *t*. Perhaps that is just as well. There was never anything particularly worthy of a literary gesture here anyway; at least the designations *u, d, s, c, b,* and *t* save space and time.

These six kinds of quarks are thought of as being distinct particles, despite the fact that they never appear alone; these six types are also referred to as being the different flavors of quarks. One can therefore have a quark with the flavor property of *up*, or even speak of a quark as having a *strange* flavor. (I am not making this up.)

All the quarks carry an electric charge. Strange, bottom, and down all carry a negative charge that is ⅓ of that on the electron. Quarks with the flavors charm, top, and up carry a positive charge that is ⅔ as large as that on an electron.

A proton consists of two up quarks and one down quark. That means its total electric charge is $+\frac{2}{3} + \frac{2}{3} - \frac{1}{3} = +1$, a charge equal to that on an electron but of opposite sign. A neutron consists of two down flavored quarks and one up quark. The electric charge is therefore $-\frac{1}{3} - \frac{1}{3} + \frac{2}{3} = 0$. The antiproton consists of two anti-up flavored antiquarks and one anti-down antiquark for a total charge of $-\frac{2}{3} - \frac{2}{3} + \frac{1}{3} = -1$; that is, it is a proton with a charge that is negative just like that of an electron. All baryons turn out to consist of three quarks.

All mesons can be explained as consisting of quark-antiquark pairs. For example, a down quark and an anti-down antiquark give a neutral pion, one of Yukawa's mesons, and the psi meson consists of a charm flavored quark and an anti-charmed antiquark. The nuclei of atoms are held together by the exchange of various mesons, which really involves the exchange of quark-antiquark pairs among the protons and neutrons in the nucleus. The process, however, is rather more complex than this, as we shall see later.

Now let us look at what holds quarks together in the pairs forming mesons and the triplets forming baryons. The force is called the strong force; the particles holding the quarks together are called gluons. There are eight different gluons. The theory behind this force is called *chromodynamics*. The reason for the "colorful" name is that there seem to be three kinds of charges necessary to explain the strong force. Because there are three basic colors, these charges have been given colors for their names: red, blue, and green. Thus we find that the strong force that at first was thought to be due to Yukawa's mesons is actually caused by the gluons. The mesons are still there, but now we see that they are themselves pairs of quarks held together by gluon forces. They are, in turn, constantly being exchanged among the baryon particles that make up the nucleus.

But let us back up just a bit. The force of gravitation pulls on all objects. We can think of gravitation as a force that has only one kind of "charge," a positive charge called mass, carried by all pieces of matter. Electromagnetic forces, however, act only on particles that carry an electric charge. Furthermore, there are two kinds of electric charges, positive and negative. In chromodynamics, the theory that explains how the strong forces act, there are three kinds of "charges," or colors, as they are called: red, blue, and green—like the three primary phosphors that produce colors on a television screen. But nature is clever. It has not only stepped up to three kinds of charges but has also embedded something like positive and negative electric charges here, for there also exist "positive" and "negative" colors. The negative colors are called anti-red, anti-blue, and anti-green.

The gravitational force is carried by gravitons, the electromagnetic force by photons, and chromodynamics tells us that the particles that carry the strong force are the gluons. Gravitons act on particles that have mass. Photons act on particles that have electric charge. Gluons act on quarks that carry color. There are red up quarks, blue up quarks, and green up quarks which come in the anticolors yellow, cyan, and magenta—the colors my former physicist colleagues call anti-blue, anti-red, and anti-green. There are red down quarks, blue down quarks, magenta down quarks, red bottom beauties, magenta bottoms, and blue tops (forsooth). All the flavors come in all the colors. True or false, chromodynamics is definitely garish!

Now there is a rule that governs how these colored quarks of the various flavors can combine to form objects we can see—that is, objects we can detect in our experiments. The objects made of quarks are all "colored" neutral. Thus objects—mesons—made of two quarks must contain a quark of any color together with an antiquark that carries the opposite color. Therefore, there can be mesons that are made out of a red up quark together with a cyan up quark, a blue down and a yellow up quark, a green bottom paired with a strange magenta quark, and so on.

The baryons must also have neutral color. The three quarks that make up a proton, for example, must carry one each of the basic colors red, blue, and green that, when mixed together, give us white. Every baryon must have one each of these colors or one each of the anticolors yellow, cyan, and magenta that, when mixed together, also give a neutral color. It is a flower garden. Hidden within the fabric of reality lies this imaginary herb and flower garden filled with flavors and colors.

But there is more. Inside these flower gardens are the gluons that hold the various quarks together to form mesons and baryons. The gluons, like bees carrying pollen, exchange colors between quarks. Each gluon can carry one color and one anticolor of any kind. A gluon can, for example, be red and magenta (anti-green). Inside a proton containing, say, a red up quark, a blue up quark, and a green down quark, a red-magenta gluon forms out of the red up quark. The red color of the up quark passes into the gluon, and magenta (anti-green) forms as the gluon leaves green color behind in this up quark. Now the gluon travels to the green down quark, where the magenta neutralizes its green color, which permits the red to be deposited to form a red down quark. It is certainly more complicated than the photon exchange that keeps an electron circling the charged nucleus of an atom.

Gluons, like bees, can also interact with one another. This interaction is simple. For example, a red-magenta (anti-green) gluon interacting with a blue-cyan (anti-red) gluon turns into a single blue-magenta gluon. The anti-red cyan on one gluon joins with the red on the other gluon to cancel out, leaving just one gluon carrying a TV-color blue, and a Xerox-color magenta.

With chromodynamics, we can explain just about everything that mesons and baryons do. However, there is still another kind of interaction that can take place, and there is yet another kind of force. The weak force has its own set of particles that carry the force and a set of effects that the force can inflict on the particles subject to it. This weak force ties together, so to speak, the leptons and the quarks.

The weak force is carried by a trinity of very heavy particles: the W^+, W^-, and Z° particles. The W^+ particle carries a positive electric charge, the W^- particle carries a negative electric charge like the electron, and the Z° carries no charge. This trinity of particles has the ability to change the flavors of quarks; for example, they can change a down quark into an up quark.

Our inventory then includes (1) the four forces: gravitation, electromagnetism, the strong nuclear force, and the weak nuclear force, (2) the quarks that form the mesons and baryons, and (3) the leptons: electrons, muons and tau particles, along with their neutrinos.

The four forces involve the graviton, the photon, eight gluons,[2] and three particles that carry the weak force. That is 13 particles.

Then there are the leptons: electrons, electron neutrinos and their antiparticles, the positron and electron antineutrino; the muon, which looks in every way just like an electron except that it is heavier, its muon neutrino, and their two antiparticles; and the tau particle, which is just a fat muon, together with its neutrino and their antiparticles. That is 12 leptons.

Finally, we have the set of six quarks, which come in six flavors and three TV-colors, together with an entire set of antiquarks having Xerox-colors and opposite electric charges. This makes a total of 36 kinds of quarks. Thus we have 36 quarks, 12 leptons, and the 13 particles involved with the four forces. Altogether that makes 61 kinds of particles. That is the "standard model."

In addition to these, there is one more particle called a Higgs boson, a particle expected to be like the quanta (*i.e.*, the chunks of energy) postulated to give the W^+, W^-, and Z° particles their heavy masses. To say the least, as Chris Quigg put it in *Scientific American*,[3] "By the criterion of simplicity the standard model does not seem to represent progress over the ancient view of matter as made up of earth, air, fire and water, interacting through love and strife."

The standard model does enormously simplify the profusion of mesons and hyperons. Still, it seems that all these particles are just some neoptolemaic theory of epicycles miraculously resurrected from the dusty pages of Roman history. Despite the magnificent science that all this represents, it sometimes seems that the true nature of reality has eluded us. There are so many particles with so many properties—charges, colors, flavors, spins, masses—that it seems as though the answers have simply been pieced together. Maybe the theory fits the facts, but there are enough knobs and dials and spare parts in the theory that it can easily be changed to fit most any set of facts!

The problem of the predicted decay of protons is a case in point. This prediction arose from efforts to reduce further the complexity of the basic theory. A unified theory would let us understand quarks as made out of leptons. After all, the standard model details how neutrons decay by showing that a down quark can decay to produce an electron and an electron antineutrino. Perhaps quarks are made out of such things. This Grand Unified Theory (GUT) reduces the total number of particles, but it also requires a few new ones. The theory predicts that protons should decay somewhat the way the neutron decays into a proton and an electron with the emission of a neutrino. In this new theory, the up quarks in the proton can decay into photons. Thus the protons in ordinary matter should decay away.

At first, it looked like the theory predicted that all the protons in all the nuclei of all the atoms should have disappeared, but then the theory was fixed. In the fixed-up theory, protons would still decay, but at a very slow rate. The experiments to test this proved to be difficult and very expensive—and they failed to find any such decay. The maximum possible half-life for the proton would have to be more than 1.7×10^{32} years. But the theory was fixed up to agree with even this lower rate of proton decay. And this points out one of the weaknesses of a theory that embraces such a large number of basic particles. Surely these particles have something to do with reality, but the ease with which large numbers of alternative theories can be created detracts from the significance of all these theories. We are left feeling that these theories may have no more to do with the real structure of ultimate reality (whatever that may be) than Ptolemy's epicycles, which, you will recall, did explain the planets' motions.

DEEPER STRUCTURES?

Physicists have begun to look even deeper into the structure of matter, looking for ways to reduce the total number of particles necessary to explain everything and ways to clear up some of the puzzling things about quarks and leptons. Why, for example, has nature given us a chubby electron, the muon, that is 207 times the mass of the electron, and an out-and-out obese tau particle, in every way just like the electron except that it weighs in at about 3480 electron masses? Why does each of these have its own similarly mysterious neutrino? The quarks, also, seem to come in three sets of particles. Such groupings suggest that these particles may have some internal structure. Perhaps if we look for things even smaller, we will find the secret of the universe. Perhaps the universe is made of "strings" of vibrating space-time.

STRINGS IN THE FABRIC?

"All this buttoning and unbuttoning." This enigmatic phrase from an eighteenth-century suicide note somehow reminds me of the frantic search for elementary particles. Finding so many possible levels of compositeness as they looked deeper and deeper into matter, physicists surely should have begun to ask, "Where will it finally end? Does it go on forever? If we search at higher and higher energies to find smaller and smaller bits of matter, will we ever find the smallest pieces?" Recently, physicists have begun looking at what they call the "Theory of Everything" (TOE) to unify all the fundamental forces, identify all the particles, explain their masses, and account for the rules governing the composite structures.

The idea is to attempt to explain all the forces of nature the way Einstein explained the force of gravitation, as being due to a distortion in space-time caused by the presence of matter. In superstring theory, this kind of space-warp force takes place in a ten-dimensional space-time (nine space-like dimensions and the usual time dimension). Particles are considered to be string-like loops in this ten-dimensional space-time, strings that can vibrate much like the pictures we imagined for the de Broglie waves of electrons in orbit around the proton. The difference is that these strings are able to distort space into exceedingly tight loops—so tight that six of the dimensions curl up so that we can see nothing of them but the effects they have on these string-like particles. But more than this, the strings themselves are nothing but vibrating distortions of space. In this TOE, the particle is just a space wave, an almost infinitely tiny vibrating knot in space just 10^{-33} centimeter small.

But what is space, and what is time? We have already seen that space and time are both relative concepts. They are not some absolute substance from which we could roll up pieces of matter. It seems almost as if in our search to understand what matter really is and what reality is all about, we reach into the very seeds of creation to find only a void. I do not want you to misunderstand me here, however. We have probed to incomprehensibly small distances and unimaginably brief instants of time. There is great achievement in all this. Science has discovered much. The engineering is wonderful, epicycles and all. And yet, as we look at this vast, elaborate structure built of layer upon layer of complex constituents, can we help but be reminded of the Land of Oz? Have we found the Emerald City? Is this what we were searching for? Is this the ultimate fabric of reality? Is this all there is?

A long way—I have come a long way now, far from that open grassy ridge west of Auburn. Those distant hills lie in the past, but the questions have remained. There was something more to her, so far back in my past, that is gone from all the atoms that once breathed the same warm air that I breathed, and I do not find any of her here in the quarks, in the leptons, or in any the superstrings of space-time. I remember a cocked head, a laughing voice, and a glance—decades away. Somewhere in our search for reality we have passed something by, something important that we no longer find amid the bits and pieces of disassembled matter—something vital that we cannot build out of these parts. There is surely something else, some piece of divinity in us, something that was before the elements, and owes no homage to the sun.

7

Many Worlds, Many Mansions

I do not know what I may appear to the world, but to myself I seem to have been only like a boy playing on the sea-shore, and diverting myself in now and then finding a smoother pebble or a prettier shell than ordinary, whilst the great ocean of truth lay all undiscovered before me.

—Isaac Newton

This is a material world, and I am made of matter, and this matter is made of atoms. Inside each atom, a nucleus; inside each nucleus, protons, neutrons, pions; inside these, quarks; inside these, perhaps, yet tinier things. This is a material world, and physics has described it—described the universe, described the matter it is made of, and developed the equations that show how it all fits together and works. With great precision, physicists probe matter to show things as small as 10^{-17} centimeter and consider times as brief as 10^{-35} second. But somewhere along the way, we have passed something by, something that is important. We have to go back. There was something we passed earlier that it seemed, for a moment, might help us find that something that we now are missing. There was something, something about observers, and there was someone long ago objecting, "God does not play dice."

The great success of quantum mechanics, of Schrödinger's equation and Heisenberg's uncertainties, and of all that has been achieved in studying the infinitesimal world of the atom has lent ever-increasing credence to the idea that the wave function Ψ actually does describe the real world out there. This wave function, this probability amplitude, really does give us a complete picture and a detailed specification of matter. It describes the complete condition of any collection of objects. Of course, physicists never attempt to write down Ψ for a whole vista of reality down to its atomic detail. Rather, they try to stick with fairly simple problems. Physicists prefer to break things down into simple pieces, but they also demand of their science that the equations completely describe how things work as assemblages. They want to know fully how things work—understanding the pieces *and* understanding the whole.

Thus Ψ, which contains pictures of all the potential configurations or states of a system, all the ways a collection of objects can be found to be—describes exactly what the object is—whether atoms or people, quarks or quasars beyond the distant galaxies. This is the complete reality of the system at any moment before it is measured, before it is observed. This Ψ is what exists; it is reality.

But all of this is contrary to the way Albert Einstein conceived reality to be. All of this great accomplishment, all the success of this quantum mechanical conception of nature, is the brilliant solution to the wave-particle duality problem, Ψ being a probability wave describing the chance of finding particles in this or that place, in this or that state of motion. But this very clever solution to the wave-particle duality problem has spawned a new and a far more perplexing problem that has continued to confound physics. It is a problem that led Einstein to confess, "I cannot bear the thought that an electron exposed to a ray [of light] should by its own free decision choose the moment and the direction in which it wants to jump away. If so, I'd rather be a cobbler or even an employee in a gambling house than a physicist."

In this statement, Einstein sounded a discordant note in the quantum symphony even before Heisenberg discovered his matrix mechanics or Max Born interpreted Schrödinger's Ψ as a probability wave. Einstein was to be proved to have been wrong. Einstein was to become, for much of his life, an anachronism swept away by the rapid developments in atomic physics with which he was no longer in tune. And yet, in this he will probably still prove to have heralded a new vision of reality. Einstein's dissatisfaction with quantum theory, his "God does not play dice" understanding of reality, is the starting point for a new physics that is only now beginning to appear. Moreover, it is a physics, an understanding of reality, entirely different from any answer that Einstein ever had in mind, an answer foreign to him and to those who opposed him. Einstein helped to create quantum mechanics and was there during its very inception, but when it finally came to fruition, he, like Schrödinger, turned against quantum mechanics and disowned what it had become. To understand why Einstein and Schrödinger rejected the quantum mechanical view, we have to understand a bit better just how Einstein pictured reality, and we have to understand what quantum mechanics had become. In order to do this, we must go back in our story, and we must go back in time.

THE BATTLE LINES ARE DRAWN

By 1920, Einstein had become the world's preeminent physicist. His intuition had painted an entirely new picture of reality, a picture that led him to make the most incredible predictions in the history of science. His momentous predictions were now paying off in experiment after experiment in laboratories around the world. His new theories of relativity rejected Newton's absolute space and time, but Einstein still pictured the universe as a great machine in the hands of a Spinozan God. Einstein pictured space and time as gently curved into a new, non-Euclidean order where objects, traveling on straight-line trajectories, curl into orbits simulating the effect of gravitational fields of force stretched out across the distances between planets and stars. Objects obey only the law stipulating that they travel along the straightest line that exists for them to travel through this curved space-time. Their motions are governed by absolute laws. There is for this motion one path and one path only. But physicists peering down into the strange interior world of atomic physics were beginning to see another motion, another behavior, in which the quantum mechanics of probability and chance permitted chaos. Whereas Einstein fashioned a cosmic engine of perfect geometric order, his fellow workers seemed to be saying that God's grand creation has roulettes for wheels and dice for gears—that photons, like bees, buzz where they please. This was not the orderly world Einstein had envisioned.

As quantum mechanics developed, everything got worse, at least from Einstein's point of view. What made the whole enterprise so maddening was Bohr's fixation on a literal conception of the probability wave as an absolute descriptor of the world. For Bohr, the fuzzy microworld of the atom was the complete story. Position and momentum were the two key ingredients in Einstein's general relativity, but in quantum mechanics, these two things are no longer mainstays of reality. In quantum mechanics, it is a fundamental principle that we can never know precisely both the position and the momentum of anything. Things are governed by uncertainty relationships that generate probabilistic descriptions of motions. For Bohr, the uncertainty relationships were not just some inability to make accurate measurements but were the very stuff of reality. For Einstein, if Bohr's arguments were correct, if the Heisenberg uncertainty relations were absolute, then general relativity had to be wrong.

As the decades passed, an enormous chasm developed between quantum mechanics and the general relativity of Einstein's creation. General relativity has been eminently successful in describing the universe as a whole, yet it seems entirely inadequate in the microscopic world. Einstein envisioned that all the forces of nature would someday be explained in terms of distortions of space-time, just as he had explained gravitation. But now physicists talk of graviton particles that carry the gravitational force—something totally foreign to Einstein's vision of physical reality. Such an idea is a mockery of Einstein's theory, and if such particles are indeed the carriers of gravity, it is difficult to see how it would be possible to keep Einstein's relativity from falling apart at its base. Is it any wonder that he would have preferred honest labor as a cobbler or even the life of a croupier than to have played the fool as a physicist in the Devil's casino?

In Berlin in April 1920, Einstein and Niels Bohr met for the first time. Planck had invited Bohr to lecture at the *Physikalische Gesellschaft* on his theory of the atom. He had hardly arrived when a lively discussion began between Bohr and Einstein. As R. W. Clark describes it:[1]

> Something was sparked off between Bohr and Einstein at this meeting, the first of a long series of mental collisions whose succession through the years was to have a quality quite separate from the impact of genius on genius. For it was Einstein who, fifteen years earlier, had first brought an air of unexpected respectability to the idea that light might conceivably consist both of wave and of particle and to the notion that Planck's quantum theory might be applied not only to radiation but to matter itself. It was Bohr who was to bring scientific plausibility to the first of these ideas with his principle of complementarity and substance for the second with his explanation of Rutherford's nuclear atom. Yet these very ideas were to create not a unity between the two men but a chasm. From the early 1920s as Bohr and those of like mind followed them on to what they saw as inevitable conclusions, Einstein drew back in steadily growing disagreement, withdrawing himself from the mainstream of physics and giving to his later years a tragic air which not even the staunchest of his friends could argue away.

In order to make this world of wave-particle duality understandable, Bohr invoked a concept he called complementarity. Never mind, said Bohr, that the wave and particle pictures are contradictory. We need them both. These two pictures are complementary to each other. In some experiments we see photons and electrons as particles; in some we see only that electrons and photons are waves. The conflict is only in our minds because we attempt to picture the world in terms of everyday images. The answer is simply that when we do an experiment to test for particle characteristics, then that is the behavior matter manifests; when the experiment is designed to show wave properties, that is the behavior that is found.

But Einstein did not like this at all. For Einstein (and for me, by the way), physics cannot be right until it is complete and consistent. Bohr was correct to the extent that we should not expect to be able to apply our macroworld pictures of particles and waves to the atomic world, but we cannot be satisfied to leave physics in a muddle of inconsistencies. Our basic ideas must be logical and complete, and they must fit the equations that so well quantify the physical world.

The ideas that Bohr expressed ran counter to all of Einstein's scientific instincts. At the Solvay Congress in 1927, with the greatest array of physicists in the world assembled, the battle flared into the open. In the words of Banesh Hoffmann,[2]

> Born and Heisenberg argued that the indeterminacy is unavoidable: that because of the absence of strict causality, the probabilities tell all there is to tell. Bohr agreed. But Einstein did not. He was unwilling to accept what his instinct rejected. He felt that the theory was incomplete. And he brought forth a succession of ingenious arguments in support of his views. Never before had the new quantum mechanics been subjected to so formidable and penetrating an attack. But Bohr and his allies, hard pressed, stood their ground. Refining their concepts in the heat of battle, they defeated the objections of Einstein one by one, and Einstein, for all his ingenuity, had to retreat. The unknowable jolt of observation could not be avoided.

On December 12, 1951, Einstein wrote these words to his old friend Michele Besso:[3]

> All these fifty years of conscious brooding have brought me no nearer to the answer to the question, 'What are light quanta?' Nowadays every Tom, Dick, and Harry thinks he knows it, but he is mistaken.

Despite Einstein's efforts to prove quantum theory wrong, Bohr's complementarity arguments carried the day. For the most part, Bohr's view of quantum theory is still held by physicists, and yet his ideas are not really theory, but rather a kind of cataloging of experimental results. Certain experiments give particle-like results, some give wave-like results. Certainly, quantum mechanics has worked too well to have been overturned by Einstein's philosophical objections. For the physicist, when the mathematics fits experiment so extremely well, the implications must be accepted. Even though Bohr's complementarity argument surely seemed as much like double talk to many of his fellow physicists at the Fifth Solvay Congress as it did to Einstein, the Schrödinger equation, with Born's probability interpretation and Heisenberg's uncertainty relations, worked. Bohr simply gave physicists a way to keep their minds at peace in this strange new wave-particle land.

But if Einstein could not prove quantum mechanics wrong, perhaps he could at least show that quantum theory omits fundamental information about reality—information that would have to be present in some future, more nearly complete theory of reality. Perhaps he could demonstrate that quantum mechanics did not provide the basis for a complete understanding of what is real. If Einstein could prove this—prove that the world is not fully described by a probability wave—then someday, perhaps, someone would be able to penetrate beyond the jitterbug picture of atomic reality. If he could at least show this, then he could make peace with how the world works, even if the final answer eluded him.

In 1935, working with Boris Podolsky and Nathan Rosen at Princeton, Einstein thought he finally had what he was looking for. These three published a paper in the *Physical Review* titled, "Can Quantum-Mechanical Description of Physical Reality Be Considered Complete?"[4] This paper was something of a parting shot against quantum theory. After this, Einstein turned his sails to chart different waters from the rest of the physics community. He would spend the rest of his life searching for his "unified field theory," which he hoped would unite his own gravitational theory with all other forces in nature and, in so doing, yield a physics that would replace quantum theory. He was perhaps searching for some divine certainty in the workings of nature to recapture the logic and order that quantum mechanics and the world had tossed aside. He was not successful in his search, but this parting shot marked the real beginning of a renewed effort to understand quantum mechanics.

In the *Physical Review* article, this parting shot, Einstein put his finger on precisely what the issue is. In this paper, Einstein saw the first light of a new understanding of reality, and yet, at the critical moment he turned back to his earlier belief about what God should do and failed to see what nature is telling us about what God is.

The ideas advanced in this now very famous paper by Einstein, Podolsky, and Rosen have come to be known as the EPR Paradox. The EPR Paradox does not maintain that quantum mechanics is in error, but it tells us just what quantum mechanics is saying about reality, and this is something horrible—and wonderful. Note what they say at the outset:[5]

> In attempting to judge the success of physical theory, we may ask ourselves two questions: (1) Is the theory correct? and (2) Is the description given by the theory complete?

They wish to show that quantum mechanics is incomplete and thus not a correct picture of reality. However, *we* are looking elsewhere. We are looking for the missing clues to the reality that quantum mechanics omits and that Einstein overlooked. Einstein, Podolsky, and Rosen attempted to prove quantum theory incomplete by proving that there exist properties that properly describe the state of any object—properties that have real values even though quantum mechanics does not ascribe such a reality to these quantities. Thus Heisenberg's uncertainty relationship says that one cannot measure both an object's position and its momentum precisely and simultaneously. In quantum mechanics, the idea of an object having both an exact location and an exact momentum has no meaning. Either can be measured to any desired precision, but both do not have simultaneous meaning as exact descriptors of an object. We may picture objects located in exact positions with exact motions, but that is not the way reality is described by quantum mechanics. For quantum mechanics, there is a limit on the precision for simultaneously measuring the *product* of the position and the momentum, a product that has a limit about equal to the value of Planck's constant.[6]

Now if Einstein, Podolsky, and Rosen could show somehow that an object does have a real position and a real momentum and that both exist at the same time, even though we may be unable to measure them, then although quantum mechanics would still describe accurately what we measure in a laboratory, it would not tell us what is *real*. If this could be proved, then Einstein would be right, and Bohr, who had won the debate at the Solvay Congress, would have been only a clever apologist for a passing fashion in physics.

In their 1935 *Physical Review* article, Einstein, Podolsky, and Rosen considered a system consisting of two particles that are permitted to interact with each other and then separate. During this interaction, the particles share momentum, angular motion, and position. As they separate, they share their total momentum, their angular motion, and a common point of origin. We assume one of the particles stays in the laboratory so that we can measure it any time we wish, while the other flies out the window and departs to the farthest part of the universe in total isolation. The Schrödinger equation can be used to describe the possible states of these two objects as a result of their original interaction. For example, quantum mechanics can describe for us their exact positions or their exact momenta. But as we already have seen, it cannot give both at the same time for either particle. Now the question is "Do these particles really have an exact position and also an exact velocity, which we are unable to measure without disturbing them, or do these quantities not exist as a part of reality at all?"

"The usual conclusion in quantum mechanics," Einstein, Podolsky, and Rosen say, "is that when the momentum of a particle is known, its coordinate has no physical reality."

However, the way quantum mechanics describes how objects behave is very peculiar. It describes reality as a collection of potential realities, a collection of pictures of the world that includes all the things we expect could be found in a measurement. If we call the two particles mentioned above A and B, then quantum mechanics will say there is a series of pictures—mathematical functions—that describe the possible conditions of *A* and of *B*. Let us say that these pictures for *A* are named $P_A(\# 1)$, $P_A(\# 2)$, and so forth. For B they are similarly $P_B(\# 1)$, $P_B(\# 2)$, . . . Now the Schrödinger equation can combine all of these pictures together into a single description of the system that is represented by Ψ. The quantity looks like this: Ψ = (first picture of A) multiplied by (first picture of B) + (second picture of A) multiplied by (second picture of B) + · · ·. That is,

$$\Psi = P_A(\# 1) \times P_B(\# 1) + P_A(\# 2) \times P_B(\# 2) + P_A(\# 3) \times P_B(\# 3) + \cdots$$

This description says that if particle A is given by picture #1, then particle B is given by its picture #1. If particle A is given by picture #2, then particle B is given by its picture that shows how it moves, and so on. This is because the two particles share all of their dynamical properties during their interaction. This means that once we measure the position of particle A, we can use this equation to find out the position of particle B exactly, without ever disturbing particle B. And if we decide to measure exactly the *momentum* of particle A, then we will know the momentum of particle B exactly, without ever having disturbed particle B.

"Thus," Einstein, Podolsky, and Rosen pointed out, "by measuring either A or B we are in a position to predict with certainty, and without in any way disturbing the second system, either the value of the quantity *P* [momentum] or the value of quantity *Q* [position]." Then Einstein and his collaborators made this observation: "If without in any way disturbing a system, we can predict with certainty the value of a physical quantity, then there exists an element of physical reality corresponding to this physical quantity." This, they believed, proved that both the momentum and the position of both particles A and B have an independent, real existence. If the momentum were to be measured on particle A, then we could say exactly what the momentum of particle B was without ever having to measure it. Therefore, B has to have a real momentum. If the position of A were measured here in the laboratory, then the position of the particle B, speeding off into the remote universe, would have to have reality also. Because we can know all this about particle B without ever disturbing it, it must have both a real position and a real momentum, despite the fact that quantum theory says it does not.

Einstein, Podolsky, and Rosen said we are left with two possibilities. First, we can assume that particle B can have a real position or a real momentum because it has such a reality for both position and momentum all the time. This would mean that quantum mechanics is incomplete, just as Einstein had said all along. Or, second, we can say that particle B does not have a reality for position or momentum until we have measured the position or the momentum of particle A in the laboratory. Concerning this second alternative, Einstein and company stated that it "makes the reality of *P* [momentum] or *Q* [position] depend upon the process of measurement carried out on the first system, which does not disturb the second in any way."

Writing on this same subject 35 years later, C. A. Hooker observed that "it is difficult to resist the conclusion that, by changing our minds about which measurement to make on one of the systems, we are able to alter the state of the other system."

On the one hand, Einstein and his colleagues presented us with the possibility of a rather classical billiard ball reality hidden beyond any reach of experiment—out of view and beyond the realm of knowing the nature of its reality. But on the other hand, if quantum mechanics does describe the true nature of reality—if Ψ is the complete representation of reality—then there exists the possibility that a choice made by one's mind, whatever that may prove to be, has an effect on physical events. Could this be what reality actually is? Is it possible that reality is not ultimately a vast collection of overly agitated billiard balls or even a world of fuzzy, hopping, and popping atoms and elementary particles? Could it be, instead, that mind actually does affect matter? Could *consciousness* possibly be a negotiable instrument of reality? With such incredible worlds at his finger tips, Einstein ends this fabulous excursion into the heart of reality with the deflating observation that "No reasonable definition of reality could be expected to permit this." Einstein had probed to the very edge of mind, but it was not the world he was looking for, and so he let it slip through his grasp.

For the time in which Einstein was writing, when the choice lay between the objective reality he understood and the complementary wave-particle duality double-think that Bohr would apply to these phenomena, Einstein's choice was a most perceptive indictment of the incomplete vision of reality that Bohr then offered. Bohr's framing of the interpretation of quantum mechanics was still couched in terms of the wave-particle duality problem—a problem that had already been laid to rest by the Schrödinger formulation but that has lived on for decades in the guise of Bohr's complementarity philosophy.

However, Bohr's school of thought, the Copenhagen interpretation (the name refers to the fact that Bohr headed the Institute of Theoretical Physics in Copenhagen during these years), has evolved, in the hands of von Neumann and Wigner, into a more mature interpretation. And this refined interpretation enables us to ask the next question in our effort to understand reality, a question that leads us to search for the nature of mind.

Einstein, preoccupied with the desire to show that quantum mechanics was incomplete, did not see that the philosophical system—the paradigm—on which he had based his physics was itself hopelessly incomplete. The world we know has conscious observers; the world Einstein described had particles moving along trajectories but no one to observe. Einstein did what many physicists have done: He assumed that consciousness and mind require no physical explanation. Physics concerns itself with questions about the nature of matter, space, and time—the building blocks of the world. Surely, physicists believe, the functioning of the brain is the domain of biology. Biology has to explain how the brain works, and it will have to do that in terms of the physical laws. With such a picture of how mind and reality work, Einstein could avow, "God does not play dice with the universe" and could believe he had at last proved quantum mechanics to be a flawed picture of reality. But in all the worlds that Einstein conceived, there was no place for any such God to exist.

THE GREATER OUR REACH, THE LESS WE GRASP

The response to Einstein, Podolsky, and Rosen was neither swift nor particularly insightful. In their answer to the EPR Paradox, advocates of quantum mechanics simply held that the representation of a physical system by the function Ψ does give a complete representation. They said that, paradox or no, there exists no hidden description of an object. The wave function is the full reality of the object. Only by measuring the system can we obtain more information about an object, and not until we make such a measurement will there be any meaning to the idea of exact position or exact speed. Einstein's argument, they maintained, rested only on a philosophical predisposition about what reality is. Thus the description of anything before measurement is, in fact, given by Ψ. After measurement, the description of the object measured is given by one of the component states, one of the pictures contained in Ψ. This is how we use the Schrödinger equation, and it is a natural, factual interpretation of quantum mechanics. This interpretation, which has come to be called the Copenhagen interpretation of quantum mechanics, represents the evolution of Bohr's ideas into a form that finally avoids the language of the old wave-particle duality problem. This language frees us of Bohr's complementarity double-talk.

Yet despite this progress toward a pragmatic interpretation of quantum mechanics, we do not entirely escape the problems that have beset our efforts to understand the reality that lies behind the practical use of Heisenberg and Schrödinger's quantum theory. It is this understanding of reality, not the practice (that is, the use of the equations) that physics is all about. We, as physicists, are trying to find out how nature really works. We can use equations for that description, but the equations must lead not simply to engineering results but to an internally complete and self-consistent picture of reality. We had thought we had a valid picture of reality in the billiard ball world of Newtonian physics. Quantum mechanics has shown us that such simple pictures are not adequate. Now we have a picture in which the processes in the world, small or large, micro or macro, simple or complex, of few objects or of many, have to be represented by this collection of probabilistic pictures—by possibilities. These possibilities, these potentialities, become actualities whenever we carry out a measurement on the system. Such a statement is fine for the laboratory, but we want a description of reality that will work for the everyday world, not just when we are in a laboratory. What does this word *measurement* mean? Well, it means *observation*. When we actually observe, or interact with, any system, then the system goes into one state. Of course, the details of how this happens may not seem important, because on the macroscopic level in our everyday world, things work essentially like the Newtonian billiard ball world. For this reason, we can have a practical interpretation of quantum mechanics that says there is a transition, a "collapse of the state vector," or, in other words, a "reduction of the wave function," that occurs when we interact with, or observe, any system.

The phrases *collapse of the state vector* and *reduction of the wave function* mean that the collection of potential pictures given by Ψ becomes, or turns into, a single state, one of the components or permitted potential conditions of the system. Ψ turns into one of the component states of which Ψ consists—a picture of the observed world that we designate by ψ_i (the subscript *i* stands for the number assigned to the picture of reality contained in Ψ that actually turns out to be the one that *happens* when we look at the world). But the terms *interact* and *observe* are not well defined in physics. One may imagine that the system initially represented by the state vector Ψ will, after a time, whirl and grind itself until it finally pops into one of the possible states as a process that happens entirely on its own. But the Schrödinger equation does not tell us this will happen, and it contains no provision for Ψ to change itself into a component state.

One might assume that this state vector collapse occurs when we disturb the system by having some measuring device interact with it in order to make a measurement on the system—that is, when the outside world interacts with it. But that is not how quantum mechanics works. When a second system interacts with the first system, all that happens is that we get an even more complicated system with more potential states: a bigger Ψ with more little ψ_i component pictures than before. The measurement device we use to look at quantum mechanical atoms is itself governed by the Schrödinger equation, and, therefore, it cannot give us anything but a more complicated Ψ that is made out of the set of pictures that represents the overall system's potentialities, but now consisting of two parts: one part for the atomic system's state in each picture and one part for the measurement device's corresponding readings.

But we do know that when we observe the system, we will see only one picture, one state, one condition. We know that state vector collapse must have already occurred or that it occurs at the time of this observation. As a result, investigators often speak of "state vector collapse on observation." But *observation* is just a euphemism for consciousness, for mind; this interpretation of quantum mechanics says that the system undergoes state vector collapse because of our mind! Moreover, there appears to be nothing else to blame for state vector collapse. Everything else is something physical and, as a physical system, must be subject to the Schrödinger equation. At best, it can only create more potential states, not remove them. Yet when we observe the system, we know that state vector collapse has occurred. We see only one state, one of the ψ_i component states. This effort to obtain an entirely practical interpretation of quantum mechanics—this Copenhagen interpretation—leads us to the incredible conclusion that mind, or consciousness, affects matter!

Physicists don't like to fool with biology or philosophy. Something must have gone wrong. Whatever is wrong, that something is called "the measurement problem in quantum mechanics." A thousand physicists have grappled with this problem and failed to come up with any satisfactory answers. They have tried to remove this "state vector collapse on observation," this intrusion of mind into the realm of physics—and they have failed. Einstein was only the first of many to have tried. And yet, this is the problem we must solve if we are to peer beyond the veil of the physical world to find the hidden workings of reality. If so many physicists have failed to solve this problem, how can we hope to find an answer? Is there any answer? Perhaps the reason why physicists have failed is that they have believed that they already know how reality works—believed too much to have paused to hear the message nature has been whispering to them.

It is January again, January 27, 1952:

I got up at 9:00 this morning. With thirty cents in hand, I went off to Sunday School and to Church. Daddy filled out those NROTC forms. I called Merilyn at 2:45. We talked about nothing till four. I tried to write a poem to Merilyn to say how much I love her—but the words could not equal what I wanted to say. It has been raining all day, as it is now.

Monday, January 28, 1952:

No homeroom teacher. Jimmy Dewberry asked me to collect pennies so he would get class favorite . . . . Mrs. Borders caught me running in the hall and sent me to Mr. Walker [vice principal, no relation]. After school, I gave Merilyn a poem I had written on the nineteenth which says that I wish her to be my wife.

She later wrote out for me a copy of the poem I had given her, the copy I have now. As poetry, it is awful. Still, it says something that you can only say once in your life in the way it was said and thought. It begins, "My maid holds the balance of my life, and may the day come when she'll be my wife." (Incredibly, the thing plumbs still greater literary depths. Out of kindness, I spare the reader!) On the same day, the nineteenth, Merilyn had written a poem to me, "Questionnaire Solitaire."

I wonder if I love him . . .
Experience has taught me to be cautious.
But can you be cautious about love?
He is in my dreams,
Good and Bad.
I dream that I declare
For all to hear:
"I love you!"

Perhaps—
If I knew his heart
I would know better my own.

I do not wish to love in vain.
Yet he does not speak
Of his heart.
So—I remain
Swinging in the balance—
First in—then out
Love's gate.

I do not wish to love in vain.
It would surely break
My heart.

—M. Z.

I called her when I got home and later we went to play tennis. Then, I went to her house (2). Tonight, I started my new and third 'book' which will be mainly poetry. P.S.: Merilyn said to me at her house, '*Ego te amo.*' I am all too nondescript lucky to have her.

WIGNER'S FRIEND AND A PARADOX BOX

Whereas Bohr championed the new view of quantum mechanics as a complete yet probabilistic interpretation of physical reality and faced down Einstein's awesome guns, it was von Neumann who revised the conception of quantum mechanics, leaving behind the wave-particle duality language, and put the interpretation into its modern form.

Von Neumann makes clear that there are two ways in which a system described by quantum mechanics can change with time. There is the usual development of a system with time, which lets us take what we know about the motions of an object right now and use that knowledge to extrapolate into the future to find out what the description of the object should be then.[7] And there is a second in which that time can change things. When a measurement occurs, there is a sudden change, a stochastic alteration in Ψ that changes it into one of the component pictures. We do not know which picture will occur, but when a measurement happens—when we observe the system—the collection of pictures collapses into one picture, one state.[8]

These, then, are the two ways in which things change as time passes. There is no mystery about the first. It is as familiar as Newton's mechanics with orbits that change the positions of planets as time passes. The only thing new is that there remain all of these possibilities that quantum mechanics allows. The puzzle has to do with the second way things change with time—as that collection of possibilities suddenly changes when we observe what is there.

Exactly what causes that sudden change in the description of the system? That is what von Neumann wanted to discover. Is it perhaps just that when we make a measurement, we find out what was there all along? Was the result there, waiting to be found, even before we looked? We will examine this question in much more detail later. But von Neumann gave the first definite answer to the question that Einstein, Podolsky, and Rosen had asked: "No, we cannot just treat quantum mechanics as ordinary chance." Quantum mechanics is not just a toss of dice that turns up one of the pictures. No, God is not playing dice with the universe. Something much stranger is going on.

But if the object or the system is really described by this series of quantum pictures before measurement, and by a single picture after we actually look at things, then, in von Neumann's words, "Just what physical event triggered the change?"[9] Von Neumann tried to answer this question. He used quantum mechanics to follow what happens when we measure something in the real world—following what happens all the way from the material events in the outside world and on into the brain and into the mind.

What happens when we try to measure an event, when we try to observe anything? What is it that changes? Schrödinger's equation for these quantum waves of probability tells us that we just get more and more possibilities. But when a message about the results of a measurement travels from the eye down the optic nerve and into the brain, it becomes a conscious thought somewhere in the brain's functions. When that takes place, then we know the system must be in one single state. Some one thing has had to have happened.

Von Neumann does not distinguish well between the brain and the mind, but that does not matter. It is clear that as far as physics is concerned, the brain, though complex, does not have in its chemical processes anything that would alter the results of the Schrödinger equation without changing what the equation itself says. Yet when consciousness enters, everything changes. The physics of Schrödinger, of Heisenberg and Einstein, of Bohr and von Neumann neither provides any explanation of consciousness nor gives us any place to put it except beyond the edges of quantum theory where the state vector becomes the picture we see.

Von Neumann traces this act of observation back to the very edge of our mind. He illustrates the way quantum theory works by giving specific examples. In one of these examples, the observer is engaged in making a temperature measurement. Here is von Neumann on the way quantum mechanics works:[10]

> We wish to measure a temperature. If we want, we can . . . say: this temperature is measured by the thermometer. But we can carry the calculation further, . . . we can calculate

> its heating, expansion, and the resultant length of the mercury column, and then say: this length is seen by the observer. Going still further, and taking the light source . . . [and] quanta into the eye of the observer, . . . we would say: this image is registered by the retina of the observer. And were our physiological knowledge more precise than it is today, we could go still further, . . . and then in the end say: these chemical changes of his brain cells are perceived by the observer. But in any case, no matter how far we calculate—to the mercury vessel, to the scale of the thermometer, to the retina, or into the brain, at some time we must say: and this is perceived by the observer . . . . The boundary between the two is arbitrary to a very large extent . . . but this does not change the fact that in each method of description the boundary must be put somewhere, . . . Indeed experience only makes statements of this type: an observer has made a certain (subjective) observation; and never any like this: a physical quantity has a certain value.

But what justifies putting a boundary anywhere? Eugene Wigner carried von Neumann's analysis one step further into an area called the Wigner's friend paradox in quantum mechanics. He asked what quantum mechanics would say if the observer that von Neumann speaks of should be a friend of his. Wigner's friend conducts an experiment and makes an observation. His friend then comes to Wigner telling him what he has discovered. According to quantum mechanics, should Wigner's friend be treated as merely another step in the sequence of interactions leading to Wigner's knowledge? Should the friend's observation of the measured result be considered just a potential state, or should we, as common sense would suggest, agree that when Wigner's friend observed the result, the result was already in some specific single state? Quantum mechanics cannot give us the answer. Surprisingly, quantum mechanics is incomplete in the face of this obvious consequence of using literally the prescription of the Schrödinger equation.

The answer would seem to be that something is wrong with quantum mechanics, that its make-up is incomplete, or that something about the act of conscious observation alters the quantum mechanical description of reality, something that plays a special role in quantum theory. Wigner has seen some of these implications of quantum theory more clearly than most physicists. In his book *Symmetries and Reflections*, he writes,[11]

> It may be premature to believe that the present philosophy of quantum mechanics will remain a permanent feature of future physical theories; [but] it will remain remarkable, in whatever way our future concepts may develop, that the very study of the external world led to the conclusion that the content of the consciousness is an ultimate reality.

And yet, even with such a strong commitment to the idea that consciousness is a part of reality, Wigner does not follow this line of reasoning any further than to say that quantum theory implies its existence. It is not surprising that he falls short of advocating that consciousness is indeed a new entity that exerts a ponderable effect on physical reality. We physicists would do this with only the greatest of reservations, with reluctance, and even with trepidation. Yet it must be considered a clear implication of quantum mechanics, to be proved or not. We must answer this piquant question that nature has put to us. But in personally speaking to Wigner about this question of the nature of consciousness, I came away with the impression that he believed consciousness does not govern state vector collapse in quantum mechanics after all, but was itself only one more phenomenology of a complex system, "like determining the entire structure of a piece of aluminum."[12] He seems to have acknowledged, but not explored, the possibility that consciousness may play a role in quantum mechanics and that, if so, "it will be remarkable."[13] Wigner's position reflects the physicist's desire to obtain a purely physical interpretation of quantum mechanics. Nearly all physicists at some time feel a desire to stand God-like above all creation and view the wheels of nature turn as one would look into the workings of a clock. But even a physicist is not a god, and the equations of quantum mechanics still fall short of perfection. Yet even in its imperfection, quantum mechanics embodies so much truth that it cannot hide what it omits. Quantum mechanics points to a new path, to an enlarged meaning, to a better physics that must come after: a physics that will be closer to the truth, closer to perfection. Quantum mechanics, even as it stands, inspires us to learn what observation and what consciousness really are.

I keep looking back, searching, like a drifter collecting pretty pieces of shell and bits of glass from a beach washed well by all of yesterday's waves—looking for memories that have slipped into the quiet past. Occasionally, I hear a voice or catch a glimpse that I had lost, and I remember the memories from long ago again. Suddenly, I seem to find myself looking up toward the garage behind her house, and then there I am inside painting a stage prop, a coming-out gateway with glittering stars and a shining diamanté moon she had designed for her sorority's winter dance at the Pickwick Club . . . . The old Pickwick in Five Points. It is gone now—all of it. The club, the building it was in, the streets, even the streets are not the same anymore.

I hear a memory speaking, teaching: "You're painting it too carefully. Don't worry about the details, they won't show when it's on the stage." A memory, a bit of broken glass, a piece of something she said and I heard, something only the two of us ever knew—and only the two of us will ever have known.

Then my mind discovers another glittering moment gliding out of the clutter of the busy, empty years since she died. The dance. A night at the Pickwick Club. A night complete with a mirrored ball reflecting pencil beams of color across the floor, streams of light walking across the dancing couples, and catching eyes in the dark, glancing, sparkling in the moment's gaiety; pencil beams meeting quickly raised glasses, flashing before climbing up to the ceiling and out into the night. And I was there, fearful of that threatening arena. I avoided the dance floor, afraid of showing my ineptitude out on that foreign terrain, dreading it like the combat zone it was. I was quite terrified of stumbling out onto that dance floor then, and the awkwardness of it all has never left me. But she had taught me—at her house and up in the garage—the steps that I then knew, and now we danced the steps, dancing to "Dancing in the Night," the melody playing in her house and now playing into the night at the Pickwick Club.

I look back, looking into my memory, looking into the next glinting splinter of glass, and I see the two of us again, a moment frozen in the glaring spotlight as we walked together through the gate we had painted back in the garage behind her house on the mountain amid the longleaf pines—splinters of glass reflecting the colors of my emotions streaming through the rails of that gateless gate, looking back into my past, trying to find something.

We stepped out the back door of the Pickwick into the alleyway and into the dark where we could hold each other, kiss, and share a moment of love. Where is that beautiful woman's life now? Where is that soft, warm, loving woman?

I look back again, looking into my memory, but if you look too long into a splinter of the past, you will see the sharp edge that cuts. I remember telling Helen; she was sitting in that old two-story, square frame house just off Highland Avenue with its 12-foot ceilings and cubical rooms. I remember sitting in the kitchen telling my brother's sister-in-law what had happened; she was a friendly soul with whom I could share my troubled thoughts then. I told her that Merilyn had died. I remember that I smiled. I had seen the same smile on her stepfather's face too as he told me about her illness—about the "trouble"—and now I understood that smile, too. We speak to one another in controlled syllables about things that have come all apart, and pain and horror well up inside to twist our words into grotesque smiles. I was telling Helen. I smiled.

INTERPRETATIONS OF QUANTUM MECHANICS

Though the measurement problem in quantum mechanics calls attention to the possibility that the observer—indeed, the very consciousness of an observer—has an effect on matter even as the observer perceives those effects, there are other avenues that physicists have explored in their efforts to understand the paradoxes in quantum mechanics and resolve this measurement problem. There are four principal explanations or interpretations, as they are called, of quantum mechanics.

Certainly, all four of these interpretations cannot be true, but each has something to offer us. Language is not adequate to describe in a few breaths the many facets of reality, so to achieve a broader understanding of nature, we need to know each of these interpretations of quantum theory.

We have already met the first of these four interpretations of quantum mechanics, the *Copenhagen interpretation*. The second has a rather fancy name, the *stochastic ensemble interpretation*. The third understanding of quantum mechanics is rather abstrusely labeled the *relative-state interpretation*, although it is more popularly known as the *many-worlds interpretation*. The fourth is mysteriously dubbed the *hidden-variable interpretation* (or perhaps *interpretations* is more accurate here). These interpretations take many forms, but nearly all efforts to solve the measurement problem can be placed in one of these four categories. Let us look at each of these in turn.

The Copenhagen Interpretation

Although we discussed the Copenhagen interpretation earlier in this chapter, it does have variant forms that are worth mentioning. Fundamentally, this way of understanding quantum mechanics accepts the original idea of a superposition of states given by Ψ that develops immediately after any interaction between two or more bodies on the atomic level. The resulting system exists, at least for a while, as a collection of possible states—as a fuzzy, unresolved potentiality—a collection of many possible things that it can become and from which what it will ultimately become is somehow chosen.

Various mechanisms have been suggested to explain how the transition, the state vector collapse, or the reduction of the wave function actually occurs. The patent explanation is that the state vector collapse happens on measurement. The trouble with this explanation is that the measuring apparatus is, after all, just another piece of physical matter interacting with the system to be measured. As we have seen, many physicists, such as von Neumann and Wigner, as well as opponents of this conventional understanding of quantum mechanics, such as Hooker and Einstein, have argued that quantum mechanics is saying that it is not the measuring apparatus as such, but rather our observation, our consciousness, our mind, that brings about this change.

Others have suggested that the change takes place naturally as we make the transition from the microscopic world to the macroscopic world. After all, is quantum mechanics not the physics of the atoms, and did not Paul Ehrenfest prove that for large objects, quantum mechanics says the same thing that Newton's mechanics says?

This is surely a good argument. If quantum mechanics agrees with Newton's equations when we get to the macroworld, then the problem will surely all go away. And Ehrenfest did prove a very important theorem about quantum mechanics merging into that large-scale physics of Isaac Newton. But Ehrenfest did not say that the macroscopic physics described by quantum mechanics is the same thing as Newton's classical physics. He only showed that quantum mechanics becomes that classical, common-sense world in certain kinds of transitions from the micro- to the macroworld. He showed how quantum mechanics, when applied to large objects, would generally yield a description like the one Newton's equations give. But many physicists have overlooked the fine print. Ehrenfest's theorem does not hold for the instruments we use to peer into the atomic world, and it does not hold for other macroworld events that depend in a strong way on microscopic fluctuations, perturbations, and contingencies that can affect our large-scale world of everyday life. Quantum mechanics must reach into the macroworld, too. If Ehrenfest really had proved that quantum mechanics becomes the same thing as Newtonian mechanics in the everyday world, then we could never even have known about quantum mechanics. There would be no quantum theory. Thus, this explanation, this way out of the measurement problem, does not work.

We have already seen that simply bringing more and more interactions and more and more complexity into play does not lead to a simpler result either. We do not get rid of the many potential states by letting our original system interact with another system—for example, a measuring device. That was the whole point of von Neumann's study of the measurement problem, where he looked at each step from the events at the atomic level all the way into the brain without ever finding a place where the Schrödinger equation would suddenly—or gradually, for that matter—converge into a single event explaining why we see only one state when we observe the world. Despite this, buzzword explanations of how quantum mechanics works are still heard coming from the lips of even some of the most distinguished of scientists.

The Stochastic Ensemble Interpretation

We have seen that quantum mechanics began as simply a statistical theory of matter. The theory initially said that matter is rather like dice in that some things are predictable and other things are not. It can tell us what photon energies could be emitted by an atom, for example, but in general it cannot predict the energy and the moment of emission for any specific photon from a specific atom in an excited state. We also run into trouble when we try to use quantum theory to visualize what exists just before we make a measurement. However, the stochastic ensemble interpretation (that is, quite literally, the "collection of random happenings" interpretation) says that such a question is meaningless. It says there is no superposition of states, no potentialities. There is simply the measurement event.

And it says each event is simply a sampling from a collection—an ensemble—of experiments done exactly the same way. All that really exists is the ensemble, the collection of results we get when we do a lot of experiments. All that physics can do is say what experiment is performed and use Schrödinger's equation to give a description of the ensemble. This ensemble does not require the existence of potentialities, according to this interpretation, but instead it claims simply that quantum theory gives a description of all the results we would see if we carried out a very large number of tests. Then, when we do the experiment, that experiment is just one event taken out of the total collection of things that will happen. Where the Copenhagen interpretation speaks about what exists before a measurement, the stochastic ensemble speaks to the question as though nothing exists except the collection of results we accumulate as we do more and more experiments.

It is probably OK to focus our attention on what comes out of the actual experiment, but what is an experiment? More precisely, what distinguishes an experimental interaction from any other physical interaction? In reality, what distinguishes it is the fact that we, as conscious beings, identify the measurement apparatus as such. It has meaning to us, whereas other events that occur in a complex interaction would have no such meaning to us. The "ensemblists" do not define measurement. That we intuitively know which processes are measurement events and which are not is so obvious that it seems unnecessary to acknowledge. As a result, the advocates of this interpretation seem to solve the problem von Neumann posed. You remember: It is the problem about where we break that continuous chain of interactions between the atomic process and the conscious brain. The ensemblists break off the sequence at the measuring device. The real issue of the measurement problem is consequently hidden in a clever shell-game shuffle. The issue in quantum theory was never about quantum mechanical events being stochastic. The problem was that a measurement no more did anything new than did the original interaction. "Measurement" is just a way of anthropomorphizing one of the interactions. If that is allowed, then one can use the stochastic ensemble interpretation to deny reality to the physical system being measured, so that one can no longer say Ψ is a collection of potentialities. But unless one can complete that task by defining measurement in some unique way, the measurement event itself must be denied existence by the same logic. If the stochastic ensemble interpretation denies any existence to potentialities and does not tell us what is new about measurement that cannot be described by the Schrödinger equation, then measurement as something that gives us the sampling does not exist either.

More than this, Wigner has offered a specific example in which an interaction, plus a measuring device that gives one result, can be turned into an entirely new experiment in which the *measuring device* of the original setup becomes a part of the experimental *interaction*—just a part of an interaction subjected to a further measuring device. In Wigner's new experiment, the arrangement leads to an entirely different experimental outcome. This is known as Wigner's double Stern-Gerlach experiment.

In the double Stern-Gerlach experiment, neutral particles—atoms, say—that have "magnetic moments" (they act like little magnets) are directed in a beam toward the two poles of a strong magnet. The experiment is set up so that all the magnetic moments of the atoms initially point up. Now, let us look at these atoms one by one as they go in, and come out from, between the poles of the magnet that lies in the horizontal plane. As these atoms enter the magnetic field, the atoms will be flipped, just as a compass needle will flip when placed near a magnet. Because the magnetic field is perpendicular to the direction in which the atoms originally pointed, they can flip into the horizontal direction, pointing either right or left. Either with or against the magnetic field, either can happen. Now we can carry out a measurement with a measuring device to find out which way the magnetic atoms flipped. To do this, we let the atoms travel through a second magnet's field that is designed to be very divergent. This causes the atoms to drift in the direction in which their magnetic moments now point, forming two separate beams. By placing detectors on either side of the atoms' paths, we can see which way each atom has been deflected and in which direction each atom's magnet points.

But there is something else that we can do. We can let the atom travel farther, past a wire carrying an electric current set up with just the right current in it to produce another magnetic field to deflect again the atoms of the two beams, bringing them back to form a single beam once more. This field will again flip the atoms in a direction perpendicular to their prior right or left directions. We can then again analyze this beam to see whether the atoms are pointed sometimes up and sometimes down or whether all the atoms now have their magnetic fields pointing up as they originally did. You might think that, because the first magnet caused them to flip randomly to the right or left (making a random 90° flip), this second magnetic field from the wire would cause them to flip randomly also, flipping them so they point up or down at random. You might think that, when recombined, they would again be randomly deflected up or down. But orthodox quantum theory tells us they will all point up. Now at one point along the way, we put a device—the divergent field magnet—to interact with the beam so as to separate out the atoms according to their magnetic alignment. That is to say, we did something very much like doing a measurement. If we had stopped there, then according to conventional physics logic, that would have had to be the complete interaction. That would have defined the ensemble, including the measuring apparatus used to sort out the left- and right-pointing atoms.

However, if we let the beam go on to be recombined by the current loop and then to a third magnet to see which way the atoms are then aligned, we would find that the second magnet, which previously defined the measurement apparatus and also the ensemble, no longer defines what the ensemble is or what a measurement is either. Hence we are forced to ask ourselves, "What is it physically that means we cannot simply treat the first magnet as the interaction and everything after that as the measuring device?" Moreover, in any experiment, if we make our restoring current loop sufficiently elaborate, we can always add on more stuff that will turn what originally looked like a completed measurement into only the first half of an interaction.

In practice, of course, we can do such things only for the simplest kinds of interactions, but in principle it could always be done. This means that we know where the experiment ends—not by means of any physical theorem, but only because we use the word *experiment* anthropomorphically. Because of this, the stochastic ensemble interpretation is a shell game that works because it sneaks the observer in by using the word *experiment* to delineate what the ensembles are. Still, we will see much later that it does help give us a piece of the answer. It does focus on the measurement event, telling us that if we could define that process, we just might be able to solve the measurement problem. The answer, as we will see, will reveal much about the nature of reality.

Everett's Many Worlds

Back in 1957, a graduate student of John Wheeler's at Princeton submitted a doctoral thesis on something he called the "relative state" formulation of quantum mechanics. In his thesis, Hugh Everett III suggested that there might be a way out of the dilemma physics had gotten itself into. He suggested that the idea in Schrödinger's equation that Ψ represents a probability wave is all wrong. As we have seen, if it is a probability, then one has to decide when the physical world makes the transition from potentialities to the real stuff. One has to decide when "state vector collapse" occurs, and that brings the onus of an observer into the sanitary world of physics. Everett proposed that instead of things going from potentialities to actual events, every one of the possible states really occurs. He suggested that every time two objects interact, *all* the possible states that might occur actually do occur. This means that every time two electrons, for example, collide somewhere in the universe, the total universe splits up into two, three, or an infinite number of copies so that there will be at least one universe that will go into each outcome that Schrödinger's equation describes. If we make that assumption—if we say that when the Schrödinger equation says there are two possible outcomes for an interaction, it is because the universe actually splits into two totally independent universes, complete with two copies of ourselves—then in one universe, one copy of myself will see the *first* outcome described by the

Schrödinger equation while the second copy of myself sees the *second* outcome. If we do this, if we simply take what the Schrödinger equation says literally, then we will not need to invent observers to bring about state vector collapse. We will not need to talk about state vector collapse at all. The measurement problem will just go away.

Everett's idea presents us with what looks like an incredible solution to the measurement problem. It appears to get rid of the problem of invoking conscious onlookers into the world. All we need is the Schrödinger equation itself, and we need have no worries about state vector collapse. It does, of course, ask us to believe that the entire universe is constantly splitting up into copies of itself. Everett's idea has gained a lot of support in the physics community, to say nothing of inspiring a lot of science fiction stories—with scenes of Captain Kirk looking on as the denizens of two parallel universes, representing good and evil, struggle to destroy their other selves while becoming trapped forever between their two universes. But of course, Everett's idea says that there could be no contact between these several universes once they have split up. There could be no way whatsoever to prove that these other universes exist. There could never be any contact between them. The universes split off, and those universes in turn split off, and the process continues in an infinite proliferation.

But Everett's interpretation says a lot more. When two electrons collide, they do not complete their interaction in an instant of time that leads to two possible outcomes. As we saw in the last chapter, the force between two electrons is caused by the exchange of many virtual photons, so every possible combination of photon interaction, including the virtual creation of electron pairs and other particles, must be included. This means that every interaction of two electrons must lead to an infinity of universes. Moreover, these two electrons do not have precise positions and speeds. The uncertainty relations tell us there are an infinity of possible initial states and, for each, an infinity of outcomes—for every pair of particles in the universe. The number of universes splitting off in one single second is not simply *nondenumerably* infinite, it turns out to be one of Cantor's alephs of superinfinity numbers. The result is that our universe of universes would be piling up at rates that transcend all concepts of infinitudes.

Mathematicians have found that there are different kinds of infinities. Whereas the infinity of the ordinal numbers 1, 2, 3, 4, 5, 6, 7, etc. and the prime numbers, 1, 2, 3, 5, 7, 11, 13, etc. are both of the same type and both just as infinite, some things differ as to just how infinite they are. Using the ordinal numbers, one can count how many prime numbers there are just fine. Even though they are both infinite, one can nevertheless proceed on the infinite task of counting them. The counting process is quite definable and orderly. But now try to count how many points lie along an inch long line. Well, no matter how you try, the effort is doomed to failure. You cannot even get started in any orderly fashion. This shows that there are at least two kinds of infinities, but mathematicians have shown that there is an infinity—a countable infinity of these types—and the designation of the types is given by their aleph number. The original work was done by the German mathematician Georg Cantor, who gave these infinities their designations in terms of the aleph symbols. As a result, they are also often called the Cantor numbers or infinities.

Despite this, many physicists would still prefer this Everett interpretation, just as long as it keeps what they feel to be the metaphysical nonsense of consciousness out of physics.

But the Everett interpretation does not, in fact, work. This has only slowly become clear to me and to other physicists. There are two major problems with the many-worlds interpretation. Let me show you what these problems are.

Let us say that we have an experimental apparatus that enables us to look at an individual atom as it radiates away some excess energy. The way we run the experiment is to press a button that flashes a high-intensity ultraviolet light source at the atom, and then we wait with our light detectors to see which of two possible photons comes out: a high-energy photon (say, of blue light) or a low-energy photon (say, of red light). Quantum mechanics will tell us just what the probability is for each of these two possibilities. Let's say that the probability for the blue photon is 0.9; 90% of the time we get a blue photon, and 10% of the time we get a red photon (a probability of 0.1). Now let us do a sequence of one hundred experiments with this apparatus. Quantum mechanics tells us we should expect 90 blue photons. Probability theory says it is unlikely that we would see fewer than 85 and less than one chance in a thousand that we would see only 80 blue photons. But if we follow Everett's idea about the universe branching at each event, then after the first event there will be two of me and two of you looking at the results in our respective universes—one in which a blue photon was measured and one in which a red photon was measured. After two experiments, there will be a blue-blue world, two red-blue worlds, and a red-red world. After three, a blue-blue-blue world, three blue-blue-red worlds, three red-blue-blue worlds, and one red-red-red world. If we are in the red-red-red world, we should be a bit surprised; quantum mechanics says there was just one chance in a thousand for three red photons to turn up in a row, but using the many-worlds theory, we have one chance in four of being in that world. By the time we have done ten of our experiments, most of the universes will contain surprised you's and me's. By the time we have done all one hundred experiments, the chance that we will be in a universe that gives results consistent with Schrödinger will be one in 1,000,000,000,000!

There are ways of patching up Everett's idea. One of these ways was suggested by Everett himself. We assume that there is a greater chance for us to take the "blue" branch than the "red" branch. This tack, however, does not work. There are two branches. Each is occupied by copies of us—or else all we have is different language, different words for this idea of potentiality. If no universe appears for one or for the other of the possibilities, then that branch was only a potential event, and we are simply paraphrasing the Copenhagen interpretation, nothing more.

There are ways to patch up Everett's interpretation that will, in fact, work. We can say that for each event there are an infinite number of blue universe branches and an infinite number of red universes and that, furthermore, for each and every red universe branching off, there are nine blue universes. This works. It raises us one notch higher in Cantor's superinfinity count. That is bad, but it is not the real problem caused by this modification. The problem is that we now no longer have a simple interpretation of Schrödinger's equation, as originally promised by Everett. The solutions to Schrödinger's equation do not have terms expressing infinities like that. We would have had to invent a special way of handling the Schrödinger equation in order for things to turn out looking the way we see them. We would have had to add new hypotheses to quantum theory that do not buy us anything that can be put to the test experimentally.

A greater difficulty comes from the fact that Everett really does not tell us where or when the universe branches. If, as is frequently done, we talk about the branching of the universe occurring when we make an experiment, when we look to see whether we have a blue or a red photon, then we are back to hypothesizing a conscious observer to define when this branching happens. Again, we have made many new hypotheses, and we have gained nothing. If, however, as is usually thought if not explicitly expressed, the branching occurs after every interaction, then we wind up predicting the wrong physics with Everett's many-worlds hypothesis. Look at the double Stern-Gerlach experiment we talked about a moment ago. Branching after the electron interacts with the first magnet leads us to expect the wrong results coming out at the end of the second magnetic field. Remember that if the universe separates after the electron passes the magnetic field produced by the electric current, then the Schrödinger equation should not contain two terms for the right and left orientations of the atom, but only a universe with a right-pointing atom and a universe with a left-pointing atom. When the atom pops out at the end, there is no longer any reason for it to right itself and become an atom pointing up. Instead, half should now point up and half should point down as we sit in one of the (now) many universes that see a stream of atoms coming out. This part of Everett's many-worlds interpretation works only if we keep the observer, or keep the idea of measurement, somehow as part of the formalism.

But why should we not actually want to keep the observer in the picture? Why should we not see the occurrence of this necessity for a conscious observer in quantum theory as something that tells us about the nature of reality? After all, conscious observers do exist. So far, nowhere else in all of science have we seen anything that gives us any hint about what consciousness really is. Maybe this Gordian knot is simply something we have created in our own minds because we cannot see the simple solution.

Hidden-Variable Interpretations

Now let us look at the fourth kind of interpretation of quantum mechanics. You may have picked up on a salient point of this whole measurement problem much earlier. Sure, quantum mechanics has had its successes, but isn't it being just a bit presumptuous to believe that no future theory will alter the basic scheme of quantum theory to explain what really lies hidden behind its outwardly random behavior? That is surely what Einstein was saying. Heisenberg gave us reason to believe that any hidden real position and simultaneous real velocity for an electron was impossible to know by present methods—but maybe tomorrow's methods will change this. Or maybe, if we could only look behind the veil of this random behavior, we would see some microscopic engine within each atom, within each particle, that generates such a complex behavior that it looks to us like a nondeterministic reality, a randomness stamped on all atomic events. If such a hidden machine exists in each particle or governs the stochastic behavior for every collection of objects, then knowing its internal workings and settings, like knowing the key to a cipher, would allow us to read nature's hidden messages. If such machines do exist within the atoms, then they must have hidden parameters, hidden variables that are like the key to some baffling cipher. Such a picture of the inner reality that might lie hidden behind quantum mechanics has been referred to as a hidden-variable interpretation.

Hidden-variable proposals began with de Broglie, who, you will remember, first showed that particles of matter (that is, ordinary objects) have a wave-like property, just as light waves have a particle nature. De Broglie's idea was that all matter exists as discrete chunks or particles that are *guided* in their motion by waves. Particles—electrons, protons, atoms—are like surfers carried along on a wave. As a consequence, their behavior is in part determined by the wave motion, but it is also in part chance behavior, because no one can say just where along the wave front the particle is "surfing."

But de Broglie's idea was swept away by another wave, the tide of quantum mechanics advancing with the equations of Schrödinger and Heisenberg and with the ideas of Born and Bohr. The idea of a hidden mechanism responsible for the statistical behavior of atomic matter seemed an unnecessary encumbrance for a theory meeting success upon success everywhere it was applied. Physicists are pragmatic to a fault. Why put aside something that works for the sake of what may be only a philosophic nicety, especially when the theory in question is working everywhere we turn. After all, de Broglie's idea only yields a more complicated explanation of quantum mechanics, with nothing to show it to be better than the Heisenberg-Schrödinger approach. De Broglie's ideas might offer a way of visualizing some hidden workings of quantum mechanics, but at the expense of reinstituting a classical conception of reality. For a physics that has progressed beyond such limitations, this alternative has not appeared attractive.

De Broglie's ideas were all but forgotten when Bohm, in 1952, again attacked the problem of explaining quantum mechanics in terms of a "quantum field," a field much like de Broglie's guide wave. Bohm, however, pictured the particles as behaving in an entirely conventional fashion. Bohm pictured these particles as following Newton's equations of motion, but he assumed something else was going on. He assumed there was a new kind of force, a rapidly fluctuating quantum force jostling the particles about. For Bohm, the quantum world was only a veneer covering a deeper, "hidden" world that could be understood without notions of potential states or conscious observers.

However, this hidden-variable conception of quantum mechanics did not work as Bohm had hoped. Physicists quickly pointed to flaws in the approach. After all, Bohm had not given us a new equation for quantum mechanics; he had only renamed portions of the Schrödinger equation. His quantum mechanical force-field, as formulated, was simply the collective effect of all the potential states of the particle. The change Bohm had proposed was no more than a relabeling. When you got down to brass tacks, his theory worked just the same quantum way. In fact, though he spoke of a hidden-variable theory in which new effects taking place on a subatomic level would be the cause of the quantum staccato, Bohm had not even provided that. Bohm's effort to take physics back to a more comfortable, classical way of visualizing atomic reality failed. But Bohm did give impetus to the idea that some additional process somewhere must be taking place. Every time we carry out an experiment to measure what the atoms are doing, we get a different answer—an answer taken from that statistical description of quantum mechanics all right, but still a different result each time. What is responsible for each measurement's particular result? Are we indeed, as Einstein had charged, merely looking into the eyes of a God who would play dice with our desires and answer our quest for truth with noise?

Although Bohm's quest was surely a good one—to advance our understanding of the causes of the quantum's stochastic nature—it was fatally flawed by his desire to return to those glorious days of yesteryear, a time of certainties when matter was matter and waves were waves. And it received the appropriate reception among the scientific community.

For Bohm to have been correct, his variables would have to have had a mechanism within each particle or guiding each particle that would determine the outcome of each atomic event. But other scientists were bent on showing that such ideas could never work. A succession of mathematicians and physicists went to work on the basic ideas of quantum mechanics to show that the concept of potential states—potentialities—could not be eliminated from quantum theory without totally altering its predictions. That is to say, if we begin with the fact that quantum theory and quantum experiments agree, then there is no way to remove the use in quantum mechanics of simultaneously existing potential states (of the kind we referred to in discussing the double-slit interference experiment), without totally changing the theory and losing agreement between theory and experiment. This seemed to be the death knell for all hidden-variable theories and an end to aspirations for any deeper understanding of the stochastic beat of nature's heart.

But the mathematical proofs were complicated and, worse, they appeared to assume exactly what they were intended to prove. The argument revolved around something called "dispersion free states." This is the idea that quantum mechanics could be written down in some new way that would not assume a superposition of potentialities but would assume, rather, that the objects or systems were in some single unknown state all along. The mathematicians and physicists working on this problem began by accepting that quantum mechanics, as it stands, represents physical systems in terms of a superposition of potential states, a collection of pictures given by Ψ. They would then prove that if that were so, then one could never find a new theory that did not contain a superposition as some part of the new formulation without altering the predictions of the theory.

Opponents like John Bell shouted, "Foul!" They claimed that any proof is invalid that starts out by assuming exactly what it intends to prove as an essential part of the theory. Because it was assumed at the outset, they maintained, it was hardly surprising that these proofs wind up showing that theories omitting the idea of dispersed states are impossible! Actually, all this was a misunderstanding. These proofs did not begin by assuming that quantum theory *had* to be in a superposition of states or *had* to be expressed by potentialities. All they did was point out that the experiments that have been done show that this is how one can represent nature. What they did next was say that quantum mechanics agrees with experiment; therefore, let us assume that the present way of expressing quantum mechanics gives correct results and then see whether it is possible to re-express these same results in a new way in which hidden variables explain the randomness we see. Thus, by assuming that quantum mechanics agrees with experiment, which so far is certainly the case, they proved there was no other way in which you could retain a perfect fit with experimental results while changing quantum mechanics into a deterministic theory. Ergo, the randomness cannot be shaken out of quantum theory.

John Bell didn't believe these mathematical proofs. He believed they were merely proving what they were assuming. He showed examples of how hidden variables could explain the results seen in many experiments, seemingly negating the validity of the mathematical proofs. But something more than Bell had to offer is needed. To negate the validity of a proof that quantum mechanics is not compatible with hidden-variable mechanisms, we require not just a single example of an experiment in which the behavior of some particle can be explained in terms of a hidden variable, as Bell argued, but, indeed, an example of a whole new quantum mechanics in which a hidden-variable formulation yields the same results as the Schrödinger equation everywhere.

However, Bohm and Bub seemed to have something that would fill the bill. They had a new quantum theory based on hidden variables. They added a new term to Schrödinger's equation that would sit there docilely affecting nothing until a measurement occurred, and then this added term would take charge. It would select one of the potential states to be the final outcome at the completion of the measurement interactions and destroy all the others, like a mathematical Pac Man eating up unwanted terms in the Ψ picture of the world. Just which potential state would occur would still be a matter of chance, but now there would be something causing the result. There would be something that would select which state would occur. The final event would depend on the value of a "hidden variable" in the added term.

But again there was a flaw in this new effort by Bohm and Bub to find a neoclassical physics. There existed nothing objective in their theory that would throw the switch to turn on the hidden Pac Man mechanism that would enable this new Schrödinger equation to select the final observed state of the system and explain the physics of measurement. The equation should describe nature. If mathematicians have to inject this term when they want the system to go into a single state and leave that term out at other times to keep the answers correct, well, then it's just a fudge factor. If one is going to cheat, surely there are easier ways to fake the result.

So it turned out that Bohm and Bub's effort to introduce a hidden variable to take care automatically of state vector collapse did not work. In fact, there were many efforts to achieve a hidden-variable "mechanism" that would still give the correct answers but would remove the seeming need to have an observer determine when state vector collapse had occurred. These efforts could have gone on and on without resolving the problem. But John Bell proposed a daring new approach. He would show how to prove quantum mechanics was actually wrong, just as Einstein had maintained all along. If Einstein had been correct in his criticism of quantum theory, if quantum mechanics is incomplete in its description of reality, there must be a way to show this, and Bell believed he knew how. The experiment that John Bell conceived is now recognized as one of the most important experiments ever done in the history of mankind.

8

The Sound of the Temple Bell

"He himself said it," and this
"himself" was Pythagoras.
—Cicero, *DeNatura Deorum*

It is a place for meditation, a place for the mind to search and to rest. It is a temple in Kamakura, framed of wood and open to the mountain air. Somewhere a gong is struck, sending its low dissonances into the zendo of the temple in Kamakura. Each monk moves on, wooden bowl lacquered black in his hand, in procession; wooden walls and wooden halls darken the path each takes. Each monk stands before the table in his place. A gong is struck. The sound breaks free, searching down the halls and out into the fields of rice, down over the hill toward the sea before it is caught by the voices chanting a thousand years of Buddhist songs and whispered words. The last syllable is spoken from the lips of the last monk. Each monk eats rice in silence. The bowls are scrubbed, the table left. A gong is struck. The sound turns the shadows and clocks the cogs that set the mind. The master reads:[1]

The bamboo-shadows move over the stone steps
as if to sweep them, but no dust is stirred;
The moon is reflected in the pool, but the
water shows no trace of its penetration.

A gong is struck. The sound touches a black koromo, brushes the temple hair, enters the mind, and mingles with a question: "It is too clear and so it is hard to see?"

Is there some hidden machinery that causes things in the atomic world to look like they jump about at random? Most physicists might have conceded that there could be something hidden going on in the atomic world—something to explain the strange nature of quantum mechanics in terms of a classical mechanical picture of reality—but in the absence of any experiment to reveal the existence of such inner workings, the whole question has seemed academic. The argument has raged like a tempest in a teapot through the pages of the physics journals, a thousand articles finely argued, but for most physicists it has seemed to be merely a philosophic squabble. As Bernard d'Espagnat, professor of physics at the University of Paris, has pointed out,[2] the exclusion of hidden variables is justified for three reasons: First, exclusion of any such hidden variables simplifies the mathematics. Why complicate matters unnecessarily? Second, the present simpler formalism of quantum mechanics predicts results that are confirmed by experiment. Why mess with success? Third, the addition of hidden variables would not explain anything that is not already explained by the present theory of quantum mechanics. So why bother? For the physicist, using hidden variables is like, well, being verbosely talkative. It strikes the mind of the physicist like redundance strikes the grammarian's ear. Conventional thought in quantum mechanics had dismissed hidden variables as superfluous—ultimately, perhaps, as meaningless. Without some way to test the existence of a hidden reality of the kind Einstein and his friends had talked about, the whole subject would be academic.

The way out of this apparent impasse was discovered by the late John Bell. I met him in Vienna at the Schrödinger Centennial Symposium shortly before he died. He was lively and quick, a carrot-topped leprechaun of a man from Northern Ireland. He had already become a legend for his strange discovery, which had changed him from one who believed quantum theory to be incomplete—believed nature to be objective and exclusively material—to one who had begun to speak about the role consciousness may play in the scheme of things.

Early in the 1960s, John Bell saw something in the Einstein–Podolsky–Rosen (EPR) thought experiment that everyone else had missed. He saw the possibility that there just might exist an experimentally testable difference between the hidden-variable picture of the atomic world and the Copenhagen interpretation of quantum mechanics. Bell felt that the EPR thought experiment just might be turned into a real experiment that would make it possible to circumvent Heisenberg's restriction on our ability to measure both momentum and position simultaneously in order to let us see beyond the veil—to let us see a real objective machinery ordinarily hidden from view.

It is not too difficult to get a sense of what Bell saw in the EPR experiment that offered the chance to solve the measurement problem. Remember what Einstein and company had talked about in 1935. They had suggested that one could carry out experiments on pairs of particles that had been produced in such a way that they would both have exactly the same values (or, just as good, exactly opposite values) of some quantity to which the Heisenberg uncertainty relationships apply. This would give the experimenter a second chance at measuring the true and hidden state of the position, speed, spin, polarization, or whatever that determines the way the atomic world works. The EPR experiment offers the possibility of making a second measurement on subatomic quantities that quantum mechanics assumed must have only a fuzzy existence—a fuzzy existence, certainly, but a fuzzy existence in which, for every possible state of the first particle, the second particle has to have a corresponding state because of all those conservation laws (conservation of spin, energy, momentum, and the like). It is, moreover, a fuzzy existence that must become concrete upon observation. All this means that if we do such an experiment on the first particle so that it goes into a definite state, then if quantum mechanics is true, that second particle must now show the result, as though it somehow has gotten the message that it now must no longer be in a fuzzy state but must take on the state that corresponds to the value of the measured quantity that the first particle took on. This second particle must take on its corresponding value instantly, no matter where it is in the whole universe.

But if we do the experiment that Einstein and company talked about and we find that quantum mechanics actually predicts the wrong result, then Einstein would be vindicated and quantum mechanics would have to be replaced by an objective theory. This would show that quantum mechanics is nothing but a fuzzy way of looking at an inner world of objects that in fact have definite positions and definite states. This would return physics to an objective, physicalistic picture of reality. This would bring physics out of its formless, twilight world of atomic tenuity. This is what Bell saw and what he hoped for.

But Bell went further than the EPR three. He showed that the requirements of objective physical realism call for things to behave in definite, testable ways. Both quantum mechanics and hidden variables can give us a world that looks random, at least on the surface. Hidden variables could even behave in the same delicate way that Heisenberg's uncertainty principle calls for. But the fact that quantum mechanics calls for the state to be made definite in both particles when either one is measured, even though the particles are separated, means that there should exist a demonstrable difference between quantum mechanics and any possible hidden-variable theory. If the test Bell proposed went his way, objective reality would be proved true. But if the results went the other way, physicists could no longer presume the existence of some unreachable, some unobservable, but nevertheless objective reality. Objective reality would have been tested and found wanting. That is exactly what has happened. Bell was right about being able to test, but wrong about the outcome. To understand what reality actually is, we have to understand how this happened.

In order to do this, there are quite a few facts that we have to muster. Taken together, these facts prove there is something fundamentally wrong with objective models of physical reality—the models we use to understand what we are, what the world is, and what it is all about. The answer[3] is 42. But first we must understand the question. We have to understand Bell's theorem.

THE BATTLE PLAN

The story gets a bit complicated, not because of intrinsic difficulties, but just because there are a lot of things we have to bring together.

We can have, as EPR proposed, an experiment in which we get a second chance to make measurements on the state of a system. But we are trying to find out whether quantum mechanics is correct in describing the world as a place where these things do not have a definite reality until the observation event.

The question in physics is whether quantum mechanics—that is, the stochastic behavior of matter at the atomic scale—could be explained by assuming some hidden mechanism, some subatomic mechanism responsible for the seemingly random behavior of photons, electrons, and atoms. Are quantum jumps that seem so random actually intrinsically random, or does something cause the randomness?

Bell turned to the EPR type of experiment to provide a way of putting quantum mechanics on trial. In the last chapter, we talked about the EPR experiment in terms of measurements on the orientation of electron spins. Now we are going to talk about another version of that same experiment, this time using photon pairs rather than electron pairs and employing polaroid filters to test their polarity. (We choose this setup because just about everybody has had some experience with polaroid filters.) The photon pair will be one in which the photons go off in opposite directions and have exactly the same polarization. This means that any polaroid filter that would let one photon pass through would also let the other pass through. Any polaroid filter that would stop one of these photons would also stop the other, allowing, of course, for an occasional exception due to imperfections in the filter or reflections off the surface of the filter. Bell's theorem concerns only one simple question: How many photons that pass through a filter set at one angle will have twins that will also pass through another polaroid filter set at another angle? In order to see how this all works, let us look at how polaroid filters work.

FUN AT THE MOVIES

Most people who have been to a 3-D movie have played with the polaroid filters in the special glasses provided. If you look through only one of these filters at an ordinary light bulb, or at most objects illuminated by sunlight, you will notice that about half the light passes through the filter. The polaroid filter has a preferred axis so that light in which the electric field waves are aligned with that axis will pass through. When the electric field is perpendicular to that axis, the light is filtered out—that is, it is absorbed. If the electric field points at some angle to the axis of the polaroid filter, then part of the light wave can pass through and part will be blocked off. The part that can get through is the component of the electric field that runs parallel to the axis of the polaroid filter. The light from a light bulb or from the sun is unpolarized; it consists of light with all possible orientations (each photon has its own orientation, and all possible orientations are present in the myriad photons that make up the light). Now let us take the left (or right) polaroid filter from the 3-D glasses and put it in front of the other filter. If we look at a light bulb, we will notice that most of the light is now blocked off. If we had scientific-grade polaroid filters, almost no light would get through. Now if we slowly turn one of the filters, we will see more and more of the light coming through the two filters until, when they are aligned (the right-eye filter perpendicular to the left-eye filter), virtually all the light coming through the first filter (about half the light that strikes it) gets through the second filter as well.

But how does this work? Let us start with what is actually a very good analogy. If we take an oven rack, stretch a string through the grating, and pluck the string in the same direction as the slits (the rods of the grating), the string will vibrate unimpededly. If we pluck the string perpendicular to the grating, the waves on the string will stop at the grating, and no motion—no wave—will propagate through. The motion is stopped by the oven grating.

This is what happens when we place one of the polaroid filters from the 3-D glasses in front of the other filter without turning them. The axis of one filter is perpendicular to the axis of the second filter. Light passing through the first filter is polarized in one direction, and when it comes to the second filter, the same thing happens to the light that happens to the vibrating string that passes through the grate; it is damped out. The polaroid filter does not let the electric wave vibrate in that direction, so the wave cannot go through. Now, in our oven rack analogy, if we pluck the string at an angle to the grating, a *component* of the wave will propagate through. The part of the string's motion that lies in the plane of polarization (parallel to the grating bars) will be free to vibrate, and the rest will be blocked off.

Similarly, when we set the polaroid filters from the 3-D glasses at an angle to each other, only a *component* of the light wave can get through. If we start with light from a light bulb, it comes from the vibrations of the hot atoms in the tungsten filament. Because these atoms vibrate at random, the light waves that are emitted come out with the electric component of the wave vibrating in every possible direction that lies perpendicular to the direction of the light beam. When this light passes through the first filter, every electric field *component* perpendicular to the axis of the first filter will be damped out. Next, the electric wave of the light beam hits the second polaroid filter. This filter lies at an angle to the first filter. What happens is exactly what would happen if we had a string pulled tightly that passed through two oven grates tilted at an angle to each other and we proceeded to pluck the string. Depending on the angle at which we plucked the string, only a part of the wave would then get through the first grate, and, again, only a part of *that* would get through the second grate. Line up both grates so that they are parallel and pluck the string parallel to the direction of the bars, and all of the wave goes through both grates. Turn the second grate at an angle to the first, and only a *component* of the wave motion gets through. That component passing through the grate is like the component of the electric wave of the light that passes through the two polaroid filters. That component of the string wave has a magnitude; its magnitude, or *amplitude*, is given by the biggest deflection that the string's wave makes. We can also talk about the magnitude of the light's electric wave at any point in space and time. The amplitude of the electric wave is the maximum value of the electric field as the wave passes any point in space.

BUT THE PHOTONS CARRY THE ENERGY

Now all this tells us how much of the electric field gets through the polaroid filters. But how many photons get through the 3-D glasses? Everything we just said was said as though light were a wave, just likes waves on water, sound waves in air, or strings plucked. But the electric waves are a holdover description of what reality is all about. The electric field is actually a special kind of wave: a probability wave. It is to the photon what Ψ is to the electron and other material particles. Now Ψ is the probability wave amplitude for particles such as electrons and atoms. Similarly, the electric field is a probability wave amplitude for photons (but measured indirectly in terms of the photon's energy).

In the case of the electromagnetic wave of light, we talk about the energy carried by the wave rather than talking about the probability of a particular photon being somewhere. We use this electric field to calculate how much energy is in the wave. Furthermore, photons are packets of energy. They carry the energy in the electromagnetic wave that is light, and the number of photons in the wave depends on the energy of the wave.

Now the energy in just about any kind of wave is given not by the wave amplitude, but by the *square* of the wave amplitude. This is so for water waves and waves on strings in oven racks, and it is so for electromagnetic radiation (light). The square of the wave amplitude determines the energy in the wave on the string, and the square of the electric field amplitude determines how much energy there is in the form of light—of photons that are there ready to go through the polaroid filters. But it is the electric field, not the square of the field, that determines how much of the wave will make it through the filters.

Look at what happens when light that passes through a first polaroid filter at an angle A then goes through a second filter at an angle B. Let us see how much light (that is, what fraction of the photons) that gets through Filter A will go on to pass through Filter B. If we let $E_{Filter\ A}$ be the electric field for the light that gets through Filter A, as shown in Figure 8.1, then the energy and the number of photons will be proportional to the square drawn on this vector $E_{Filter\ A}$, as also shown in Figure 8.1. Now, taking the component of $E_{Filter\ A}$ parallel to the axis of Filter B will give us $E_{Filter\ B}$, which is the magnitude of the electric field coming out of Filter B, as is also shown in Figure 8.1. A square drawn on this is proportional to the energy and the number of the photons that get through Filter B. The small square in Figure 8.1 that connects $E_{Filter\ A}$ and $E_{Filter\ B}$ represents the energy and the number of photons absorbed by Filter B.

What all this means is simply that (1) only a component of the wave can make it through a polaroid filter, (2) the energy of the light that gets through is given not by the wave amplitude but by the square of that wave that gets through, and (3) the number of photons that get through is given by the energy—that is, the square of the amplitude. I have said all this in a way that I hope makes each *item* logical, but logical or not, these facts have been established by innumerable experiments. The law in physics relating the number of photons that pass through the second filter to the angle between the two filters is called Malus's cosine-squared law.

Now, we have to come up with Bell's theorem in some fashion that will retain the central facts of the theorem without imposing too much mathematical encumbrance. This will be a bit tedious, but it is necessary to see just what physical ingredients determine what stops a photon at a filter and what lets it go through. The reward for all this effort is coming up soon.

If we have twin photons created at the same time and place and with the same polarizations (see Figure 8.2), the chance that the second photon will get through its filter, if the first one gets through its filter, also depends on the angle between the two filters. You just might want to ask yourself, "How does the second photon know what the position of the other photon's filter is?" Think about it for a while. We will come back to this question later, but this little puzzle is the whole Bell theorem problem in a nutshell.

Look again at the three squares in Figure 8.1. What we have constructed is the standard figure from the Pythagorean theorem. The sum of the areas of the two smaller squares on the right triangle equals the area of the larger square. Thus we have a simple graphical way of showing how many photons (in any time interval) will pass through a pair of polaroid filters at any angle—and also how many will be stopped. The square on the hypotenuse is the number of photons coming in,

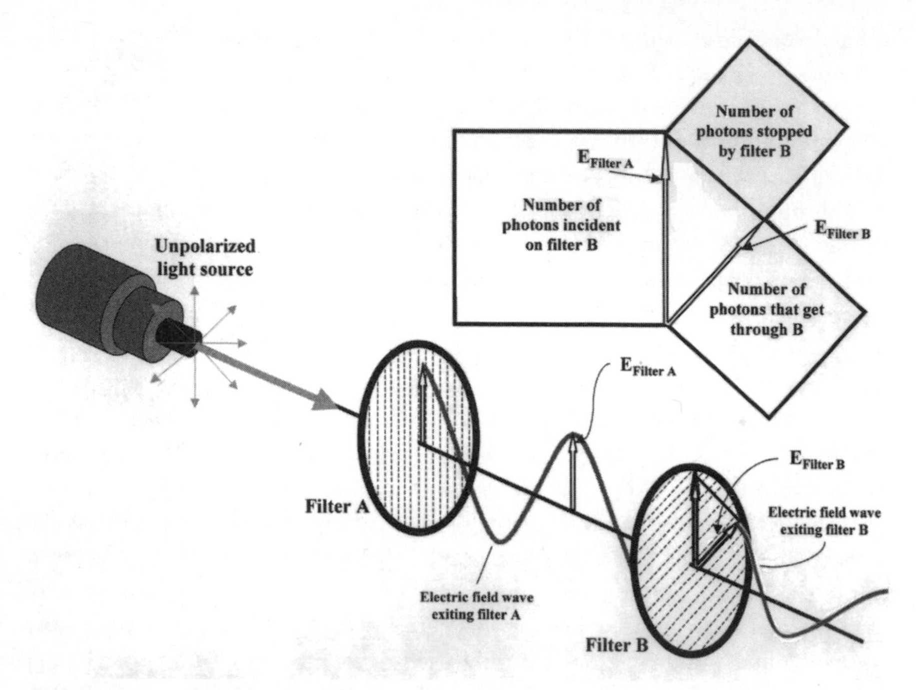

FIGURE 8.1 A beam of light is shown being polarized on passing through polaroid Filter A. The emerging light beam is polarized parallel to the filter's axis, as shown. When the beam of light reaches the next filter, Filter B tilted at an angle to Filter A, only a component $E_{Filter\ B}$ of the light's electric field $E_{Filter\ A}$ that got through Filter A, can pass through Filter B. Thus a portion of the photons will be stopped by Filter B. How many photons get through and how many are stopped depend on the electric fields and can be determined by using the graphical procedure shown, where squares are constructed on the lines representing the electric field components $E_{Filter\ A}$, and $E_{Filter\ B}$ and on the perpendicular line connecting them. This procedure also applies in Figure 8.2 for the twin photons.

and the other two squares on the right triangle represent how many photons pass through and how many are absorbed. This is actually what quantum theory tells us the photons do.

The photons carry the energy,[4] but they do not carry that energy as though it were attached baggage on the electromagnetic train. The photons come in stochastically. The electromagnetic fields tell us only about the probabilities of individual photons. Quantum mechanics tells us that each photon acts independently and probabilistically. When we pass a light beam through a polaroid filter, we are actually cutting down on the chance that any given photon will pass through.

Now there is a subtle but curious aspect to the way these photons behave. If we think of the photon as having its own polarization angle, then we must recognize

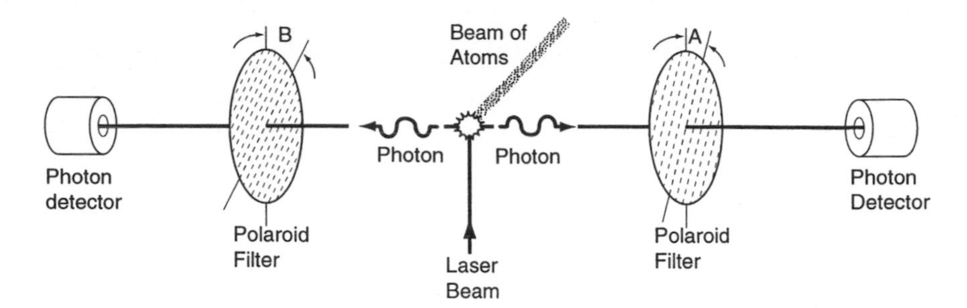

FIGURE 8.2 Experiment designed to test Bell's theorem. At the center, a beam of atoms crosses a laser beam. Atoms in the beam absorb single photons from the laser and reradiate a pair of photons in opposite directions. According to quantum mechanics, the fraction of photons with twins going through Filter A that will pass through Filter B depends on the angle between the axes of the two polaroid filters, A and B. How does the second photon know, so to speak, which way Filter A is pointing? Bell's theorem assumes that they cannot. (*Note*: Only the A and B orientations of the polaroid filters are shown here. The C orientation occurs when, in a second run of the experiment, B is set at a new angle C.)

that when it reaches the filter, that filter does not simply block it if the two axes are different, but instead the photon "makes a choice" as to whether it will go through or not. Whenever the photon reaches a filter at an angle A, then it sort of calculates out the square of the cosine of A, uses that to calculate the absorption probability, flips a coin or whatever photons do to make a random choice consistent with that probability, and then either proceeds through the filter or gets itself absorbed. But if the photon goes on to another filter, a filter set at exactly the same angle as before, it must act as if it "remembered" what it did at the last filter. There is nothing strange here if *we* remember Bohr's rules about complementarity—we, as opposed to the electron as an objective mechanism. We can talk about electric fields and waves and polaroid filters setting the total energy that gets through the filter, and then we can go down to the individual photon and say that any given photon is just a packet of that energy. But if we try to tell the story in a consistent way entirely in terms of this photon, it does seem as though the little critter makes its stochastic choice when it reaches the first filter and afterwards has to remember what it just did. It is as though the little guy carries with it a plan. The word *plan*, of course, is merely a way of summarizing how each photon goes about satisfying all the rules and regulations placed on the photon's behavior by the laws of physics. The concept of the photon having a plan will let us look in detail at how it is possible for nature to follow the laws of physics. This in turn will let us put the question of the nature of space, time, and physical reality to the ultimate test.

THE CAT IS RED, THE DOG WAS BLUE

Now we are going to see what John Bell's theorem has to say about such creatures as these little photons. Bell's theorem is all about just what that little photon plans to do when it meets up with a polaroid filter and how it shares its plan with its twin in the EPR experiment we talked about earlier (see Figure 8.2). Both photons, created with the same polarization at the same time and place, will always respond in exactly the same way to the same filter. If one photon makes it through a polaroid filter turned at a given angle, its twin brother will too.

Now let us see what quantum theory tells us must happen if the twin's polaroid filter is turned at an angle to the filter that its brother went through. According to quantum mechanics, it has to pass through or fail to pass through with a probability that depends on the angle between the two filters. It has to do this irrespective of what angle its brother's filter is set to. It is as though when the first photon gets to its filter, when it finds out the angle at which that filter is set, it sends a message to its twin to tell it what the angle is and whether or not it is going to pass on through itself. Then that twin, when it arrives at its filter, measures the angle between its filter and the angle its brother said was the setting for his filter, and then figures out from the cosine-squared law what to do. Quantum theory demands that the photons behave in this manner even though physics also tells us that there is no way to send such a signal between photons and, moreover, no way to send it fast enough. If this were the only way to explain what happens, it would be as though the twins had to be in telepathic contact.

Bell believed that it could not work that way. In fact, anybody reading what was just said should think that quantum mechanics is crazy and that the world cannot work that way. Bell believed there had to be some hidden plan, some hidden variables, that determined what each photon would do when it encountered a filter. He believed the plan made the two photons look as if they were correlated—behaving in unison—but he believed this was nothing but appearance. Bell believed that any such plan had to have been created when the photons were created. Their plan had to have been arranged before the two photons separated, at the very latest. There is no way that the photons could send signals at the last moment when they got to the filters. At that point, they would already have to be carrying their plan.

Obviously, if the photons can send messages back and forth—if state collapse is global, to use the language of quantum mechanics—then any kind of outcome would be possible in experiments with polaroid filters. But if the photons cannot send any such message—if state vector collapse is entirely a local process that can affect only the photon actually being measured—then the correlation between what this photon chooses to do and what its twin does must be limited to the information it could carry in its hidden variables—in the plan that he stuck in his suitcase when he and his brother left the home where they were born as little ripples!

So imagine that there is a plan. The plan tells what each photon will do if confronted by any polaroid filter or any series of polaroid filters. This plan permits correlated behavior, and it permits stochastic behavior. For any given pair of photons, the plan can call for anything. Now for the big question: Is it possible for such pairs of photons to formulate a plan that would reproduce all the statistical results we see in experiments and meet all the requirements of quantum mechanics? Can they satisfy Malus's cosine-squared law, for example? Can this be achieved without changing our conception of space and time, without introducing what Einstein called "spooky action at a distance" signals between the photons, without things happening that look like these photons can send telepathic messages? The fact that both photons follow this plan gives us a chance to make two measurements to see what the plan is. Bell showed that there exists an almost trivial limitation on any such plan. This seemingly trivial restriction, however, turns out to say that quantum mechanics predicts one outcome experimentally, and *any* hidden-variable theory in which the particles have only the information they carried with them when they "left home" predicts an entirely different outcome. Bell had found that sharp cutting edge that Einstein had longed to find.

To follow Bell through his theorem making, we are going to assume that each pair of photons has a plan. Now let us look at how we are going to find out something about what this plan looks like. Figure 8.2 shows the setup that we will be considering. A laser beam is used to excite atoms into emitting a pair of photons that will have the same polarization. The photons travel off in opposite directions, each passing through a polaroid filter, and each, if it is not absorbed, passing into a detector so it can be counted.

The polaroid filters are set at different angles. In Figure 8.2 we show the filters set at the two angles labeled A and B. Two additional experimental runs are to be carried out with the filters set at angles A and C and at angles B and C. The particular angles chosen are to be selected so as to maximize the difference between the predictions of quantum mechanics and of Bell's theorem. We have already seen what quantum mechanics tells us is the way the photons will go through filters; now let us see what Bell's theorem tells us about the numbers of photons that go through both A and B, both A and C, and both B and C, or about the numbers that will go through A and be stopped by B, and so on, assuming only that these photons cannot have any spooky telepathic contact to help create correlations once the two photons have separated.

If Bell and Einstein had been correct, what would happen in each case would depend on what kind of plan each pair of photons had. For example, one pair of photons could have as a part of their plan that if they encountered a filter at angle A, or B, or C, then they would pass through the one at angle A, stop at the one at angle B, and go through the one at angle C. (They might also have a plan about filters set at other angles, but any other such plan will not be relevant to the experiment we are considering.) The two particles give us two chances to check on the plan, but only two chances. What we need to do is check on all three, somehow. The reason is simple enough. We need to check on what the photon pair does at each of the three angles, because one of these really serves only as a reference angle. That means we still need to look at two more angles simply to find out how the behavior of the photons changes as we change the angles. So we will be looking for information about the photons' plan at each of three angles, even though we can make measurements to discover what that plan is on only two of these angles for any given pair of photons.

Now let us see how we can do this with the experimental data we collect with the apparatus shown in Figure 8.2. First of all, let us just consider those photons that will encounter the filter that is placed at angle A. All the photons having an internal code A^{yes} will pass through the filter, and all those with a code saying A^{no} will stop at the filter. If we count the number of photons that pass through the filter when it is at angle A, then this will give us a number that we will call $N(A^{yes})$. We will designate the number that do not pass through as $N(A^{no})$. Now let us look at the case having both Filter A and Filter B as in Figure 8.2. Let us call the number of photons that pass through both filters $N(A^{yes}B^{yes})$. For the number that pass through A but not B, let us use the symbol $N(A^{yes}B^{no})$, and so on.

We will be talking about a photon plan that has $A^{no}B^{yes}C^{no}$, and those with A^{yes} $B^{no}C^{no}$, and every other combination. Since our measurements can check out only two of the three angles, we will also have to deal with cases where our information is limited to $N(A^{yes}B^{no}C^{I\ don't\ know})$. Next we will gather our data, fill a book with that data, and then use that data to check to see whether Bell's theorem is true or not. But we almost forgot something!

WHAT'S BELL'S THEOREM, ANYWAY?

In all my efforts to explain what quantum mechanics requires, and what kind of experiments we are going to do to test Bell's theorem, I almost forgot to say what Bell's theorem is. Bell's theorem is very simple. I will explain in a moment, but first, just one last digression. It's a nice one. It's about writing a book!

Let us talk of writing an odd and very long book. It is a book of a million sentences, and each sentence has but three words. The first word is the subject, and it is the word *dog* or the word *cat*. Either word; it does not matter. The second is the verb, either *is* or the word *was* can be chosen in each sentence, so long as every sentence has a verb. And then each sentence in this book will have to have a predicate modifier: the word *red* or the word *blue*. It is an entire book of these sentences that are created randomly—or not. Let us give the book a name. Let us call it *The Guide.*

Now let us ask a simple question. If one counts their number, will there be more sentences in *The Guide* that say "Dog is blue," or will there be more that just say *something* "is blue"? Well, if there is even one sentence that says "Cat is blue," then there will be more of the second than of the first. What is more, this is true regardless of whether the subjects, verbs, and modifiers are selected at random or not.

Let us ask another, harder question. If we add all the sentences that say *something* "is blue" to all those that say "Dog was" *something*, how many will there be? Will there be more or fewer than those that say "Dog" *something* "blue"? If you think about it, you will see that

$$N(\textit{Something is blue}) + N(\textit{Dog was something}) \geq N(\textit{Dog something blue})$$

where the symbol ≥ means "greater than or equal to."[5]

And that is Bell's theorem!

Let me explain. Each of these words has been used to stand for one of the options the photons have to pass or not to pass through a polaroid filter turned to one of three angles A, B, or C. If the two photons decide that they will go through a filter at angle A, then *The Guide* would use the code word *dog*. If they decide to stop for angle A, then the code word is *cat*; the mnemonic *dog* stands for passing through Filter A. At the filter with angle B, the word *is* designates passage; *was* indicates that either photon will stop. For angle C *red* means go and *blue* means stop. Thus the sentence "Cat was red" means either photon would stop at a filter at angle A or at a filter at angle B but would pass through a filter at an orientation C.

Thus our mnemonic Bell's theorem from above really says,

$$N(B^{yes}C^{no}) + N(A^{yes}B^{no}) \geq N(A^{yes}C^{no})$$

All that Bell's theorem says is that even if the photons prepare a plan before they separate, a plan that details everything they might do under any circumstance when they encounter polarizing filters, there are limits on what they could possibly achieve in the way of their correlations. All that Bell's theorem says, when it is done, is that plans always have their limits. When we make our plans, you and I, and we part company, things can still go wrong, even for the best of plans.

Bell's theorem tells us that if we do experiments with polaroid filters set at the angles A and B, B and C, and A and C, the number of photons that make it through A but not through B added to those that make it through B but not C will always have to be bigger than the number that make it through A but not through C. Otherwise, the little devils are acting as though they are in telepathic contact, and we know no physicist in his right mind would accept that. It would even be silly to think such a result would come out of an experiment to test Bell's theorem. It would be silly, that is, unless that happened to be what quantum mechanics says happens. In that case, we should want to know what really happens in an experimental test of this theorem. We should want to know whether photons act sensibly and carry around a plan—which would mean either that quantum mechanics is wrong or, if quantum mechanics turns out to be correct, that we would be faced with the problem of figuring out how these photons are sending their telepathic signals!

Well, does quantum mechanics call for photons to violate Bell's theorem? From what we have learned about photons already, it is now pretty easy to answer that question. This is why we went through all that tedious business about polaroid filters and Pythagoras. Look at Figure 8.3, and refer also back to Figure 8.1. In Figure 8.1, we saw that if a polaroid filter at an angle A passes certain photons, whereas one at orientation B stops it (or, referring to Figure 8.2, if one of the photon pair passes through the filter having an orientation angle A, whereas its twin is stopped at the filter at angle B), then the number will be proportional to the square constructed opposite the angle between Filters A and B. This square opposite the angle between Filters A and B, therefore, stands for $N(A^{yes}B^{no})$, which we notice is the second term in the Bell's theorem inequality.

Now, in Figure 8.3 we have drawn lines that indicate the polarization axes for the three settings of the filters in the experiment shown in Figure 8.2. These lines point in the directions A, B, and C, which we will take to be set at small angles to one another since that makes it easy to see where Bell's theorem differs from what quantum mechanics calls for. Now if we check Bell's inequality and check Figure 8.3, we will see that each term is represented by a square. Bell's theorem tells us what we should find, if Bell was right and Einstein and Podolsky and Rosen—and Bohm, for that matter. Figure 8.3 shows us graphically what quantum mechanics calls for. The square constructed opposite the angle between Filters A and B is proportional to the second term in the Bell's theorem expression, as mentioned already. The square constructed opposite the angle between Filters B and C is proportional to the first term in the inequality, and the square opposite the angle between Filters A and C in Figure 8.3 is proportional to the third term. It is quite obvious that the square constructed opposite the angle between Filters A and C is much larger—nearly twice as large—as the sum of the other two squares. This is exactly the *opposite* of what Bell's theorem calls for. Quite obviously, quantum mechanics calls for results that would violate Bell's theorem.

So who is right? The only way to resolve this conflict is to see what the photons themselves actually do. The experiments have been performed. The verdict is in.

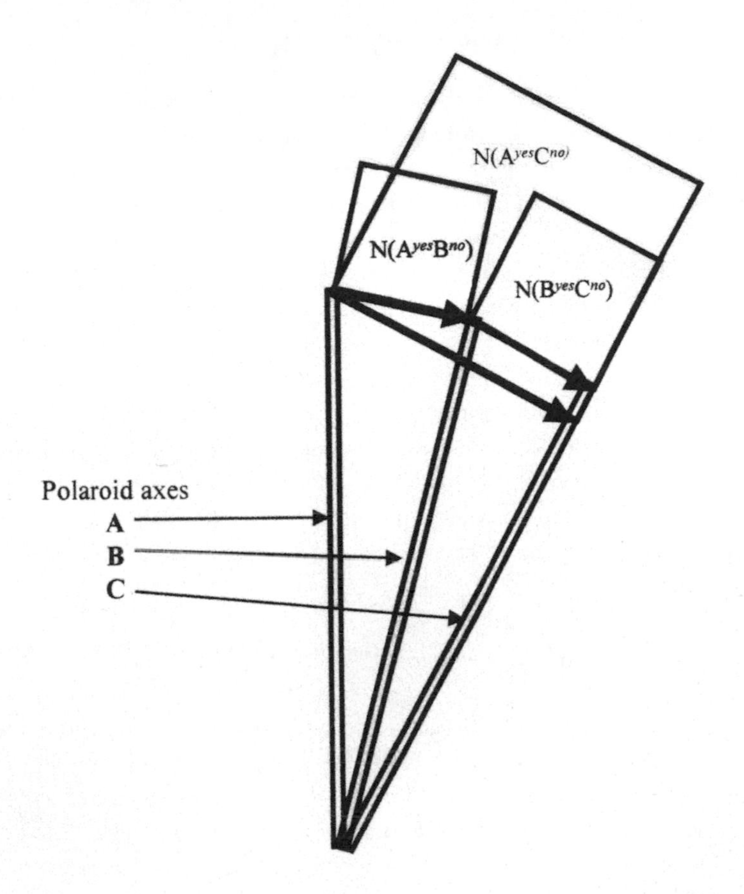

FIGURE 8.3 These three squares show the number of photons that will pass through one filter while their twin is stopped by the second filter according to quantum theory. The lines A, B, and C represent the orientations of the polaroid filters used in the experiment in Figure 8.2, C being a second run of the experiment with B set at a new angle C. $N(A^{yes}B^{no})$, for example, tells us the number of photons passing through Filter A at orientation angle A while their twins are blocked by their filter turned at orientation angle B. The figure is constructed assuming that the experiment is run until the number of photons that make it through the first of their filters is the same in each case.

The photons do things the way quantum mechanics calls for things to be done. This means nature violates Bell's theorem. The photons do not carry around a plan—an awkward enough hypothesis in any case. Instead, the photons seem to use telepathy! Of course, they don't use telepathy, but they do something that is, whatever it is, just that bizarre. Let us take a look at what the experiments revealed.

CHECKING IT OUT

Doing the experiment turns out to be just a bit more complicated. The problem is that it is awfully hard to do experiments in which we look for photons that do not go through one of the polaroid filters. Back in the Bell's inequality expression, we talked about the number of photons that go through Filter A but have twins that fail to go through Filter B. What we need to do is change the experimental setup so that we are talking about measuring photons passing through Filter B at the same time that the opposite member of the pair is passing through Filter A. If we set things up this way, we can put a lot of fancy coincidence circuitry into the detectors that will keep us from counting stray photons. With the coincidence circuitry, we will count only those events in which a photon goes through A at the same time—to a billionth of a second or better—as one goes through B. In that way, we can be sure that any pair of photons detected actually started life together as a pair.

In order to set things up this way, all we need to do is to rotate the second polaroid filter through an angle of 90°, a quarter of a circle. If we do that, then Bell's inequality will read

$$N(B^{yes}C'^{yes}) + N(A^{yes}B'^{yes}) \geq N(A^{yes}C'^{yes})$$

where C' and B' simply remind us that we will have to rotate those filters through 90°. The argument we used in Figure 8.3 still works, but the labels on each of the squares have to be changed from $N(A^{yes}B^{no})$ to $N(A^{yes}B'^{yes})$, and so forth.

Another way to make the change is to talk about photons that are emitted with polarizations at right angles (at 90° to each other). This means that the photon twin will always pass through a polaroid filter that would stop its brother photon. If we use photons like that, then Bell's theorem takes the form

$$n(B^{yes}C^{yes}) + n(A^{yes}B^{yes}) \geq n(A^{yes}C^{yes})$$

Here, I have used the lower-case *n* just to indicate that we are to do the experiment with photons that are cross-polarized. By the way, this last form for Bell's theorem is also the equation to use if we do the experiment on the spin of protons or electrons, because there, the only thing that is convenient to use in such an experiment is pairs of oppositely aligned particles.

In 1969, while at Columbia University, John F. Clauser got the idea of using an apparatus like that shown in Figure 8.2 to test Bell's theorem. Bell himself had not described in any detail how to do a practical experiment. Kocher and Commins at Berkeley had already set up an apparatus to look for inconsistencies in quantum mechanics, but without obtaining any conclusive evidence. Clauser went to Berkeley hoping to use their equipment. A student of Commins, Stuart Freedman, worked with Clauser to set up the experiment using a beam of calcium atoms irradiated by ultraviolet light to produce pairs of violet and yellow-green photons that have correlated polarizations. In 1972, Freedman and Clauser published their results, showing that nature agrees with quantum theory: The photons violate Bell's theorem.

In 1973, Holt and Pipkin repeated the experiment using mercury atoms producing violet and green photon pairs, which are emitted with their polarizations at right angles to each other. They got results that seemed to agree with the limit imposed by Bell's theorem. In 1976, John Clauser at Berkeley and a separate group, Edward Fry and Randall Thompson at Texas A&M, repeated the Holt-Pipkin experiment using mercury atoms and improved on the accuracy of the prior test. Both showed conclusively that the photons behave the way quantum mechanics says.

Since then, experiments have been done using very high-energy photons, gamma rays produced by positron-electron annihilation. In 1974, Faraci, Gutkowski, Notamigo, and Pennisi, at the University Catania in Italy, obtained results in agreement with Bell's theorem, but theirs was recognized to be a marginal result. Repeated at Columbia University in 1975, this experiment yielded results in agreement with quantum theory. In 1976, Lamehi-Rachti and Mittig, at Saclay Nuclear Research Center near Paris, carried out an experiment using pairs of protons, ordinary pieces of matter. This experiment also gave results in agreement with quantum theory.

Although a couple of these experiments seemed to support the hidden-variable hypothesis that Bell's theorem tests, in each case when the experiments were refined, the more careful tests gave results that unequivocally agreed with quantum theory. Despite all of these results, there remained one possible "out." Perhaps the polaroid filters generate some field that makes their presence felt at a large distance. If so, then the second photon could be responding to a local effect produced by the distant photon's filter. Alain Aspect, Jean Dalibard, and Gerard Roger, at the University of Paris-South in Orsay, France, however, ruled out this possibility. It was necessary to acknowledge that the tests of Bell's theorem prove quantum mechanics to be correct and prove Bell, Bohm, and the EPR trio wrong. In his conclusive experiment, the direction of the polaroid filters was set by acoustically driven optical analyzers capable of switching the polarity in a time much shorter than the time required for the photons to travel through the apparatus. As a result, there was not even time for any signal to go from one photon to the other or from one filter to the other before the photon had already passed through the polarizers and into the detectors. Yet the photons still behaved as though they knew what had happened to their distant photon twins.

For photons, their property of polarization violates Bell's theorem. Measurements show that particle spin also exhibits this strange violation of Bell's objective reality. But it all goes much further. If we return to the original paper by Einstein, Podolsky, and Rosen, we find that any of the other pairs of quantities for which quantum mechanics exhibits an uncertainty relationship could equally serve as the basis for such a test, and the results now make it abundantly clear that quantum mechanics is right and Bell was wrong. We are forced to conclude that momentum, position, energy, and temporal intervals must exhibit this same nonlocal behavior that has already been established for the property of particle spin and for photon polarizations.

So the photons do not carry some kind of plan that lets them mimic quantum theory. No set of hidden variables inside each enables them to behave in an apparently stochastic way. Instead, we must realize that something about the way we had imagined the world to work must be wrong. Bernard d'Espagnat says that this "discovery discredits a basic assumption about the structure of the world."[6] He adds that "it was possible until a few years ago to believe in an independent external reality and simultaneously to regard Einstein separability as a completely general law bearing on that reality." But since all objects ordinarily regarded as separate have at some time in the past interacted with other objects, in some sense everything constitutes an indivisible whole. He says, "Perhaps in such a world the concept of an independently existing reality can retain some meaning, but it will be an altered meaning and one remote from everyday experience." Objects—all objects—that once have been in contact remain forever in intimate touch, forever entangled, no matter where they go, no matter how far in space they seem to be separated from each other.

. . . And the temple bell sounds for evening meditation.

IS THE MOON THERE WHEN NOBODY LOOKS?

who knows if the moon's a balloon, coming
out of a keen city in the sky

—e.e. cummings, "pity the busy monster"

What does it all mean? We have covered some heavy stuff. It may take a while to sort out all the pieces. I hope, though, that I have adequately emphasized that these tests of Bell's theorem show there is something wrong with the usual idea of objective reality. Just how great the hiatus is can be seen in an article by David Mermin in *Physics Today*,[7] (the *JAMA* for cyclotron jockeys). Mermin is director of the Atomic and Solid State Physics Laboratory at Cornell University. In 1985, he wrote an article explaining the consequences of this "spooky action at a distance" that comes out of the experiments showing that Bell's inequality is violated by nature. Quoting Einstein, he titled his paper "Is the Moon There When Nobody Looks? Reality and the Quantum Theory." He says, "Anyone who's not bothered by Bell's theorem has to have rocks in his head. . . . The EPR experiment is as close to magic as any physical phenomenon I know of."

But where do we go from here? To answer this, we should first realize that the outcome of the tests of Bell's theorem does not tell us something new but rather puts an end to something old. It puts an end to those ideas that inspired all the efforts to reinterpret quantum mechanics that we talked about earlier. It puts an end to the false conception of reality that motivated Bell to attempt to explain away quantum mechanics with local hidden variables. It undermines a conception of reality that most people—especially physicists—have subscribed to for generations. It discredits the idea of an absolutely objective reality made up of localized, discrete events and bits of discrete matter. It is the death knell of the false doctrine of materialism. Madonna may be a material girl, but this is not a material world. That, Madonna, is what Bell's stuff is all about!

We must remember, too, that Bell's theorem was only the last gasp of the concept of local objective reality. The literature of quantum mechanics was already replete with demonstrations that this concept is false. One after another, between 1935 and 1970, physicists and mathematicians had told us, had in fact proved, that quantum mechanics is totally at variance with such ideas of a materialistic world. They had proved that the experimental data on which quantum mechanics was based could not be satisfied by any reconstruction of quantum mechanics in which local objectivity—that is, conventional materialism—was assumed. But with the tests of Bell's theorem that have shown nature and quantum theory to be in agreement, while showing the mechanisms assumed in Bell's theorem to be in disagreement with nature, physicists are at last beginning to acknowledge, albeit grudgingly, that quantum mechanics demands that we see reality in a new way. What is that new way?

Remember the ideas we used to obtain Bell's theorem in the first place. We said, for example, that we could make statements like:

$$N(\textit{dog is something}) = N(\textit{dog is red}) + N(\textit{dog is blue})$$

In other words, we assumed that even though we could not observe what these photons would have done at C because we have to destroy them to see what they do at A and B, still we believed they might contain some plan about what would have happened at C. We assumed that an object that would show a blue or red code if we looked at it must have a blue or red code even if we do not look at it. That is, we assumed objects have an independent objective existence.

But quantum mechanics does not make that assumption. Quantum mechanics assumes that the object is somehow both blue and red—that it exists as a combination of two mutually exclusive things until we observe it. Thus, in order to resolve the problem about Bell's theorem, we could reject the assumption of objective reality and instead accept that the act of observing things somehow affects the reality of the things we observe. This is the first way out of the dilemma posed by the failure of experiments to satisfy Bell's inequality. We can assume that objects are not independently real but instead depend in some way on the observer.

The derivation of Bell's theorem also assumes that one photon cannot "know" what is happening to its remote twin. The derivation assumes that the only way one photon can behave in a fashion that is correlated with its twin is if a plan was established when the two photons were originally formed. We ruled out the idea that a signal could be sent from one photon to the other so that one could respond to the polaroid filters in concert with its twin. More than that, the experiment by Aspect and his associates in Orsay made it impossible for any signal to go from one polarizer to the other or from one photon to the other in the short time available. Any message would have had to travel faster than light and, in so doing, violate Einstein separability—the idea that no influence can travel from one object to another faster than the speed of light. Hence, in deriving Bell's theorem, we assumed that any correlation had to be due to a plan that the photons already had. We excluded the possibility of any nonlocal, global, or faster-than-light influence. This, then, is a second avenue out from under the restriction that Bell's theorem imposes. We can say that nonlocal, or global, or faster-than-light influences are possible; we can drop Einstein separability from the description of reality. We should note, however, that almost every aspect of physics now depends on this nonlocal *entanglement* assumption being true, from the elegantly proven effects of the fine structure constant to the harsh fireball that swept away Hiroshima.

There is also a third way around Bell's theorem. We assumed, in order to derive it, that it is reasonable to draw conclusions from the experiments we perform. We agreed that even though we can never truly repeat exactly the same experiment, still we can repeat it closely enough so that we can abstract the laws that govern the universe. We assumed that we need not look at every physical phenomenon as each being a totally singular event. We assumed instead that given any change in the way a physical system behaves, there must be lawful relationships that can be discovered through experimentation. We can, of course, drop this third assumption. In that way, we can turn the simple physical experiment involving the change of one variable (the angle of the polaroid filter with its resultant effect on photon correlation) into an uninterpretable jumble of data in which conclusions cannot be drawn about causes, despite the fact that the regularities are plain to see. That is the third way out.

Thus we have made three assumptions about the nature of reality in order to obtain Bell's theorem. At least one of these must be wrong. Bernard d'Espagnat puts it this way:[8]

> This world view is based on three assumptions, or premises . . . . One is realism, the doctrine that regularities in observed phenomena are caused by some physical reality whose existence is independent of human observers. The second premise holds that inductive inference is a valid mode of reasoning and can be applied freely, so that legitimate conclusions can be drawn from consistent observations. The third premise is called Einstein separability or Einstein locality, and it states that no influence of any kind can propagate faster than the speed of light . . . . The[se] three premises . . . form the basis of what I shall call local realistic theories of nature.

But it is exactly these three assumptions forming the basis of Bell's theorem that now must be questioned. One or more of them must be false.

At a point in the history of intellectual endeavor when the strides of modern physics have propelled the mind to the far reaches of the universe, to the finest grit of matter's parts, to the very dawn of all existence, that suddenly we encounter doors that are blocked by our very reason—beyond which there seems to be no path to take and no waiting footfalls to tread into a future understanding of reality. But something must lie ahead. If we are to advance, if we are to understand what is faulty in the picture of reality that physics has painted, we must be ready to set aside one or more of these notions about reality. We must adjust our understanding and prove that the change we have made yields a better picture of reality. We must set aside the Escher picture of reality that spins the mind with complementarity word soups that do not hold internal consistency. We must give up local hidden variables and faster-than-light Bach chants and replace them with a new image, one seen in a Gödel vision: a conception of reality that is consistent, that is more compelling, that works, and that leads, as you will discover, to a surprising conception of reality. The old image of reality has faded away.

WHICH PATH IS THE WAY?

D'Espagnat gives us three alternatives: Reject objectivity, reject Einstein locality, or reject deductive reasoning based on the results of experiments. Which do we reject?

Obviously, if we reject the last of these, we are finished with the development of any further understanding of reality. To reject our ability to draw conclusions via experimentation would be to say that reality cannot be fathomed. Only if no other course were possible should we accept this choice.

This seems to leave us with two options: Reject Einstein separability (locality) or reject objective reality. Physicists, being steeped in the background, training, and tradition of the scientific explanation of facts about the objective world, have little predisposition to give up objectivity. If forced to choose, they would almost universally look to the more mechanistic explanation of the Bell's theorem results in some violation of locality. Better to yield just a bit on Einstein's view of locality than to abandon the commitment of science to objectivity.

But it is the truth we seek. Timidity is sometimes not the way to resolve such difficulties. The Ptolemaic system had to be changed by *De Revolutionibus*, phlogiston by recasting fire itself as the turbulent chaotic movements of matter. If we give up Einstein separability, we will be left with an incredibly complex situation. Because all particles came out of the original fireball that created the universe, all particles have been in contact. If we try to resolve our dilemma over Bell's theorem by saying the photons send faster-than-light messages back and forth, we are left with a picture of each piece of matter sending messages back and forth between every other piece of matter—messages that somehow have no associated mass or energy and no particles as carriers but that still affect the recipient particle. And because of the complexity of the measurement interaction (because of the fact that it is never easy to say which interaction is a measurement and which just a part of a series of interactions, as we saw in the case of Wigner's double Stern-Gerlach experiment), the particles themselves must take on a complexity in the way they sort out all these messages that would make each about as complex as an observer. It is not clear that abandoning Einstein's conception of locality would lead us to a workable new mechanics that would be explained in terms of these hidden variables. It seems that something even more drastic may be required to solve the problem. Perhaps if we examine closely the original question that instigated the measurement problem, perhaps then we will be able to find the solution to this problem.

BACK TO THE FUTURE?

We got into the measurement problem in the first place because quantum mechanics seemed to be saying that observers really affect the things that are observed—because quantum mechanics tells us that physical properties have, in general, no objective reality independent of the act of observation. As Pascual Jordan, one of the major early contributors to quantum theory, put it,[9] "Observations not only disturb what has to be measured, they produce it . . . . We compel [the electron] to assume a definite position . . . . We ourselves produce the results of measurement." Maybe the way out is to accept the fact that hidden variables, in any usual sense of the term, are not going to solve the problem. Maybe we need simply to accept the fact that the observer (the consciousness, for *observer* is just a euphemism for this concept) actually exists as a participatory constituent of physical reality. Maybe the way out of the Bell's theorem problem and the measurement problem in quantum mechanics is to stop denying the obvious answer. Quantum mechanics requires that we take into account the fact that conscious observers exist as unique entities, as a part of the total reality of the world. What we have to do now is find a way to understand what consciousness itself is.

Here in this examination of what Bell's theorem is all about, we have seen quantum mechanics forcing physicists to acknowledge consciousness as a reality affecting the material world. But why should the answer be so hard to accept? Consciousness should have long since been the topic of reasoned scientific study, and yet it has been largely ignored.

Bell's theorem was an effort to escape this obvious conclusion about quantum mechanics, and that effort failed. It failed because it was an attempt to design a universe that would leave out consciousness. The way out of our difficulty, the path we must take now, is to try to understand what was previously rejected. We must recognize that objective reality is a flawed concept, that state vector collapse does arise from some interaction with the observer, and that indeed consciousness is a negotiable instrument of reality. Our entire conception of reality must now be rethought. We stand at the threshold of a revolution in thinking that transcends anything that has happened in a thousand years. Now the observer, consciousness, something self-like or mind-like, becomes a provable part of a richer reality than physics or any science has ever dared to envision.

And let us recognize at the outset that this is not some obscure effect, some baroque brain-born notion that pops from the corner of the physicists' laboratory when some odd apparatus is wheeled out to demonstrate nonlocal magic. Einstein, Podolsky, and Rosen, in writing their pivotal paper, spoke principally about position and momentum; Freedman, Clauser, and Aspect and his colleagues ran their tests on the polarization of light; Lamehi-Rachti and Mittig measured the spin of protons, the nuclei of hydrogen atoms. All of these investigators spoke about rather simple and idealized configurations of matter—simple interactions of elementary particles. But nothing except the level of complexity changes if we go to the larger collection of matter in the real world. We cannot follow such complexity in order to show just where and how Bell's theorem gets violated in the interactions among people, for example, but this same nonlocality, this transcendence of space, must be manifest even there. That is the way physics is. We could never measure the leaves of all the trees blowing in the wind to show that Newton's laws apply there. We know that Newton's laws govern the leaves because we have carefully measured the effects and tested the laws where they are best demonstrated. And this is how we know that the conscious observer must touch everything: We have seen it where it is best seen.

This revised view of reality in which we see the observer and the consciousness as central to reality itself is as significant as if we had found the key to the soul. Perhaps that is what *has* been found here. Our quest for the fabric of reality has brought us from religion to science. But that science, when asked to show us reality, has caused us to look into a mirror to see what we are. It is the last place that science would have chosen to search, yet now we must look there into that image we see within ourselves. Now we must find out what the observer is and find out what threads consciousness weaves in creating the quantum mind.

THE NEXT STOP

I need to clarify something (and high time, you may say!). I need to clarify a particular point about Bell's theorem and the direction we have begun to take here. To review very briefly: We have found that quantum mechanics itself seemed to require an observer in order to make the Copenhagen interpretation work. Others, including Einstein and Bell, tried to show quantum mechanics (or at least the Copenhagen interpretation of quantum mechanics) to be wrong, and this attempt led to the tests of Bell's theorem. Those tests showed instead that any effort to picture the physical world in terms of local and objective reality (independently existing objects) would yield results that are inconsistent with the known behavior of atomic phenomena. What we have done, therefore, is to say that maybe the original idea—the Copenhagen interpretation, which tacitly assumes a conscious observer to affect the things observed—is correct. We are exploring the possibility that the solution to the Bell's theorem paradox, if you will, lies in understanding what the observer is.

I am not saying—and this is the point I wanted to clear up—that introducing consciousness into physics is the only possible way of dealing with the Bell's theorem results. I believe that all other efforts to patch up the physicalistic ideas will lead either nowhere or to some epicyclic morass. Such confused efforts to develop nonlocal hidden-variable theories have already been looked into—so far, entirely unsuccessfully. Examples of such efforts assume nonlocal hidden variables that "know" when a measurement is being made. But those dedicated to a physicalistic view of the world will have to explore those avenues. I have looked at this interpretation and found it wanting. It had already been examined and discredited by hundreds of the most capable physicists before Bell came on the scene. For me, Bell's theorem is the final nail in the coffin. But as I said, Bell's theorem does not prove that consciousness plays a role in physical processes. We are going to prove that, not by looking further into Bell's theorem or further into quantum mechanics, but by searching out consciousness itself, by savoring the fruits we gather as we discover what consciousness is.

◈

Nearly 20 years have passed since I last looked through the few things of hers that I still have. And yet, though she is gone, when I see these things—particularly when I see her signature, the way she would write her name—I know somehow instinctively that something of her is still here now.

I am not talking about the paper. I am not talking about the fiber that has been marked by ink. There is something that has had character, that has expressed that character even in these few strokes of her pen. There is something there that persists, something that I experience, something that I know whenever I see her name, whenever I hear her words or hear her name whispered in my mind. There is something . . . .

9

A Golden Brocade

And if you gaze for long into an abyss, the abyss gazes also into you.

—Friedrich Nietzsche, *Zur Genealogieder Moral*

In Kamakura, the temple bell sounds a dull, dissonant reverberation full of questions instead of harmony, questions that hang in the air like the sound and go nowhere. Hui Hai says, "What do you suppose is not the Buddha? Point it out to me!" Hui Hai says, "If you comprehend, the Buddha is omnipresent, but if you do not awaken, you will remain astray and distant from him forever."

It was not until recently that I found Zen Buddhism, timeless and sparkling, hidden amid its oriental foil. I had never had a moment to look into this Zen. I had other places to go and no time for Zen. I had no time for something that had for its meaning only the trappings of antique pedagogy, a remote absurdity that had no handle on my relevancies. I had other things to discover and my own places to go. I had no time for Zen. I had no time for Zen because I was spending my time finding other answers. But as I developed answers, answers that needed to be shared, I found that certain critical ideas were very difficult to communicate. Things that should have been the easiest ideas seemed oddly to be the most difficult. The more simply they were stated, the more clearly they were put, the less others seemed to grasp them. I was puzzled why this should be. One day, not too long ago, however, I picked up a little book about Zen written by Daisetz Suzuki, and our minds met across the gulf of cultures and the chasm of time between us. Here was someone who understood the beginnings that must be understood before one can pass to the bridge.[1]

> A monk once went to Gensha and wanted to learn where the entrance to the path of truth was. Gensha asked him, 'Do you hear the murmuring of the Brook?' 'Yes, I hear it,' answered the monk. 'There is the entrance,' the master instructed him.

In the previous chapter, we talked about Bell's theorem and proved that nature shows us a world in which local objectivity does not work. We found that the concept of the observer, of consciousness itself, must be recognized as an instrument of existence. We found that the alternatives to the Copenhagen interpretation of quantum mechanics did not work so that now we have to understand that strange interpretation and somehow make *it* work. We found that consciousness is something real, and now we are obliged to discover what that means.

What is consciousness? What is Zen? To take the next step, you will have to grasp, in the space of this chapter, the fullness of Zen thought. You will have to read—and understand—the words of the eighth-century Ch'an master Ch'ing-Yuan:[2]

> Before you study Zen, mountains are mountains and rivers are rivers. While you study Zen mountains are no longer mountains and rivers no longer rivers. When you have obtained enlightenment mountains are again mountains and rivers again rivers.

You will hear one ask, "What is the sound of one hand clapping?" and immediately answer by touching this page, so that no one else will understand. Yet it is actually all very simple. All we are talking about is just the *observer*. All we are talking about here is *consciousness*. When you know what consciousness is, then you know Zen. And koans are at once clear.

Suzuki talks about the enlightenment experience. Sometimes it takes many years to achieve that state. Some who go to the temple to learn the truth of Zen never achieve satori. For others it takes years, decades, a lifetime. I had assumed that I merely needed to say a few words and my audience would know—would follow. They did not know what consciousness is other than as something that has to do with the brain being awake. I, for myself, did not know this Zen Buddhism. I only discovered many years later that my experience had involved this "thing" sometimes pretentiously called enlightenment. After learning of others' experience in searching for satori, I realized why my audience had not understood what I was saying. I had assumed, without realizing it, that they were enlightened people. Educated, knowledgeable, widely read, and surely capable people—they were all of that, but enlightenment is about something else.

I remember the moment when I had the revelation, the experience that finally let me catch hold to the consciousness experience in such a way that I have never since let it go. The enlightenment experience had been there. Indeed, it is always there waiting for you to take hold of it, but my knowledge of physics, my training in that discipline, like the habits we all form in our everyday life, made it difficult for me to catch hold of what I already had. The revelation permitted me to catch hold. I was walking in an open field across from the Physics and Astronomy building at the University of Maryland down toward the main road that passes through College Park on a summer afternoon in 1966, when in one moment the problem was suddenly solved. I would still have to do all of those things one must do to show just what that answer was, but suddenly I had reached to the source. The river would flow tranquilly on, but now the flowers were blossomed red.

BY THE GRAVE OF MAD CAREW

At least Hofstadter admits it.[3] He says, "I'm not sure I know what Zen is." He says, "In a way, I think I understand it very well; but in a way, I also think I can never understand it at all . . . . To me, Zen is intellectual quicksand—anarchy, darkness, meaninglessness, chaos . . . . Yet it is humorous, refreshing, enticing." He keeps a watchful eye on the admonition: Those who know do not speak, those who speak do not know. Hofstadter comments on some of the koans the Zen master Mumon (which means "no-gate") wrote in the thirteenth century, seemingly with the one intent of keeping Hofstadter—and everybody else—from ever knowing what he was talking about. The koans appear to be totally opaque. Here is an excerpt from one of the koans that Hofstadter also quotes.

Koan:

A monk asked Nansen: 'Is there a teaching no master has ever taught before?'

Nansen said: 'Yes there is.'

'What is it?' asked the Monk.

Nansen replied: 'It is not mind, it is not Buddha, it is not things.'

Mumon's commentary:

Old Nansen gave away his treasure-words. He must have been greatly upset.

Puzzling. Hofstadter believes that the koans are intentionally idiotic. He says that perhaps they are meant to show how useless words are, as though they are used merely to "break the mind of logic." To further prove his point Hofstadter gives an even more perplexing koan:[4]

The student Doko came to a Zen master, and said:

'I am seeking the truth. In what state of mind should I train myself, so as to find it?'

Said the master, 'There is no mind, so you cannot put it in any state. There is no truth, so you cannot train yourself for it.'

'If there is no mind to train, and no truth to find, why do you have these monks gather before you every day to study Zen and train themselves for this study?'

'But I haven't an inch of room here,' said the master, 'so how could the monks gather? I have no tongue, so how could I call them together or teach them?'

'Oh, how can you lie like this?' asked Doko. 'But if I have no tongue to talk to others, how can I lie to you?' asked the master. Then Doko said sadly, 'I cannot follow you. I cannot understand you.'

'I cannot understand myself,' said the master.

Hofstadter comments,[5]

> If any koan serves to bewilder, this one does. And most likely, causing bewilderment is its precise purpose, for when one is in a bewildered state, one's mind does begin to operate nonlogically, to some extent. Only by stepping outside of logic, so the theory goes, can one make the leap to enlightenment.

Hofstadter goes on to explain what is wrong with logic, to explain what he believes to be its impediment to enlightenment. To answer this question he has posed to himself, Hofstadter tells us what he believes enlightenment to be. He says, "Perhaps the most concise summary of enlightenment would be: transcending dualism." And "dualism is the conceptual division of the world into categories . . . [h]uman perception is by nature a dualistic phenomenon."[6]

But I think that in this Hofstadter has entirely missed what Zen is about. If there is anything we should know about Zen, it is that perception is not dualistic. The problem is that our perception, our thinking, our natural outlook is to see everything as one thing—matter.

My friends and I go to lunch together. One says, "Let me see what I will have to eat." I tell him, "You cannot eat." I get an irritated look. We have played the game before, and he knows this is not a question of permission, so he doesn't respond with "Why can't I?" He just looks annoyed and goes on to peruse the menu. Of course "he" cannot eat. Only his body can. It is as we see in the series of paintings by Kakuan where the revelation that brings about enlightenment is represented by the image of an elusive bull—first he must approach the bull to catch a glimpse. He must first see that he is not the body. He must first see that there is a dualism, that there is an observer. He must first catch a glimpse of his own consciousness if he wishes to know enlightenment.

In Zen, one seeks enlightenment. I

Alan Watts in his book *Three*[7] gives us Huang-lung's (A.D. 1002–1069) *Three Barriers*, a kind of Buddhist S.A.T. exam for monks consisting of three questions and answers:

> *Question:* Everybody has a place of birth. Where is your place of birth?
> *Answer:* Early this morning I ate white rice gruel. Now I'm hungry again.

> *Question:* How is my hand like the Buddha's hand?
> *Answer:* Playing the lute under the moon.
>
> *Question:* How is my foot like a donkey's foot?
> *Answer:* When the white heron stands in the snow it has a different color.

Notice the simple logic of each answer. "Where is your place of birth?" The body may have come from Atlanta or Rome, but you came into being this morning when your consciousness began. You have memories that go back before this, but you began then—and more than this, you are born each moment in each moment's hunger, each moment's joy, each moment's sight of the visions around you.

"How is my hand like the Buddha's hand?" You can only know how if you see the hand not as an object but as an experience that is inseparable from the entire sentient moment. Your hand is like the Buddha's hand if you see your hand *doing*, being your conscious experience of the moment.

"How is my foot like a donkey's foot?" Again, you must see this foot as a part of all experience, all transformation. The answer does not lie in the words, it lies in the experience that is the totality of being. The consciousness that is the true reality is the same thing whether it looks upon a foot that looks like another form or looks upon a colorless bird that becomes another experience when it stands in the colorless field of snow. The idea of Zen and the reality that is consciousness being are simple. It is an immediate part of you.

Hofstadter says, "At the core of dualism, according to Zen, are words—just plain words. The use of words is inherently dualistic . . . a major part of Zen is the fight against . . . words." But is there a problem in achieving monism? Our materialistic thought is immersed in monism. We believe in physicalism through and through. That is why the outcome of Bell's theorem is so at odds with our common-sense understanding of the world. The problem is that one must break away from that monism into a true dualism before one can reach this new monism of Zen. The koans are there to tell us something that is very simple and quite direct. The koan points us first of all to the fact that there is something to us that is more than the objects, more than the things that others regard us to be. We may say, "Let me see what I will have to eat," because we have thought of ourselves as objects. This is exactly why Doko's master, in the last koan, could say, "I have no tongue." Before we go further, we must first see that the monism of objective reality is a false doctrine. The koan is there to teach in clear language something that is at first very difficult to understand, but which, once understood, becomes clear and easy to grasp. It is something that becomes obvious so that eventually, as we go out from the temple to "mingle with the people of the world . . . , the dead leaves become alive."[8]

In our first koan, where we met Old Nansen, we were told that he must have been upset at having given up so much of his truth. Indeed, his answer does give away the teaching of Zen. True teaching, the teaching of that which is not so easy to teach, is taught best by just pointing to what is true, by pointing to the thing that is not thing. It does not deny things. It is not nihilistic. It points to the *non*-thing that is, nevertheless, all things at once. By affirming that it is something and yet that it is not any of the things that fill any of the categories that can be expressed objectively in any way, we point to what consciousness is, even though it is not mind, not Buddha, not things, and not consciousness. It is not that mind of mine, not my consciousness, not things. But it is consciousness and it is mind, and it is all things. As soon as we know consciousness, and as soon as we know Zen, we know that it is dead leaves, the red flowers, the river flowing—things. Suddenly we see that all things are actually one kind of non-thing, thus-ness, as the Zen master would imperfectly say, to keep us from knowing that there is an objective word for what this is, to keep us from knowing, to keep us from misunderstanding.

The koans often do try to direct us away from logic and from words. It is done for a very good reason. Hofstadter, seeing this pervasive nonlogic of Zen gives Zen his own new designation. He calls it an *ism*. It is not a bad name. But unfortunately, it is misapplied to Zen. Hofstadter defines *ism* to be "an antiphilosophy, a way of being without thinking." He says, "the masters of ism are rocks, trees, clams; but it is the fate of higher animal species to have to strive for ism, without ever being able to attain it fully." But are the rocks masters of Zen? Do they have this something that the Zen master should so wish to devote his life to knowing? Do rocks have this thing that we have that underlies the totality of what we are? Do they have what we experience? Do they possess what the dead leaves are that brings them to life? Do they have consciousness, and do we even know?

"When you study Zen, mountains are no longer mountains and rivers no longer rivers." Ch'ing-Yuan is telling us that when we study Zen, we will think that Zen is an ism, as Hofstadter has, but "When you have obtained enlightenment, mountains are again mountains and rivers again rivers." This theme, that when we have caught the Kakuan ox—the eternal principle—we return to the world, runs all through Zen Buddhism. Hofstadter's ism is caught up on a mentalism, and he—and all of us—must first let go of that analytical approach to seeking truth before we can see what is already right before our eyes—what is even closer than that which is before our eyes.

When Doko's Zen master said, "There is no mind, so you cannot put it in any state. There is no truth, so you cannot train yourself for it," he was not saying that nothing exists; he was not saying that we must become like a rock or that we must sink into oblivion. He was not even saying there is no mind and no truth. He was saying that when we begin to analyze, we begin to think these word objects are the pieces out of which we can construct the observer. Zen is not reached by manipulating word concepts, just as consciousness is not to be found among physical objects. Zen is not approached by a state of mind; it is mind, but it is not my mind; it is not your mind; it is not God's mind. It is the consciousness which is first before all these things, and it is consciousness that is none of these things. Doko's master says, "But I haven't an inch of room here." Here, when you have reached Zen, is the mind that is not measured in inches, so there is no room, no space, to put monks into.

"To suppress perception, to suppress logical, verbal, dualistic thinking"[9]—is that the essence of Zen? Is Zen holism? Hofstadter, seeing Zen as antiphilosophy, believes it to be a way of being without thinking—like a rock. His holism is the absence of analytic thinking, letting the mind go blank, or fusing all concepts into an undifferentiated whole. But it seems Zen is there when I think, when I differentiate, when I analyze:

> A master was asked the question 'What is the Way?' by a curious monk.
>
> 'It is right before your eyes,' said the master.
>
> 'Why do I not see it for myself?'
>
> 'Because you are thinking of yourself.'
>
> 'What about you: do you see it?'
>
> 'So long as you see double, saying "I don't" and "you do," and so on, your eyes are clouded,' said the master.
>
> 'When there is neither "I" nor "You," can one see it?'
>
> 'When there is neither "I" nor "You," who is the one that wants to see it?'

Hofstadter reacts as follows:

> Apparently the master wants to get across the idea that an enlightened state is one where the borderlines between the self and the rest of the universe are dissolved. This would truly be the end of dualism, for as he says, there is no system left which has any desire for perception. But what is that state, if not death? How can a live human being dissolve the borderlines between himself and the outside world?

In this explanation, it seems that Hofstadter sees you, me, and himself as something that, at its base, is data processing. To look for Zen means to quiet the machine. To achieve Zen is to turn the machine off so that there is no system left that has any desire for perception. Hofstadter is left with a Zen that is perfect only in death. And that is where he leaves us.

But is that the way? Is not the way "right before our eyes"? And, I might add, is it not right before our living eyes? So long as we say "I" or "You," we show that we have not understood what our own self is. The Zen master says, "When there is neither 'I' nor 'You,' who is the one that wants to see it?" Do we know whom it is that wants to know? He is not saying there will be no one. He is pointing to the source, to the desiring reality seeking to know itself, to consciousness that is everything, both "You" and "I."

Judging by their writings, it seems likely that most, if not all, Western writers who have sought to explain Zen have never mastered Zen. Perhaps some of the Zen masters have not either. Perhaps many of those Zen masters who seem to emphasize only the inability to use words and logic to grasp Zen never went beyond the stage of clearing out logic. Perhaps they never went beyond the formula responses to the koans.

> Hyakujo wished to send a monk to open a new monastery. He told his pupils that whoever answered a question most ably would be appointed. Placing a water vase on the ground, he asked: 'Who can say what this is without calling its name?'
>
> The chief monk said: 'No one can call it a wooden shoe.'
>
> Isan, the cooking monk, tipped over the vase with his foot and went out.
>
> Hyakujo smiled and said: 'The chief monk loses.' And Isan became the master of the new monastery.

The chief monk's answer was good, it was very good. But Hyakujo saw that it was formula. The least of the monks, the cooking monk, gave a new expression to the Zen-being of the vase. He saw the vase as the experience and shared that experience. Hyakujo recognized the danger of promoting those who learned the formula, the correct answers to the koan. There are no correct answers. There are no wrong answers, either. There are those, however, who try to give us the formulaic answers. They have only discovered the footprints of Zen.

Those who have mastered Zen are usually wise enough that they do not write about it. These wisest of masters know already what I have learned only so slowly, that knowledge of Zen cannot be imparted in a physics lecture in Denver. I remember in 1972 speaking about the question of consciousness in that seminar for the Physics Department. I tried that day to build a bridge between physics and consciousness. The audience had absolutely no conception of what I was talking about. The knowledge comes in the twinkling of an eye and only slowly, even though all have it already. Hofstadter was honest with us. He did tell us to begin with "I'm not sure I know what Zen is." I only wish I were free to say this. But my purpose is not simply to tell you about Zen. Zen, as you will discover, is only a means to an end. We must go beyond the Buddha. To go beyond, we must pass by the Buddha.

> Tozan said to his monks, 'You monks should know there is an even higher understanding in Buddhism.' A monk stepped forward and asked, 'What is the higher Buddhism?' Tozan answered, 'It is not Buddha.'

I will try to tell you what Zen is. We have already begun, of course, but we have only begun. I know that as some of you read this, you will say to yourselves, impatiently, "Well, of course," and the simple idea will have escaped you. But some of you will suddenly exclaim, "Well, of course!" Enlightenment, that elusive experience of satori, will have grasped your very soul. You will have that seed that you must have to go beyond Buddha to understand the meaning of "observer."

To go beyond Buddha, one must first know Buddha, one must first know Guatama's truth. With that knowing, one can then discover that he was wrong. Only if you do this can you ever know the true Buddha, the Buddha that Gautama never knew.

All of you, I am sure, even some of those who have found enlightenment, will think that in saying these things I am playing with words—only being figurative. But I am not. I am giving you my literal words. It is the only reason I am writing these things, so that by the end of this book you will discover that these things are true. Old Nansen must have been upset.

Merilyn wasn't at school that Monday, nor the next day nor the next. She was quite sick. Her doctor came by her house to give her penicillin shots that Monday and the next several days. In class, things continued as usual. In English, I gave a book report on *The War of the Worlds* by H. G. Wells. I should read it some day. As usual, nothing happened in debating class. After school I called Merilyn, but she couldn't talk to me, which upset me. On Tuesday, I saw Violet; "She said Merilyn had told her that she would not let me come over to her house last Saturday. This made me mad the rest of the day. I had not even asked Merilyn if I could come over on Saturday. I started not to call Merilyn later that day because of these two things, but because I thought I might be wrong and because I had a lot to tell her, I flipped a coin. It was heads, so I called her." I told her what Violet had said. She told me she had not even talked to Violet since Saturday. All my concerns had been over nothing.

By the end of the week, Merilyn was better. By Sunday, she was well enough for us to go to Trinity Methodist together. But Merilyn was not quite so cordial, not quite so warm as she had been before. The sickness still lingered and still colored her feelings. For a moment, I felt the shadow of loss, a presage of what the year was to bring. For a moment, for a moment only, and then it was gone. Beauty and tranquility came back into my young days.

A NEW KOAN

Let us say at the outset that consciousness cannot be defined. I will show you what I mean by this and why it cannot be defined. Then we will go on to say what consciousness is. Briefly, however, the consciousness that we speak of is the immediate experience, that sentience of the mind, by means of which everything we see and feel has reality. This is that inner source of reality that moved Jaynes to write[10]

> O, WHAT A WORLD of unseen visions and heard silences, this insubstantial country of the mind! What ineffable essences, these touchless rememberings and unshowable reveries! And the privacy of it all! A secret theater of speechless monologue and prevenient counsel, an invisible mansion of all moods, musings, and mysteries, an infinite resort of disappointments and discoveries. A whole kingdom where each of us reigns reclusively alone, questioning what we will, commanding what we can. A hidden hermitage where we may study out the troubled book of what we have done and yet may do. An introcosm that is more myself than anything I can find in a mirror. This consciousness that is myself of selves, that is everything, and yet nothing at all—what is it?
>
> And where did it come from?
>
> And why?

And yet it is so much more than this. Not just the inner eye that glimpses its selfish thought, but the moment's whole world painted across the canvas of sensations: The red polish of a woman's nails and parted lips that mark the moment in forgotten sensations, immediately sensed, immediately known without reflection, without comment to others or even to one's own conscience. It is the full wide screen of living: the sounds of laughing, the sounds of children, of music, the clashing noise of striking metal, the quiet turning page, a distant horn's intrusion into the corner of the mind, the hum within a silent room. Consciousness is not just the hidden me that struts or cowers opposite this curtain of life before the eye, but the luminous veil itself as well. The glint of light that, streaming through the shade, blinds the sight. The barely noticed hand that holds the edge of the book, the glass across the room unnoticed, out of focus, but *there*, rounding the edge of the uncanvassed mind. It is the blue that suffuses your sight as you look into a cloudless autumn sky. All these things of the sentient moment, all the feelings of knowing happiness or joy, hate or pain, longing, love, frustration, fear, or loneliness, and all the thoughts that fill your mind with schemes or reflections—all are your moment's being. These things and everything are consciousness. These things are not the objects of our outer existence; they are the consciousness that is you or I. These things all together are the question, the puzzle of consciousness that is everything and yet cannot be found in the scattered parts of an ended life. This is consciousness.

This is the consciousness of philosophy. This is Descartes's *cogito ergo sum*, correctly thought and wrongly stated, as he confounded the conscious knowing with the brain's data-processing thought—as he confused living, thinking, and sentient being with consciousness. *I think, therefore I am*, that thought of consciousness, that correct idea, but the misstatement of Western philosophy's modern founding that embodies its greatest coherence and its clearest confusion and leaves me no choice but to explain the inner reality that you already know.

Consciousness is not thinking. Consciousness is not thinking about one's consciousness. It is not self-reflection. Consciousness needs no words and needs no things. Those born blind or deaf and mute, they are as conscious as you or I. A fly blankly staring at a red table cloth in a red room will have redness consciousness. A man sitting on the beach at Waikiki, eyes closed, mind thoughtless . . . even after six months will still have consciousness.[11] In fact, he will be consciousness. He will be the consciousness.

Consciousness has nothing to do with "consciousness raising." It has nothing to do with becoming aware of some new experience or activity, although that can be the subject material of the conscious experience. Consciousness is not perception, or wakefulness, or attention, even though these are all so closely linked that a week's arguing and discussing may not be enough to tease apart and illuminate these related ideas.

Perception is properly a technical term in psychology. It refers to the brain's imagery and conception of its environment resulting from the information processing of input stimulus signals. *Wakefulness* is a physiological term that properly refers to a mode or state of the brain's functioning, the mode of functioning when we are awake. Of course, we are ordinarily conscious when awake and, except for dreaming, are ordinarily not conscious when asleep.

Attention is also a technical term. It refers to that part of the brain's perception that constitutes the brain's primary information-processing load at a particular time. Our conscious experience seems to follow the attention set to a great extent. But again, these are different concepts, different subjects.

The term *conscience* is obviously quite distinct from consciousness, but even here confusion often arises. My reading of some of the teachings of Siddhartha Gautama suggests to me an earlier mixing of the concepts of consciousness, conscience, and thinking—conscience as something that was at once all three of these in an undifferentiated sense, which required Gautama to speak much more obliquely in order to distinguish the idea of consciousness that I am talking about from the common usage of the word *viññāna* of the Pali language, in which the earliest Buddhist texts are transcribed from Gautama's native Magadhī.

Consciousness is the blue of the sky; it is C#, the taste of sweetness as it fills the mind, the smell of gardenia, the pain of love that is lost, the experienced murmuring brook as it is, the moon reflected in the pool. Consciousness is also the experience of images, ideas, words, and thoughts that play on the mind as we read a novel, as we remember the past, or as you now read these words. It is the feel of this book's cover, the texture of the paper, the weight of the volume, the space that separates your eyes from the black type on this page. But consciousness is not thinking. Consciousness is the carrier of conscious thought. It is what is left when thought is removed, when the mind is stilled. When thought is entirely gone, the consciousness is still there in the void experienced by the Zen master, the *sunyata* as the channel capacity of consciousness (as we will see later on). Consciousness is far more than simply thinking about any one subject, including one's self. It is far more than thinking about or with words. Consciousness comes before words—or what is a thesaurus for?

Consciousness can be compared to the luminous image on a television screen. The set is the body. The electronic circuitry is the brain, and the image is the consciousness that lights up the set. The image is not the picture tube, not the phosphorescent screen, not even the light radiating from the screen. Of course, there is nothing else there, and in fact the image exists in our mind, which is why it serves for this analogy. The light that is emitted by the screen of the television set can be said to radiate from a physical image. It is physical, and the glowing phosphors provide the physical correlation to this image. Of course, the television set has no consciousness; the image is not the television's consciousness. It is our consciousness of that image that gives it meaning and only as an analogy to the consciousness. But the mind of the body is the "image" our brain creates. That consciousness, that image, is who we are.

Consciousness cannot really be defined. These words merely suggest analogies, and you have been asked to examine your own experience to see whether you recognize what we are talking about. The reason why we cannot define consciousness is that definitions, true delineations, require objective demonstration, and here that is not possible. We cannot objectively define "red," for example. For a person blind from birth, how would one define red (the experiencing of redness)? We use the word, and yet we can never know that another person's experience of red is the same as our own.

Suppose there is an ice cube lying on the floor, melting. Does the ice cube feel pain? If you have understood what I mean by consciousness, you will recognize that this question is neither frivolous nor answerable. There is no measurement or observation that we can make on this object that will tell us whether it, like a person, has consciousness. The ice cube does not look like a person. It does not answer, "yes," when we ask if it is conscious. We cannot program it to respond to interrogations as though it were conscious, the way we can program a computer. Yet it could be conscious, or it may be that it is not conscious. And there is something that this statement means, once you understand what consciousness is.

FINDING THE CENTER

And yet consciousness is much more than all this. Consciousness is all things in totality. Consciousness is reality. Consciousness is the word on this page and the word that represents the thought. It is this book in your hands. Consciousness is the child playing beyond the window and across the street, consciousness is the trees and the houses, and consciousness is the beyond—it is the very distance you see that your mind deceives you into believing separates you and that child playing out there.

Consciousness is the feel of things. It is the roughness and the hardness of a brick. It is the strength of a steel rod, the sound of the temple bell. It is space and time—the sentience of being that separates its contents from its own contents while holding them together as one thing. Here, put your hand on the wall of your room and feel what you yourself are. See the man across the room. That is what you are, and as he sees you, you are what he is. The lines in his face, the fineness of his hair—these too are you. That is reality.

You have never touched a brick, never touched a wall, never held a book, never seen a word on a printed page, never held another, never touched a lover's lips. Those, if they exist at all, lie beyond you, and you can never reach out a hand to touch them. That is reality.

If to all this sophistry you say, "Well, of course I know that the objects of the external world are displayed on my mind's eye—in the brain—and so I have no primary experience of objects," then you have failed to understand what is being said here. Realize: all those concepts, all those arguments are part of that same experience, that same reality. And there is something more you should realize. We have seen that there is no objective world. The tests of Bell's theorem have undercut that logical avenue. That objective reality is no longer there. Now you should realize that it is a whole new reality that must be discovered. You should not think that the brick is an external object with an internal image in your brain. You are the brick. If this comes as a wondrous revelation, changing all your understanding, then you have just touched satori. You have discovered enlightenment, the consciousness of consciousness.

The Zen master says to you, "It is pouring rain now. How do you stop it?" You should know that you can; you should know how. You should say, "I will turn my head" or "I will shut my eyes." Hyakujo will give you no monastery for these answers, but perhaps he will not strike you either, for you must say this without saying "I." You should walk away or say "The redbird is black." Then you will know that the murmuring brook is the entrance to the path of truth. You should see that simply holding up a finger or simply saying "Good morning" to a friend is an inexpressibly deep experience.

Suzuki in his chapter that denies Zen is nihilistic gives an example that would seem to deny his very thesis:[12]

> 'I come here to seek the truth of Buddhism,' a disciple asked a master. 'Why do you seek such a thing here?' answered the master. 'Why do you wander about, neglecting your own precious treasure at home? I have nothing to give you, and what truth of Buddhism do you desire to find in my monastery? There is nothing, absolutely nothing.'

Later, Suzuki gives us another view of Buddhism in the Zen literature:[13]

> Seikei (Tsing-ping, 845-919) asked Suibi (T'sui-wei): 'What is the fundamental principle of Buddhism?' 'Wait,' said Suibi, 'When there is no one around I will tell you.' After a while Seikei repeated the request, saying, 'There is no one here now: pray enlighten me.' Coming down from his chair, Suibi took the anxious inquirer into the bamboo grove, but said nothing. When the latter pressed for a reply, Suibi whispered: 'How high these bamboos are! And how short those over there!'

And of course you should realize that when the bell sounds, the sound, the bell, the monastery are all consciousness, all one thing that has no thing. "When both hands are clapped, a sound is produced: listen to the sound of one hand." If you hear the sound of one hand, you have grasped reality. Pause. Know what you are.

> Buddha said: 'I consider the positions of kings and rulers as that of dust motes. I observe treasures of gold and gems as so many bricks and pebbles. I look upon the finest silken robes as tattered rags. I see myriad worlds of the universe as small seeds of fruit, and the greatest lake in India as a drop of oil on my foot. I perceive the teachings of the world to be the illusion of magicians. I discern the highest conception of emancipation as a golden brocade in a dream, and view the holy path of the illuminated ones as flowers appearing in one's eyes.

And this too is reality.

AT THE EDGE OF ETERNITY

The walls of the room are not mental images of objective objects. Such a characterization does not do justice to the extent to which everything about the walls derives from consciousness and is consciousness. The wood paneling of the walls around me, the pattern and grain of the wood, the feel, texture, color—are all consciousness. The odors of the room; the objects, the lights, and windows; and the trees, buildings, and hills beyond are all consciousness. And the space. The fact that space is also a part of the fabric of reality embroidered in consciousness seems so incredible and yet so clearly so. Space is mind. And mind has the space of the whole universe spread out before us.

Mind and consciousness, not body. Body is an element of consciousness, and because we have shown that objective reality does not exist, so too the objective body must be more illusion than real. Body depends on consciousness, and consciousness therefore does not depend on body. If consciousness is not conditioned on body, then that reason for conceiving of consciousness as experienced must be given up. Mind is eternal. Heaven, nirvana, eternity: These are space.

It is to these heights that pure consciousness propels the mind. It is a vision of transcendent reality. And yet Zen, where it does do so, goes too far in having us reject logic. It gives us a truth, a truth we need, and it gives us answers to unasked questions. It points us in the right way (if we have not lost logic and curiosity and a desire to know what Zen itself is) to understand that which lies beyond the Zen. Zen is not the fulfillment but a plateau. When we have come fully to understand Zen, we do not understand Zen. And indeed the Zen master would say there is nothing to understand.

But there he is wrong at the very moment he has attained his understanding of Zen! There is something beyond. And to achieve knowledge of what lies beyond Zen, one must attain Zen: Zen, free of logic, free of words, free of symbols, free of discourse and counterargument. Consciousness, free and pure, understood and caught, held within the palm of one's hand. This must be achieved if you are to see beyond Zen to the meaning of consciousness, to know the fabric of reality—the knowledge of which lies quite beyond Zen.

You will note that we have said nothing about any mystical experience. Some of you may find it dissatisfying that the term *enlightenment*, as it has been used here, does not refer to some sudden, overwhelming vision of an indescribable glory—a vision from the threshold of heaven. People do have these experiences. When they do, they often know at once just what I have talked about here. Suddenly, they know that consciousness exists as something in its own right, for now they have seen it where it exists quite apart from the objects of the external world. They have seen music and heard the lilies sing. They have tangibly felt the oneness of all creation. They have seen wonders stretch out into heaven before their eyes.

And yet this enlightenment experience may seem to you to be something less than the satori experience of Zen. Where is that inner authority that comes with enlightenment? It is there. That authority, that conviction comes from the recognition—by being—of what consciousness itself is. This is satori; this is enlightenment. One does not have to be struck by a bolt from heaven or see blinding lights. One need only see now, for perhaps the first time, what the most simple piece of conscious experience really is and discover authority in one's own self. And that, for the existence of consciousness, is authority enough. One need only see the redness of the bloom or feel the coolness of the spring's water to know that these are things that lie quite beyond the realm of material science. The blinding light of the nirvana experience is a wonderful gilded lily. Enjoy it if it comes, but we only need enlightenment, and that is satori.

But there is something beyond the Buddha, and beyond satori. Now our science can deal with the nature of consciousness. When our fathers were boys, merely understanding the propagation of a neural impulse was still an age away.

Now we can see a multifoliate philosophical complexity behind the basic facts of quantum mechanics—asking. Once, Newton's mechanics alone rolled the planets around the heavens, and the mystical states were our only way to know other realities. But we don't need such states now to step into the presence of a greater light. We only need to understand the nature of consciousness.

THE MAGNETIC MIND

Let me tell you a story about a professor of mine, Ernst J. Öpik, who would look out at me with his one good eye and somewhere else with his mind. I could relate many stories about this incredible man, particularly about his days in Russia at the time of the revolution, during World War I, and later of his days inside Hitler's Germany as Allied bombs, carrying both hope and desperate fear, fell feet away and killed those he loved. I would like to tell you of the tears I saw as he told me of these things a world ago. But these stories would take us too far away from our goal. Maybe later, maybe another time. Right now, I want to tell you about his science, the science of this professor of physics and astronomy at the University of Maryland at that time and this director of the Armagh Observatory in Northern Ireland.

What made this man's thinking so marvelous was his ability to take all the known facts about something—the most tenuous and most disparate specks of experimental observation—and shape a whole new understanding of his subject. He worked mainly in a subject area that had only the faintest hints of solid facts: astronomy, planetary astronomy, the physics of the solar system—the moon, comets, meteors, asteroids, the sun, and the astronomy of the stars beyond. It is a beautiful science that creates wonders of knowledge out of whispers of data.

I knew Ernst Öpik and worked with him just before and during the earliest days of the space program, days when only a handful of satellites had gone up and the moon was still a distant goal. Öpik was already quite elderly then. I remember the last seminar I took from him. The subject was the nature of comets. In a previous seminar on the subject of Jupiter, he had collected all the data available about Jupiter: things like how starlight fades from sight when occulted by the disc of the planet. He turned this into a complete picture of the planet: its atmosphere, its interior, the dynamics of its Great Red Spot.

In this seminar on comets, he was doing the same thing, collecting the scattered bits and pieces to construct a picture of what comets are. I remember that as he got into the subject, as he began to piece together what was going on in the comets, he began to talk about magnetic fields. The hairs went up on my skin. Others sensed it too. Comets are just blocks of ice; they are not magnets. I thought, we all thought, "The old man has gotten too old."

I went to see him in College Park back in 1969, shortly after Neil Armstrong had walked down the ladder onto the surface of the moon as a TV camera and half the world watched. He was delighted, ecstatic, as he spoke to me, his one eye always looking somewhere else, smiling broadly as he spoke in his strong Estonian accent. He had just received a copy of a book about the moon. It had just gone to press. The book had chapters by all the great names in lunar science—Whipple, Urey, Gold, Shoemaker, and, of course, Öpik. As he showed me his chapter, he smiled even more widely and said, "All the others had to rewrite their chapters, all of them. I did not have to change one word." When the astronauts landed, they found a moon that was just what Öpik had described, just what he had been describing for years. I had worked with Öpik, and it was also something of a victory for me as well, for what they found there was what I had described in my few papers on the subject, even to the size of the grains of dust on the moon's surface. Öpik had taken all the data obtained over many decades searching out the moon through telescopes on earth and had pieced together just what the moon had to be.

When Halley's comet plunged back toward the sun from the inner fringes of Oort's cloud in 1986, the nations sent space vehicles to look at and probe this passing visitor from the solar system's ancient past. Space vehicles were launched from Russia, Japan, and Europe. As these vehicles flashed past their tiny target a hundred million miles away, they measured the magnetic fields about this 10-mile ball of dirty ice, and incredibly, there they were. There were the magnetic fields that Öpik had seen with that other eye of his that could always see into the heart—that saw with the light of reason into the hidden corners of reality.

If we will look with that same eye of reason at all the facts and let those facts carry us where they will, then we will be able to go beyond the present limited understanding of how the world works. We can do this because we know that somehow, ultimately, all the facts must fit smoothly together into a complete vision of reality.

I brought up Öpik simply to show that such logic can carry us a very long way toward a greater understanding of the things we see. This kind of visionary logic can show us things even about consciousness, about Zen. You must know what consciousness is, and you must carry that knowledge along as we take the next step toward an understanding of reality. In order to know Zen, we have had to suspend logic, and we have had to grasp that which is beyond grasping so that the logic of that knowledge that violates logic—the logic of quantum mechanics—can take us to the next plateau. Suzuki, in talking about Zen, tells us something about its logic that seems to speak to the logic of quantum mechanics:[14]

> We generally think that 'A is A' is absolute, and that the proposition 'A is not-A' . . . is unthinkable.

In this, Suzuki tells us that words destroy Zen and that logic kills mind. In this knowledge of Zen, words are symbols that substitute for the experience of reality, so the telling of ideas gets in the way of achieving Zen—in the way of consciousness.

But this does not remove logic, and it does not remove words. If you know where the Zen handle is, and if you can take hold of it, then the logic will no longer interfere. There is the Zen of the words and the consciousness of the logic that lets us go to a higher level. These must be the words of experience that come before the words of the symbols we speak. The consciousness of the idea comes before the words, and it can exist within us as our consciousness without words.

But where is that handle that will hold Zen? I remember the moment when I had the revelation, the experience that finally let me catch hold of the consciousness experience in such a way that I have never since let go. As I was walking across that open field at the University of Maryland, knowing consciousness but still trying to discover how it could fit with my knowledge of physics, the answer came to me. The answer was simply "Do it." That is the handle by which you can take hold of Zen and carry it wherever you wish to go, even into logic. In Zen, we hold consciousness like a flame. We hold it in the palm of our hand, and we gaze with our eyes lighted by the glow of enlightenment. As we hold that flame, we know the reality that transcends our common reason. In that light, we can understand the contradictions of reality and ask the questions that open the next door, but we cannot grasp the flame.

"Do it." What this means is simple. Do the physics of consciousness, just as if it were anything else in science or in physics. Find its properties and fit it in. When you stop philosophizing and begin to do it, the confusion begins to fall away and the answers come. Physics lets us grasp the flame. You will see what this means as we go along. That is what we are going to do next. You will see how this "do it" approach enables us to take hold of the whole thing and piece reality together into a new cloth.

BEYOND BUDDHA

Now what did I mean when earlier I talked about going beyond Buddha? What did I mean when I said that first you must understand the truth of the Buddha before you can know that the Buddha was wrong? Buddhism points us to a particularly intense and clear understanding of what consciousness is. It points to a realization that consciousness experience is reality. It is what I am. It is what you are. It points to a realization that this book in your hand is your actual being. This book at this moment as you hold it in your hand is in fact you because that is the content of your consciousness at this moment. Not just "content" but, in fact, the full being of your existence. You are nothing else.

But this is only a beginning. When you understand fully what this means, you have hold of a profound experience and knowledge. Scientists will say, "How do you know this is true?" The Zen master will smile, and perhaps—perhaps—he will say, "If you know Zen, then you know." But here the Zen master has reached only a half-truth, and the scientists have failed to understand the question that they themselves asked. The Zen experience as an experience cannot be wrong, but it may be no more than the experience of truth as a conscious experience. The experience of Zen as an experience—as data to be understood—cannot be wrong. Data is data. It is not right or wrong, as such. However, we can, and often do, think that the data we have means one thing when it means something else.

The scientist who asks the Zen master, "How do you know this is true?" recognizes the fallibility of our thinking faculties; he recognizes that people can have such an intense experience that even an absurd conviction can cause him to reject everything else he has believed. This false conviction can even wind up taking control of the person's life. People can err woefully. But the scientist has failed to recognize that his own question is the wrong question. The scientist should say, "Show me this data." His question should be "And what should we conclude to be the meaning of the data you have?"

The data means something simple. It means here is something that science has not understood. It means here is something that is not quarks or photons, not magnetic fields or space as such. Science is incomplete and must be greatly expanded if it is to meet the challenge of this data.

But as we will discover, the Buddha's truth is also wrong. It is wrong because it is incomplete. Rather than an end point, a final achievement of enlightenment, it is a stop along the way. This enlightened understanding of consciousness is only a piece of data. It is a piece of data that lets us take the next great step in our search for truth, in our effort to understand the fabric of reality. And beyond? Whole worlds await.

From the book called the Shaseki-shu, we find this story by the Zen master Muju:[15]

> A university student while visiting Gasan asked him: 'Have you ever read the Christian Bible?'
>
> 'No, read it to me,' said Gasan.
>
> The student opened the Bible and read from St. Matthew: 'And why take ye thought for raiment? Consider the lilies of the field, how they grow. They toil not, neither do they spin, and yet I say unto you that even Solomon in all his glory was not arrayed like one of these . . . . Take therefore no thought for the morrow, for the morrow shall take thought for the things of itself.'

Gasan said: 'Whoever uttered those words I consider an enlightened man.'

The student continued reading: 'Ask and it shall be given you, seek and ye shall find, knock and it shall be opened unto you. For everyone that asketh receiveth, and he that seeketh findeth, and to him that knocketh, it shall be opened.'

Gasan remarked: 'That is excellent. Whoever said that is not far from Buddhahood.'

10

Satori Physics

Day ends, night comes
And stars in the indigo night
Twinkle in myriad points.
The earth, robed in black velvet,
Catches a fleeting glimpse
of their splendor.
Night wanes, dawn breaks,
The candles of the night flicker
And are gone.
The sun, fiery and red,
Harnesses his golden horses
For the western ride.
Dawn becomes day.
The sun brushes the dew-drops.
They sparkle, then vanish
Before his sweeping rays.
The wind whispers in the trees—
God is at hand.

—Merilyn Ann Zehnder

Now let us look back for a moment. Let us look back at the ending of day, the coming of night, and then the dawn. We have seen the conception of our world as objective reality develop as a replacement for primitive religious myths. We have seen logic and science replace superstition, and we have seen the secularization of religion. We have seen the growth of rationality. We have seen this pragmatic rationality stimulate the birth of science. We have seen the sciences yield a progressively more detailed understanding of the cosmos until that knowledge has reached every end and edge of the vast universe, probing into every crevice of its parts. Science is like the eyes of the tiger in the forests of the night, always searching deeper, always looking farther, always burning bright. And there, probing the fearful symmetry of nature, we find mirrored in that world our own image, observing from the other side of reality—ourselves—not as creatures of the forest, but as the fire in the eyes.

Out of the confusion of primitive myths, a few philosophies have garnered several enduring facts. Out of these scattered facts, Newton built a system of physical laws. Out of these laws and from the deft tools of science, succeeding generations of scientists have built a universe of quarks and quasars, of atoms and galaxies, of tissues and planets. We have found a universe not built of the four elements—earth, water, air, and fire—but held together by the four forces—gravitation, electromagnetism, and the strong and weak nuclear forces—forces that seem to have boiled out of one original common material in the first instant of the universe's birth.

But strangely, so very strangely, as we have looked deeper into the nature of this reality, we have found that the reality we observe depends on where we stand to look at it. By moving, as Einstein has shown, not only our vantage point changes, but the things we look upon as they move past us change as well. The space and time and matter we see are changed by our own condition of looking at the world. We have found that nature reveals herself in her symmetries that are no more than reflections of the ways that we understand the world. And we have found that *our* limits to observe the world are nature's own limits to its very existence. We have found, in relativity and in quantum mechanics, that the very reality of the universe is bound to our own experience of it. And we have watched as tests of Bell's theorem have melted the isolation of objective reality into reflecting pools that mirror the images of those who observe.

We have looked at quantum mechanics. We have looked at how it has been pictured by Heisenberg and by Schrödinger. We have seen the arguments of Einstein, Podolsky, and Rosen and looked at some of the arguments of Bohr, Wigner, von Neumann, Bohm, and Bell. And in all this, we have seen at times, here and there at the corners of our glance, the edges of our own mind curling out of the objects of our reality. We have seen that we as observers play a role in the make-up of this world we live in.

Finally, we have come back to the beginning, back to ask what we are ourselves. We have come back to look at our own pure consciousness—to see our own reality, to look upon the fabric of conscious existence, of sentience. We have found a new reality, a new substance, a "matter" that physics has yet to probe. But if we are going to know the whole of reality, we must master the science of consciousness.

We usually think of science as a collection of experiments designed to test our hypotheses and to build the theories that represent our knowledge of reality, but science itself has been a great experiment. Its successes have affirmed confidence in an orderly world, and these successes have tended to make us believe in a totally materialistic reality. But the tests of Bell's theorem have shown that this concept is wrong. This material for the fabric of reality turns out to be torn. Patching this fabric with a nonlocal stitch is not how science should work. When the theory is worn out, we should throw it away, even when we must discard the whole cloth. The logic in physics that proved to be in error must be rejected, and we must return to those ideas that worked. Thus it is back to the Copenhagen interpretation with its concept of the observer, but now with a better idea of just what this "observer" is about. Remember that those who endorsed the Copenhagen interpretation of quantum mechanics were vindicated by the outcome of the tests of Bell's theorem. They were the ones who correctly predicted that the tests of Bell's theorem would confirm quantum theory.

Unfortunately, the mote in the eye of those who followed the Copenhagen interpretation has been that they have not been able to admit that the Copenhagen interpretation depends on understanding what this observer does. Physicists in the Copenhagen camp have been too satisfied with their successes in most areas to recognize the incomplete status of the Copenhagen interpretation itself. They have not said what this "observer" is. They have not said how such ideas can be made complete and internally self-consistent. They have not been willing to explore the meaning of consciousness, which is the distinctive characteristic of the observer that might help us understand how state vector collapse is caused by observation.

MINDLESS SCIENCE

It is a quest as old as the paintings of Lascaux; indeed, it is a quest as old as mankind. The existence and nature of consciousness have been the central questions of Western thought for centuries, the basis of Descartes's philosophy. It is the point where we must begin if we are to answer questions about the meaning of life, the nature of soul, or the existence of God.

But the modern view, held by most scientists and philosophers alike, is that the mind has no independent existence. It is the view that only the material world exists. Such views of man's nature have ancient roots. Lucretius, founder of the School of Atomism, concluded,[1] "The nature of the mind must be bodily, since it suffers from bodily weapons and blows." His views on the nature of mind are as modern as his theory of the atomic structure of matter.

For Thomas Hobbes, living at a time when Hans Schlottheim and Joachim Fries were fabricating the earliest mechanical automata of wires and gears, the comparison was even more pointed: "For what is the heart, but a spring, and the nerves but so many strings," and concluding with "That which is really within us is . . . only motion."[2] The idea of mind entirely disappears in such views.

Benedict Spinoza, however, viewed mind and body as one substance: existing as a part of the universal substance, at once the matter of which the universe is composed and the embodiment of God. But Gottfried Leibniz viewed mind and body as being comprised of "true atoms" that he called monads—particles each existing as individual beings, combining both material and mental properties, exerting force, endowed with perception and apperception, and moving under the supreme, preordained control of God.

Neutral Monism

In more modern times, the development of pragmatism under Charles Peirce rejected not only Cartesian dualism but even the basic question of mind, focusing instead on the pragmatic nature of thought in terms of operation and control. William James elaborated and refined this philosophy into the doctrine of neutral monism, in which the substance of the world is reduced to pure experience. Mind and matter become the same entity, the stream of experiences rather than states of a mind substance. The doctrine of neutral monism has been advocated by such neorealists as Bertrand Russell and Alfred North Whitehead. In his *A History of Western Philosophy*, Russell writes,

> While physics has been making matter less material, psychology has been making mind less mental . . . . Thus from both ends physics and psychology have been approaching each other, and making more possible the doctrine of 'neutral monism' suggested by William James' criticism of 'consciousness.' The distinction of mind and matter came into philosophy from religion, although, for a long time, it seemed to have valid grounds. I think that both mind and matter are merely convenient ways of grouping events. Some single events, I should admit, belong only to material groups, but others belong to both kinds of groups, and are therefore at once mental and material. This doctrine effects a great simplification in our picture of the structure of the world . . . . There remains, however, a vast field, traditionally included in philosophy, where scientific methods are inadequate. This field includes ultimate questions of value; science alone, for example, cannot prove that it is bad to enjoy the infliction of cruelty. Whatever can be known, can be known by means of science; but things which are legitimately matters of feeling lie outside its province.

This quote not only reflects Russell's ideas but also corresponds to a considerable body of scientific opinion; however, there is a serious inconsistency. The great simplification that this doctrine of neutral monism effects is that the whole of existence can be expressed as a single thing in the form of a single expression or equation. Indeed, the effort of physicists to consolidate the physical equations into the simplest representation, which has been going on since the days of Newton, is largely responsible for the concept of neutral monism. If this doctrine is true, everything that exists is derivable from a single equation. Value judgments, feelings, and morals are either derivable or do not exist as anything meaningful. Russell gives neutral monism his blessing for the sake of its aesthetic appeal, but he seems unwilling to do without the concept of value judgments as something having an existence so basic that it rivals physical existence. Somehow, what the physical laws allow us to do must be tempered by the addition of moral laws or "values."[3]

To find such an inconsistency in Russell's advocacy of neutral monism is not particularly important. This inconsistency could easily be removed by eliminating Russell's "vast field," which includes these "ultimate questions of values." The point is that however desirable monism may be in the pursuit of a pure and objective science, we are left feeling that such a philosophy must ultimately provide offer us only an incomplete picture of reality. These feelings cannot provide a satisfactory basis for a dualistic philosophy, but they help us realize the direction in which science must develop if we are to achieve a complete picture of reality. And if we are to solve the great questions of moral values that a materialistic science has now so clearly magnified, science must be enlarged; it must go outside the bounds of this neutral monism. A proper solution requires examination of this question of conscious existence within the context of physical existence as defined by physics, neurophysiology, psychology, and related scientific disciplines.

Russell's arguments should be contrasted with an argument that Churchland makes against the dualistic conception of reality. In the last chapter, we stressed the need to understand consciousness as the experience and not just as a philosophic idea so that it would be possible to recognize philosophical errors. Churchland, like Russell, wished to prove that the correct philosophy is, in principle, monistic. But Churchland seeks to disarm the introspection evidence that could serve as the source of Russell's "vast field where scientific methods are inadequate . . . ," and his "questions of value . . . matters of feeling . . . outside its province." Here is Churchland:[4]

> The argument from introspection is a much more interesting argument, since it tries to appeal to the direct experience of everyman. But the argument is deeply suspect, in that it assumes that our faculty of inner observation or introspection reveals things as they really are in their innermost nature. This assumption is suspect because we already know that our other forms of observation—sight, hearing, touch, and so on—do no such thing. The red surface of an apple does not look like a matrix of molecules reflecting photons at certain critical wavelengths, but that is what it is. The sound of a flute does not sound like a sinusoidal compression wave train in the atmosphere, but that is what it is. The warmth of the summer air does not feel like

> the mean kinetic energy of millions of tiny molecules, but that is what it is. If one's pains and hopes and beliefs do not introspectively seem like electrochemical states in a neural network, that may be only because our faculty of introspection, like our other senses, is not sufficiently penetrating to reveal such hidden details. Which is just what one would expect anyway. The argument from introspection is therefore entirely without force.

But Churchland's statement is quite misleading. "The red surface of an apple does not look like a matrix of molecules reflecting photons" because that is not what it is. "The sound of a flute does not sound like a sinusoidal compression wave train" because that is not what it is. "The warmth of the summer air does not feel like the mean kinetic energy of millions of tiny molecules" because that is not what it is. If we wish to determine the structure of the object that reflected into the eye the photons of wavelength 6000 to 7000 Å to stimulate the neural events that resulted in the look of red, then our science can do that, and it can, indeed, show us the matrix of molecules at the surface of the apple. But these two are hardly identical. One is the initial train of events that gave rise to a neural stimulus; the other lies somewhere beyond that neural stimulus. In their basic nature, those two are totally unrelated, just as the electron microscope is entirely different from the gold-coated microbe that we view with its aid. The fact that one device interacts with another does not make the two identical. The sound of the flute that Churchland analyzes is not the sound that is the experience of our conscious hearing. That was Huang-lung's second barrier, you may recall. Question: "How is my hand like the Buddha's hand?" Answer: "Playing the lute under the moon." Clever Churchland is trying to play a koan game in reverse. With it, he hopes to take one who has glimpsed his own reality and steal it away with illusions of analysis and hypothesis. The sound of the flute is indeed a compression wave train in the air, but the sound you experience is not a compression wave train, and it is not located in the air.

"If one's pains and hopes and beliefs do not . . . seem like electrochemical states," it may be, in fact, that such a concept is an entirely false hypothesis. Churchland gives no scientific data to prove that such is the mechanism of consciousness, nor has anyone else. In fact, if you were to ask, he would likely have no idea how you would prove such a notion scientifically. No neurophysiologist has proved such a hypothesis. In fact, no one has even formulated such a notion explicitly and scientifically. The phrase "electrochemical states in a neural network" is only a scientific allusion to "something (not yet known, not yet understood) about the brain." Our pains and hopes do not *seem* like electrochemical states because they *are* not electrochemical states. They are what they are. We wish to understand how that real thing is related to the rest of reality. It is good that we know something of the structure of the apple and something about light, but if asked a question about something else, we should not use that knowledge to offer an irrelevant answer just because we know that answer. We should wait until we know the answer to the question that has been asked. It is by giving wrong answers to the right questions, and by looking for answers where the street light is brightest rather than looking where we know the answers lie hidden from view, that has caused philosophy to deviate from an earlier recognition of just what the problem of the nature of consciousness is.

Wittgenstein

Just how far philosophy has deviated from its earlier dualistic doctrines is perhaps best exemplified by the work of Ludwig Wittgenstein. His writings show just how great is the impediment to establishing a science of consciousness. His work epitomizes efforts to expunge the concept of consciousness as a distinct substance from logical discourse. He is concerned with the semantic questions associated with the problem of other minds. He uses language as a measuring instrument to probe the question of the existence of consciousness, and his conclusions, in a way, are strangely pertinent to our present task. Wittgenstein concludes that there is no place in an objective language for the existence of the subjective "I."

We will find later that this point is basic to the treatment we will use, for if this were not so, then, ultimately, physical instruments could be used to measure directly the presence of consciousness, making it a physical thing and proving the so-called identity theory of consciousness in which consciousness and the physical state of the material brain are taken to be exactly the same thing. But if the subjective "I" is removed from the language to make language objective, one no longer has a justification for making a statement like "I feel pain." One can only state, "There is pain." Wittgenstein says of this that although there can exist information that is not known, all legitimate knowledge is, at least in principle, public. What does this mean?

I see another person touch a hot object, and he reports his experience. Had I no previous experience with hot objects, I might be surprised when I touched the object to find that there is more to it than was conveyed by that verbal description. If I were to continue to reason, as does Wittgenstein, that there is no legitimate private knowledge, then because other persons do not convey, in a verbal message, anything even similar to what I experience when my (objective) hand touches a hot object, I must conclude that other persons are completely different creatures from me (objective). What I call pain they do not communicate. The knowledge assumed public is not conveyed. Thus I (objective) am unique. The formerly objective world collapses—the world suddenly becomes entirely subjective. I must doubt all of my former objective knowledge about existence, and then

I arrive at the beginning of Descartes' doubt. Descartes' doubt proceeds of necessity from precisely this premise that is basic to Wittgenstein's philosophy.

Gilbert Ryle's Ghost

Gilbert Ryle claims that the mind-body problem is nothing but a category mistake. Ryle compares this category mistake to that of treating both "rocks" and "Wednesdays" as the same kind of thing. Both rocks and Wednesdays exist, but whereas rocks are material, tangible objects, "Wednesday" is just a word that describes the way we classify events. Wednesdays do not exist as fundamental constituents of reality, but only as mental constructs. Ryle argues that whereas the brain is an object, mind is merely a way of talking about the brain's functioning. Ryle tells the story of a peasant in India who, on seeing a locomotive for the first time, asked what makes it move. The peasant was told in detail about the fire box, the steam, the pistons, and the crank shafts. At the end of this explanation, the native said, "Yes, but what makes it move?"

Ryle says that the Indian was looking for the ghost in the machine, and Ryle charges that exactly this same error is made by those who search for the mind in the brain. Ryle sees the world as made up of material objects. According to Ryle, "Minds are not bits of clockwork, they are just bits of non-clockwork."[5] Ryle's arguments are compelling, but he fails to come to grips with the mind-body problem itself. He gives examples of category mistakes and claims the mind-body problem is of this same type, but he fails to prove that his arguments apply to the mind. He feels that merely pointing out that category mistakes can be made, and then telling us that the world consists only of material objects, is sufficient to prove his case. But it does not work. First of all, there is something that is the subject of the satori (enlightenment) experience we discussed in the last chapter. It was to get this fact firmly in mind that we prefaced this subject with that chapter. Even when one achieves the void experienced in meditation, there is still the empty mind—no rocks and no Wednesdays, but something, nevertheless, that exists in its own right. But more than this, we can see that Ryle's idea fails to be satisfactory for even a cursory look at the problem. It would be quite possible for consciousness to be identified with electromagnetic phenomena, say, in which case it would be real, as real as rock, and yet different from ponderable mass. Possible, but as we will see later, it is not actually correct. This example just serves to show that there are categories of real things other than rocks that do not make the Wednesday category mistake.

Ryle's error lies in his assumption that the world is made up exclusively of "rocks," as it were—of physical matter. He views the world from the perspective of the nineteenth-century physics from which he takes his "ghost in the machine" parable. But that physics is gone. The idea of an objective reality has turned out not to work. The strange things of quantum mechanics, the state vector and the question of the role of the observer in quantum physics, did not exist in that nineteenth-century physics from which Ryle drew his stories. For this reason, Ryle does not have the predisposition or the knowledge to understand how the "ghost" could be brought into mechanics. The ghost is there; physicists have already had to come to grips with this Ψ-ghost. But Ryle does not know how to find it. Ryle would look at an atom just the same way he sees the locomotive and say, "I don't see any Ψ there." Ryle does not know the new physics. He does not appreciate the problem of the observer in quantum mechanics. Yet the question of the observer strikes at the very foundations of physics. That Ψ is there in the physics that describes the world we live in; and the observer, with all the mystery that it entails, is there, too. It is something as strange as any ghost. Alas, the "ghost" never dreamt of in nineteenth-century philosophies is at the heart of twentieth-century quantum physics. Ryle's well-turned phrase is simply out of date.

The Identity Theory

We mentioned the identity theory a while back. The identity theory posits that consciousness and the neurophysiological activity of the brain are identical. For example, pain is a particular neurophysiological state of the brain, and *pain*, as the conscious experience, is merely the word we use to designate that state of the brain. Consciousness is simply the sequence of such neurophysiological states. This is another way of saying that the problem of consciousness is merely semantic.

The identity theory can be developed in two ways, into two schools of thought. One of these is physically based: The world is just as classical physics sees it, with real physical objects. We are examples of such objects. That, of course, is just what Ryle was saying, and it is just what people such as Hilary Putnam and Stephen Pepper argue. The alternate view is that taken by Whitehead, among others. It argues that the physical theories are true except that all the elements of physical equations, such as the ideas of space and time, are actually "sensory." In physical theory, objects exist. Our sense organs, which are objects, interact with these other objects, and as a result, the neurophysiological consciousness experiences them. But in Whitehead's view, the concepts of space and time are aspects of consciousness experience. All the quantities in the physical equations are, in a way, sensory, and the chain of logic goes the other way. Physical objects are, at least in part, made of the stuff consciousness is made of, which he believed was also the stuff of the material world—somewhat of an "objectified" mind material.

Both of these approaches, however, fail to satisfy minimum requirements for an understanding of consciousness. In the first version of the identity theory, consciousness is a name given to a collection of objects undergoing a particular process. The neurophysiological state of the brain can be defined operationally, and likewise we recognize or define a corresponding term, *consciousness*. Identity theory says these two are one and the same.

We can always say this, of course, but how do we decide whether two things, two terms that have been defined, refer identically to exactly one and the same object? The answer is simple: All complete definitions of any single object must be logically equivalent. Here we use the term *definition* to signify any means for the recognition of an object (or class of objects). Thus we recognize what is meant by the term *pain*: We feel it. We feel pain whether we know the definition of a neurophysiological state or not. Is the process employed for the physical determination of a neurophysiological state involving the measurement of velocities and positions of the material of the brain—including levels of activation, locations of atoms, electron positions, and so on—identical to the means by which we recognize or feel the sensations of pain? Of course not. We would rather say, "When I feel pain, it is possible to discover that neurophysiological state number 27 exists." Were these two conditions identical, this would be a tautology. Until someone can show how this statement can be converted into a tautology by some set of logic operations on the subject or predicate of the statement, the identity theory will remain unproved. For my part, I think it is clear that the identity theory fails this test.

There is a further argument against the identity theory: It is not a serviceable concept. It does not permit us to ask reasonable questions. According to the theory, if John has the defined neurophysiological brain configuration, then he is in the particular conscious state, because these two statements are construed as identical. Now, let us assume we build a computer out of wires, resistors, and computer chips that manifests the same behavior as John. Since this collection of circuit boards is surely not the same as the neurophysiological state of John, we cannot identify the computer as being conscious. We cannot ask the reasonable question "Does this machine experience consciousness as we do?" Consciousness has been identified with the particular neurophysiological state. No more general definition is available to the identity theory, so it is not possible to remove what may be a trivial requirement for consciousness—namely, the presence of neurons.

Functionalism

In his book *Mind and Matter*, Churchland suggests that the theory of mind called functionalism solves this problem with the identity theory. According to functionalism,[6]

> The essential or defining feature of any type of mental state is the set of causal relations it bears to: (1) environmental effects on the body, (2) other types of mental

> states, and (3) bodily behavior. Pain, for example, characteristically results from some bodily damage or trauma; it causes distress, annoyance, and practical reasoning aimed at relief; and it causes wincing, blanching, and nursing of the traumatized area. Any state that plays exactly that functional role is a pain, according to functionalism.

As Churchland states,[7]

> Functionalism is probably the most widely held theory of mind among philosophers, cognitive psychologists and artificial intelligence researchers. It characterizes mental states as essentially functional states. It places the concerns of psychology at a level that abstracts from the teeming detail of a brain's neurophysiology. Functionalism is very much like behaviorism except that it extends its categories beyond external manifestations of behavior, to catalogue the internal behavior of the organism.

But although functionalism may succeed as philosophy, it fails as science. It fails to give us any reason why one complicated set of interactions going on in the brain leads to vivid consciousness, whereas other activities of the brain that are as functionally viable as mind states—as viable as those that involve pain, love, or the experiencing of redness—involve no consciousness at all. Most of the vast data processing operations of the brain that involve visual and auditory pattern recognition, for example, fall into this category. Functionalism succeeds well as a philosophy because it superficially catalogs away most of the examples we have that provide us with data about what consciousness is, yet without doing the science that should be done to solve the obvious questions: "Why does this or that functional activity of the brain give rise to conscious experience?" "What are the special processes that are going on that give rise to a state of consciousness?"

Elsewhere, Churchland argues against all dualistic theories, stating that they have failed to explain anything about what mind is, but here we find that exactly this fault afflicts functionalism. At the very point where one should begin to ask, "What relationships?" functionalism is telling us to simply catalog these relationships as being identical with mental states. It is so easy to anthropomorphize; for the philosopher or the scientist to envision one set of processes he knows involves consciousness as having consciousness, and so easy to forget all the counterexamples. Churchland's anthropomorphism about the relationships he sees as important in behavior keeps him from recognizing that the neural activity sustaining these causal relations in the brain physically represents only a small part of the activity of the brain. There are networks of chemical processes going on in the brain to sustain the individual cells, to provide oxygen, burn sugars, synthesize proteins, and replicate DNA. This vast array of unseen activity in the brain—which is also carried on throughout the body—while perhaps no more complex than the most complex mental activities of the brain, is, nevertheless, surely far more complex than the activity of the brain that supports a simple conscious experience, such as that filling the dullest mind involved in simply staring at a blank red wall. And yet, that redness is experienced as consciousness, whereas all the other complex activity of the brain is not experienced at all.

The functionalist says of this, "Well, of course, DNA replication is not the same functional activity as the brain activity that goes on when we stare at a red wall." This comment, however, betrays the bankruptcy of functionalism. Functionalism explains nothing; it simply catalogs. And even worse, it catalogs against nothing, for like the identity theory, it denies that these experiences of redness or pain can be anything but those functional relationships. Though it is clear that we should be asking just what functional process is critical to the occurrence of a conscious state, functionalism has (as philosophies often wrongly do) defined away a problem that obviously must be solved by means of analyses that are entirely different from those made available by that philosophy.

What should be done is to look into the brain to determine just what physical process is going on that correlates with the occurrence of consciousness. Armed with the answer to that problem, we would find it easier to say if that correlation were identical to consciousness, or consciousness were a property of the physical correlation. But if we do this, we will no longer have either the identity theory or functionalism, for these are no more than philosophical semantics games intended to leave the impression that there is no consciousness problem in the first place. They are designed as a balm, substituting for the difficult task of developing a science that would bring real understanding to this subject.

In 1952, I was already deep into my flights of "scientific creativity," for lack of a better term. The diary shows it all. On Tuesday, when I called Merilyn after school, I told her that I was building—yes—a "time machine." I have now, fortunately, no memory of what I could possibly have had in mind, but I am convicted by my own entry in the diary. The diary, surely for reasons of national security, does not record the workings of the device, but it does mention certain vital parts of the machine. The words *Tesla coil* appear with increasing frequency.

Merilyn, however, was more practical than I. She said simply, "That's impossible." I am sure you can imagine how she said it—"That's imPOSsible!?" with a questioning tone to show that she had doubts about my sanity. Perhaps she added "You're silly. What are you playing with now?" to offer me a way out. "That's impossible."

Women are like that. Women throughout all ages have said that. When men have said, "I'm working on a time machine," they have always said, "That's silly. That's impossible." That, of course, is why there are no time machines!

But I could imagine strange and wondrous devices—things as yet unseen in the pages of *Mechanics Illustrated*, or even in my favorite, *Popular Science*. My eyes were wide with the wonders the future would bring. I could imagine bizarre new experiments. My diary records my wondering what would happen if two high-frequency, high-voltage currents passed through the same point in space at right angles to each other. What strange new effects might occur in some fleeting moment in time? What strange effect might be visited in the space around such a macabre event? Perhaps a beam of positrons might radiate out and . . . "maybe it would make a fine ray gun! Bed at 12:08 A.M."

"Wednesday, March 5: In Auditorium today Merilyn was talking to me, and some teacher made her move." Merilyn and I had been making plans for a stunt we could pull in the variety show planned for the debating club's activities. Merilyn had just said she was planning to sing "Saint Louis Blues" just like she did at Barbara Riddlebaugh's party. I said, "I'll dress up like a tramp and come strolling through the auditorium pulling a rope that goes on and on while you're singing." I still remember that; I remember her leaving as some warmth being taken away from me. "After she moved I talked to Tommy about Tesla coils." Later, "I went home and called Merilyn. I went over to Homewood. I met Jimmy Yeldel, who had some firecrackers. I used some to make a marble gun." I put a firecracker into a half-inch steel pipe that had a hole drilled in the cap at one end and rolled in a marble. I lit the fuse and ran. Blaough! Primitive, but no less dangerous, dangerous in all directions. "I bought a magazine and I called MAZ when I got home." The next day I talked to Whitson some more about building a Tesla coil. During "8th period I told Mr. Newman that I wanted to come walking in dressed like a bum during Merilyn's act in the variety show." Mr. Newman was a good man. He was one teacher who still seemed to remember being just a kid in school himself. He got a kick out of the idea. "Do it. Just don't say I said you could!"

Later, "Tommy and I went to town to a radio parts shop." We were looking for things to make a Tesla coil. Who knows what time machine might yet be built, but it would have to have a Tesla coil right in the center! That night I called Merilyn, and we talked and we talked and we talked into the night.

Friday evening, Merilyn and I made more plans. We talked about our house, and we drew plans. We planned a ranch-styled house, L-shaped with an observatory built into the house so we could see the night sky, and stars, and the galaxies of stars beyond the stars. And there was going to be a German shepherd and . . . children. I don't remember exactly what we said, and I didn't write it down—but we talked about that as well; we made plans.

HIGH PRIESTS AND PHYSICISTS

Physicists have one thing in common with psychologists: They have always considered consciousness to be outside their domain. Physicists assume that if any aspect of consciousness is objective, it should simply be the concern of neurophysiology or maybe of that social science, psychology. If it is not something absolutely objective—and for the physicist that has always meant everything that exists—then the physicist has no interest in it. Of course, things like animals, brains, nations, and money exist as actual physical entities, but for the physicist their peculiar characteristics are left to biologists, psychologists, sociologists and, well, physicists do take a remarkable interest in the last one, quite beyond their own field of expertise. The physicist's interest extends only to making sure that physics provides an understanding of how all the parts of things work so that, in principle, the total functioning of any collection of the basic constituents can be deduced. He is not interested in the detailed behavior of, say, a bird, but he does feel that the principles determining its flight, the mechanical laws that determine how it moves, the physics of fracture that may affect its bones, and the laws that determine how any of its constituent molecules, atoms, or elementary particles behave is within its province. The physicist is prone to seeing himself as the high priest of science, and with some justification. He is inclined to pass out to others the more detailed tasks. The biochemist determines the chemical make-up, the neurophysiologist the circuits of the brain, and the psychologist investigates the animation that the brain gives the biologist's beasts.

But consciousness does not appear to be in any of these parts. Caught up in the glitter of the computer craze, physicists have been known to liken consciousness to the software of the brain. One almost universal belief of physicists is that computers, if sufficiently large and if properly programmed, can be made into conscious beings. Thus physicists too are, for the most part, believers in functionalism, where the mind, though not the hardware of the brain or body, nevertheless does lie in the pattern of its functioning and in the implementation of its programs. The functionalist does not think that the mind causes the neurons to fire or the brain to behave in any way but as a machine. As Paul Davies puts it in his *God and the New Physics*,[8] "The computer is simply a lot of circuitry, and anything it can do is determined by the laws of electricity. Its output is an automatic consequence of its following predetermined electrical pathways." For the functionalist, the mind is like

> the computer solving equations, making comparisons and decisions and arriving at conclusions based on information processes, *i.e.*, pushing ideas round. So it is possible to live with two different levels of causal description—hardware and software—without ever having to grapple with how the software acts on the hardware. The old conundrum of how the mind acts on the body is seen to be just a muddle of conceptual levels. We never ask 'How does a computer program make its circuits solve the equation?' Nor do we need to ask how thoughts trigger neurons to produce bodily responses.

Davies finally concludes:

> Functionalism solves at a stroke most of the traditional queries about the soul. What stuff is the soul made of? The question is as meaningless as asking what stuff citizenship or Wednesdays are made of. The soul is a holistic concept. It is not made of stuff at all.
>
> Where is the soul located? Nowhere. To talk of the soul as being in a place is as misconceived as trying to locate the number seven, or Beethoven's fifth symphony.

Marvin Minsky, cofounder of M.I.T's Artificial Intelligence Laboratory, has this to add:[9]

> Consciousness is overrated. What we call consciousness now is a very imperfect summary in one part of the brain of what the rest is doing. The real problem is that people who ask 'Could a machine be conscious?' think that they are. They think they have a pipeline to what's happening in their minds. That's not true. People scarcely know how they get ideas at all . . . it makes putting consciousness into machines easy, because I don't think it'll take very much. For a machine to solve very hard problems, it's going to have to have a brief description of itself. When there is a better theory about how certain parts of the brain summarize what's happening in other parts, then we'll understand it and be able to make machines do it.

But consciousness is not a "doing"; it is a "something." It is not thinking; it is the existential being that has thinking as its subject. How could Minsky believe that computations in one part of the brain, or a computer, could have sentience—feelings and experience—that happens just because its input data comes from other parts of the brain in spite of the fact that those parts themselves, doing the heaviest load of the data processing, have no conscious experience whatsoever? But we have heard all this before. This functionalism is Ryle's "category mistake" argument. We have already seen that such a concept fails to recognize the existence of something more than behavior. Ask yourself, "What is it that the functionalist says that is objectively different from what the behaviorist says?" We have already seen that the reality the Zen Buddhist finds beyond the void is the stuff of consciousness that still exists when the program is put to rest—that is, when thought is stilled. And most importantly, we have already seen that the efforts of physics to describe the world in such purely objective terms fail when put to the test.

The physicist calls upon such ideas as software to explain consciousness because he knows that nothing like consciousness is part of his definition of mass, of electric charge, or of leptons. He uses fanciful terminology, invoking color, charm, and flavor to talk about quarks, not because such psychic characteristics are part of quark particles, but because they are an amusing language that has absolutely no relationship to the physicist's picture of the constituents of matter. By time, the physicist does not mean that hour of waiting that the introspective psychologist would have meant. If we look at the domain that the equations of physics represent, we can see atoms, living bodies, stars, and even the universe. We can see the motions of rivers, the movement of horses, and the universe unfolding from some early Big Bang in the first instant of creation. But if we approach what is in those equations exclusively in terms of those ideas physicists have put there, we will see that there are some things that are missing and that cannot be derived from the things that have gone into those equations. The equations have positions and intervals, quantities and forms, and they describe responses. But feelings are not there, nor is pain, C#, or the colors we see in the budding red rose. "Motives" are there, but emotions are not. Conscious being is not in these equations. If consciousness is to play its role in physics, it must be included in its own right, on its own terms. Quantum mechanics may beg the question of what the observer is, but making length be the conscious experience of length will not change the Schrödinger equation or answer our questions about the measurement problem. Instead, it will be necessary to introduce something new into physics on its own terms. This is how it has always been in physics when we have wished to understand something totally new. This is how we must do things now.

When Isaac Newton wished to study the phenomenon of gravitation, he did not look for the hooks and springs that would make it work, as Robert Hooke had argued should be done. Instead, Newton introduced the force of gravitation as its own phenomenon, on its own terms. When Faraday wished to study the effects of electrostatic charge, electricity was studied as its own phenomenon, as a new subject of physics with its own laws and mechanisms and reality. Success in understanding gravitation was not achieved by ignoring the phenomenon, in some vague expectation that it would be explained ultimately as a by-product of the study of hooks and springs, and the successful development of electromagnetism did not come by waiting for someone to derive the behavior of electric charges from Newton's gravitation.

An understanding of the nature of consciousness, which we need if we are to understand reality, must be actively sought. We must treat it as though it were a distinct phenomenon worthy of study. That bold step must be taken for science to advance its understanding of consciousness and with it advance the understanding of who we are ourselves. As surely as gravitation cannot be reduced to strings and springs, nor electromagnetism to Maxwell's spinning wheels, so too consciousness cannot be understood without recognizing that it exists on a wholly distinct level in science. Consciousness that gives meaning to our concepts of matter cannot be understood in terms of those old concepts of matter. Moreover, leaving out something fundamental in formulating the basic physics of reality is bound to lead to a physics that has an epistemological problem.

To understand what part consciousness itself plays, it must enter upon the stage; it must play its part. We must listen to its dialogue with that of the other players. We must see each actor act out his own part, or we too must fear our own Virginia Woolf.

REFLECTIONS OF THE OBSERVER

As we have seen, the tests of Bell's theorem have made it clear that we will have to deal with the observer and as such, with the role of consciousness even though efforts in this direction by the physics community have been slow in coming. Almost every aspect of modern physics reflects the fundamental role that observation and, ultimately, consciousness must play in understanding the basic physics of the world. Concepts that at first sight should play no role at all in describing the laws of material objects turn out to be basic principles. The meaning of simultaneity might be important to an observer, but why should the way objects interact depend on that observer's perspective? Should an ambiguity about which of several events took place first alter the basic way in which time brings about the decay of radioactive atoms? Incredibly, relativity tells us it should. Should our inability to distinguish one atom from another affect how these objects behave thermodynamically? Incredibly, we find that it does. In the thermodynamics that has developed out of quantum mechanics, we have found that this indistinguishability makes a measurable difference in the way matter behaves. Should the symmetries we see in nature play a central role in determining the equations of physics? Again, in elementary-particle physics, the answer turns out to be "Yes." In each of these cases, the answer in physics has turned out to be "Yes." In none of these cases are we forced to resort to a conscious observer to find reason enough for the particular innovative principles these questions have revealed to us about nature. And yet each marks a step in the same direction, a step toward recognizing that what affects our ability to observe reflects nature's most basic rules and determines her most basic structures. Each is a step toward recognizing that we must accord the conscious observer a place in the total picture in order to understand how things work.

All of these discoveries have shown, little by little, that consciousness is a fundamental part of reality. The quantum mechanical description of a world of state vectors with state selection caused by observation moves us still closer to seeing the observer as co-participator—I am tempted to use Constantine's word *consubstantia*—in all physical phenomena. Bell's theorem has moved us a step closer as we have discovered that nothing "out there" has an entirely independent existence. Little surprise, then, that limits in observability should reflect the basic structure of physical reality. Little surprise, then, that symmetry should be so important in physics, or that the lack of distinguishability of individual atoms should affect the way matter behaves, or that the coordinates of the observer should determine the physical laws of relativity. How basic is this observer to the nature of reality? How much does mind create, or is pursuing this idea even the correct path to follow? Only by doing one thing that is entirely new in science can we hope to find the answer. Only if we assume that the consciousness of the observer exists as a legitimate subject for scientific scrutiny can we hope to understand what we mean by *observer* in the first place. Only in this way is it possible to show what consciousness really is.

In the past, science has woefully ignored the issue of what consciousness is and so has come up empty-handed in its efforts to deal with these questions raised by the measurement problem in quantum mechanics. The entire meaning of the new physics experience will be lost if we do not succeed in understanding consciousness now.

Bell's theorem destroyed the foundations of our old objectivity—destroyed the basic underpinnings of objective reality. It showed us that we cannot fully understand nature as made up of an indefinite Ptolemaic succession of layers of elementary particle. It has undermined that classical self-assurance in objective reality to the extent that physicists must actually ask themselves,[10] "Is the moon still there when nobody looks?" And yet a new world of understanding lies within our present reach if we will only grasp the clues and let them lead us where they will.

More importantly than all of this, we have looked at consciousness itself as the central experience, as the ineffable knowledge of Zen Buddhism. We have looked with the masters beyond emptiness (*sunyata*) and nothingness (*nasti*) to discover that there is a reality distinct from the physicist's insistence on physical processes and physical objects. It is something real, yet beyond. It is something that has qualities other than those of the material world, as we have previously used those words to build our scientific edifice. Understanding it calls for new brick, fresh mortar, unused trowels, and masons who see with the inner eye. Now we must build a science of this most elusive substance.

DUALISM

Consciousness is something that exists in its own right and has its own identity. It is distinct from all other objects, processes, energies, and realities that physics or science as a whole reveals. We must accept, at least tentatively, a kind of dualism.

The reason for this is simple. Physicists do not mean, by any of the constructs in the physical equations, either directly or indirectly, anything that constitutes the substance of what is meant by the term *consciousness*. Because no such property appears in the equations—equations that provide us only information about how quantities vary relative to one another—they do not have the power to let us discover anything fundamentally new. Therefore, if consciousness is not in the equations to begin with, it cannot be derived from those equations. We can grapple with understanding our Zen mind with our analytic brains just so long before the effort to achieve such abstraction overtakes our mental capabilities.

As a result, we are forced to bring consciousness in as something distinct, something new. We must do what science has always done when there is a new thing to study. We will create a new category, a new place, and a new way of thinking about this subject that has been with us unsolved for so long. We will set consciousness aside as being a distinct phenomenology. As a consequence, we begin with what looks at first like a conventional dualistic philosophy. What I propose, however, should be recognized as something different. It is a dualism that consists of the concepts and principles of material phenomena on one side and consciousness as something entirely different on the other. And yet the two distinct things interact. We look for the laws and mechanisms of those interactions. But that is not the end of our search; it's the starting point. We will treat consciousness in this fashion until we can better understand what it is and how it properly relates to those things in nature that science already understands well. Maybe we will find that this dualism dissolves into some single replacement for both materialism and consciousness, or perhaps we will find this dualism to be the most effective description of the relationships we discover as we go along. Our dualism, therefore, is of a most modest sort—a kind of pragmatic dualism. We assume nothing about it, and discard it if and when it has outlived its usefulness.

Why do this? The reason is simple. Physicists deal with physical quantities, by which they mean those things that are measurable. Consciousness is not measurable in the way physicists use the term. You cannot make a measurement on an ice cube and by that measurement determine directly and incontrovertibly that the ice cube feels pain when it melts. We certainly believe that it does not, but that is nothing more than an anthropocentric bias. If it is not physically measurable, and if we are going to keep sharp the powerful tools that physics has given us so that we can discover the nature of reality, then consciousness has to be put in a category of its own. Doing that is tantamount to saying we are going to treat consciousness as though it is not physical, not physical the way the things of physics are physical. We will shortly see that we are forced into this dualism just to get our investigation of the nature of consciousness under way.

The classic philosophical controversy between dualists and monists when put in a scientific context, especially into the context of the new physics, takes on a totally transformed significance. Monism and dualism are not antagonistic concepts in the development of an understanding of reality. It is true that all physical models in the past have been framed by their creators with the clear intent of explaining reality in objective or materialistic terms. But while that must now give way to some new, larger conception of nature that embraces consciousness, the ways in which physics has gone about the business of formulating its pictures of reality have served too well for us now to regress to the imprecise devices of philosophy. In physics, philosophies become embodied into basic equations: into a system or set of equations, or indeed into one single equation that describes, precisely and numerically, the whole universe—ultimately, it is to be hoped, all that exists.

In its purest realization, physics becomes one equation. That equation is the word that immaculately posits all reality, and that word becomes the perfect monism. But this monism would be meaningless without the parts of the equation whose equivalence form the perfect dualism: balanced, counterpoised, symmetrical. There is no way to describe reality, whatever it is, that does not at once demand and satisfy such a relationship of parts and wholeness. If philosophy could have surrendered mind, physics still could not have structured reality out of matter alone. Such a conception is an error of eighteenth-century thinking, a thinking in which the fundamental nature of energy, space, and time had not yet been recognized as constructs, as pieces of reality as basic to what reality is as is matter.

Thus consciousness enters as a new fundamental construct. It is a new actor on our stage that at this point enters unannounced, almost devoid of any descriptive characterization, yet a character who must be present, who must exist—a character drawn out of the audience to stand on stage, an actor who must participate in the functioning of the universe and play his role in some grand equation about to be spoken.

But though consciousness must be treated as a real quantity if any progress in our understanding is to be achieved, we must acknowledge a basic characteristic of this new thing. We must recognize that characteristic of consciousness that lies at the heart of science's difficulties in handling the issue at all. We must recognize that consciousness itself is fundamentally nonphysical! Such an admission at the start would seem to end everything—certainly any hope of building a science—before we even start. But let me explain what this actually means, and let me tell you how we will get around this seemingly intractable impediment that such a concept imposes on science.

All things in science—all physical things, including all the constructs we use in physics—are measurable, or constructed of measurable things. Space (distance relationships) is quantifiable. Time is measurable. Speed is constructed from distance and time measurements. Mass, force, energy, numbers of stars, and the existence (yes-no measurement) of a galaxy at specified celestial coordinates can all be measured. A book, for example, is a physical object. It has mass, and it can be weighed; it has a position, and that can be measured.

And yet this is something we cannot do for consciousness. We can ask the meaningful question "Does an ice cube feel pain when it melts?" We know what the question means, and yet we cannot answer this question in any physical way.[11]

It seems we can answer such a question when the subject is another person. But there we make a judgment, a leap of faith, that those who look like ourselves and act somewhat as we do also have the same feelings, the same consciousness substrate as we. We cannot, however, enter their minds to feel their feelings. We do not know if their "red" is our "red" or if their sound, their taste, or even their perception—their conscious experience of space or time—is the same as ours. Indeed, these likely are not quite the same. Even colors we see ourselves differ from one eye to the other for most people.[12]

Questions we might ask another in order to determine if they have consciousness like ours, or to find out if the quality of their conscious experience is the same as our own, are quite pointless. All the phrases we might use to refer to internal states are communicated entirely in terms of externally defined words. This is why Zen avoids words as expressions of the Buddha nature. To know Zen, to know consciousness, is to know one is beyond the domain of words. We have the word *consciousness*, but it is difficult to communicate just what even this means to another person unless that other person has experienced on his own the same questions about his inner nature. And if that person has not, you might as well try to communicate with a tree stump. The ideas will not have the same meaning when you speak to him that they had when they formed in your mind. Even simple ideas, such as red, coldness, and C#, are externalized and objective when we learn their meanings. Ask someone what red is, and he will show you a red object and most likely he will mean by that that red is the category in which the dye belongs. We saw just how easily Churchland identified the external redness of an apple as being a matrix of molecules and substituted reflected photons for the experience of redness in the mind. But ask the Zen master what red is, and he may pat your head or splash water (because these are, in fact, closer to what redness is).

Characteristics of conscious existence, such as pain or redness, cannot be measured directly by the use of any measuring device known to science. Even using an electroencephalograph to measure brain wave activity does not tell us anything about the actual presence of pain or any other characteristic of consciousness. Only if one measures his own brain waves, might he find that an experience of pain always accompanies a particular set of brain wave patterns in his own brain and, further, if he finds that in other people when inflicted with pain in a similar way that these same patterns occur, only then is he ready to make the logical leap of assuming that because that person is somewhat like himself, then that person feels pain just as he does. But he has no more measured pain when he does this than he measures pain by looking at an anguished face. What electric measurement would tell us that the ice cube feels pain? No scientific measurement can tell us; no tool of physics can probe that deeply.

And yet it is the ability to use physical tools to make direct measurements on objects that defines physicality. If it is real and physical, we can measure it—gauge some characteristic of it directly. We cannot do this with consciousness, so it is nonphysical. Yet no one can deny that one's own pain is real. Thus we reach a basic postulate about consciousness:

Consciousness is real and nonphysical.

Such a characterization of anything goes against our conventional wisdom that all real things must be physical. Moreover, it seems that by making such a postulate, we have placed consciousness beyond the pale of science. I will show you shortly, however, that we have just taken the first real step toward a scientific understanding of consciousness. We have seen others search for the ghost in the machine, not knowing what they were looking for. They have sought answers to the question of consciousness that would have been only rearrangements of material processes. To understand consciousness, we must first understand that this something has its own existence that cannot be reexpressed as a combination of other factors or objects. Consciousness is not so many atoms. It does not consist of photons or quarks. Neither is it molecules spinning about in the brain. Consciousness is something that exists in its own right.

And yet consciousness does interact in some way, by some connection, with the things of the physical world that physicists measure, know, and describe. If we are to approach the issue of the nature of consciousness in a methodical and scientific manner, we must lay out what is not a part of the phenomenology (physical processes), lay out what is a part of it (consciousness), and then find the connection between the two. This is not philosophic dualism, and it is not Cartesian doctrine. This is simply how science should go about dealing with new phenomena.

Of course, Descartes did try to do this. He was a brilliant scientist as well as a genius in philosophy. He set for himself the greatest problem, the grandest task. And yet in his day, there was no hope that he could succeed. What could not prosper as science eventually became a philosophy of stagnant ideas. His incisive perception of the central issue of philosophy has been turned into a misconstrued proposition for philosophical gyrations and semantic games: "*Cogito ergo sum*."—an often repeated quote that has ended within itself.

We have a new phenomenon. It is, nevertheless, a part of the whole cloth of reality. How does it tie in to the physical cloth? It seems like a shimmering color that floats somehow above the fabric. Where is its connection to the rest of reality? These questions imply something about how we must search for an answer, and they imply something about the phenomenon—namely, that we seek to understand some *one* new constituent of reality, and we seek to find its *one* connection with the rest of reality.

To put this another way (the way the physicist would ordinarily express it), we are dealing with something fundamental in nature. Consciousness is something that we cannot understand by describing it in terms of something more elemental. We cannot resolve any of the issues related to the nature of consciousness by specifying the atoms it is made of; there are no monads of consciousness, no more than a physicist would think of trying to explain mass by specifying the atoms it is made of (Higgs bosons not withstanding). Nor would the physicist try to explain electromagnetic phenomena in terms of the elements of the periodic table. But the physicist has had to say how electromagnetic phenomena and the mass of material objects interact. In doing this, the physicist has introduced the fundamental connecting link: electric charge. No, we are not going to define a fundamental unit charge of consciousness. That would be to repeat the most frequent, flagrant, and ignorant error that some commit in their efforts to cope with new questions. They attempt to apply approaches that have been successful in the past without understanding the unique characteristics of the new problem and without introducing a single new idea. It is like trying to solve every new problem in physics by proposing a 5-, 6-, or 11-dimensional space in a pale mimicry of Einstein and Minkowski. Each phenomenology has its own nature, its own questions, its own path to understanding. What we need in the case of consciousness is to answer the obvious question "What is the fundamental physical quantity that links consciousness, a thing in its own right, to material phenomena?" This is the essential idea. There is some link between physical processes and consciousness, a single fundamental link. And this leads us to the second fundamental postulate that we will need in order to understand what consciousness is:

> *Physical reality is connected to consciousness by means of a single physically fundamental quantity.*

These two postulates—"Consciousness is real and nonphysical" and "Physical reality is connected to consciousness by means of a single physically fundamental quantity"—organize the problem and point to a methodology. They say what we should and what we should not try to do. They tell us that we should not attempt to explain consciousness as being something else—as being made up of neurological processes, for example—and they say what the first step should be in our effort to solve the mystery of consciousness: Find the connection between consciousness and the rest of our body of scientific knowledge. When we have done that, we can begin to probe more deeply into the structure and the nature of the interaction of consciousness with other parts of reality. Ultimately, we will find out whether consciousness is a part of the cloth, thread, weave, pattern, dye, color, or form of the fabric of reality. Ultimately, we will be able to look past these aspects of reality and see something more: the loom and the weaver.

One might think that in these postulates we have already presumed our philosophy. Watch what unfolds. You may be amazed!

11

Looking for the Emerald City

Do I dare
Disturb the universe?
I have known them all already,
known the evenings, mornings,
afternoons,
I have measured out my life
with coffee spoons.

—T. S. Eliot,
The Love Song of J. Alfred Prufrock

Our goal is to discover the ultimate nature of reality. But in that search, we are still looking for bedrock. We are still looking for that secure place to stand so that we can find truth and know reality. We have passed by the flickering lights that long ago cast primitive creeds into the minds of ancient man, and we have paused, watching science categorize the world into charts and parts—slices of objective reality that almost work. We have plumbed the depths of reality's finest particles and looked into the eye of the quantum world's observer, searching—perhaps to find something of ourselves there—searching for the source of it all. And the search has led us here, trying to probe the stuff that our own minds are made of. We still have a long way to go, and, for good or ill, we must travel alone in our quest for the Emerald City beyond the conscious mind.

In the last chapter, we proposed two postulates about consciousness. These postulates are not restrictions on our thought, nor do they limit our ultimate understanding of consciousness. These postulates merely give us a starting point. If they are wrong, we will find out that they do not serve us well and they will be changed. But it is impossible to develop knowledge or a science about something without spelling out what we are dealing with. With these postulates, however, we now have a foothold. We do not wish to redo all of science, nor redefine the definitions of the physical world in order to blend consciousness into an imagined understanding of what matter is. We do not wish merely to absorb consciousness into some presupposed philosophical nonproblem. That would not give us factual knowledge, but only a substitute belief system. Instead, we intend to learn exactly what consciousness is and how it works. We have introduced consciousness as a new "element" of reality. We have "defined" it by pointing to what we mean by consciousness, and now we must find out where it fits in physically and find out how it works. That is all that is intended by the postulate that consciousness is nonphysical. If later we find that we can measure consciousness the way we would measure mass or time or some physical object's form, we will toss the postulate aside and replace it with whatever we have found to be more serviceable.

But quite apart from whether or not consciousness is distinct from the category of physical objects, we know that consciousness—our own stream of conscious events—does contain information about physical processes. Consciousness is affected by matter or by events in the physical world. So we must find out how contact is made between those events of the physical world and consciousness. More to the point, we must discover what goes on in the brain—if that is mind's proper home—that is associated, item by item, one for one, with the occurrence of consciousness.

THE YELLOW BRICK ROAD

Sherlock Holmes, in Sir Arthur Conan Doyle's *The Sign of Four*, remarks, "How often have I said to you when you have eliminated the impossible, whatever remains, however improbable, must be the truth?"

We too look for specifics. Out of the whole world of causes, we search for the truth to show Watson. And like Holmes, we must eliminate the impossible. We have a whole world of facts that physics and science have put together to cover the data of physical reality. What we need is a cutting blade that will let us peel away this apple to find the core causes, the mechanism of consciousness. What we need are the tools that will enable us to look at this world of facts with that razor eye of a Newton or an Einstein so that we can discover the pieces that form *theory*.

I emphasize the word *theory* because it is so much misunderstood, so often misused to refer to almost any nascent idea as though it were synonymous with *hypothesis*, *conjecture*, or *notion*. *Theory*, as the word is intended to be used in science, refers to a conception of the nature of some set of phenomena that fits all the facts and that has survived the tests of time and experiment. Experiments give us raw data. Theory, and only theory, gives us understanding.

Now let us look for that point of contact between mind and brain. Let us look for the bridge between these two worlds. Where is the point of contact? In *The Tao of Physics*,[1] Fritjof Capra writes,

> The central aim of Eastern mysticism is to experience all phenomena in the world as manifestations of the same ultimate reality. This reality is seen as the essence of the universe, underlying and unifying the multitude of things and events we observe. The Hindus call it Brahman, the Buddhists, Dharmakaya (the body of being) or Tathata (suchness), and the Taoists, Tao.

The physicist calls it the fundamental equation, and physicists have come close to showing that their depiction truly embraces more of reality than any other reach of intellect or grasp of mind. This fundamental equation, still elusive as to its final form, but very well approximated in the equations of quantum theory and relativity theory, embodies the physical world we wish to link to consciousness. Because consciousness is nonphysical, contact with the physical must be achieved by means of some part of that physical reality represented by the equation. We should look at this equation to see what its components are and then use these as our list of fundamental constituents by which consciousness must be mediated in order to interact with the physical world.

Using Schrödinger's equation and Einstein's general relativity as our guide, we can enumerate all the fundamental things that make up physical reality as science presently describes the world. This will give us a kind of shopping list to guide us as we look for the way consciousness might be coupled to the physical world. For each item on this list, we ask, "Is consciousness tied to or associated with . . . ?"

1. Everything? Is consciousness tied all at once to all of physical reality that the Schrödinger equation deals with?
2. A single mass point or particle—an electron, a quark, a neutrino, a gluon?
3. Space?
4. Time—or possibly the space-time of relativity theory?
5. One of the four forces in nature—gravitation, electromagnetism, the weak nuclear interaction, the strong nuclear interaction?
6. Ψ, the state vector in Schrödinger's equation?"

These are the fundamental things of physical reality. Every physical thing in the universe must consist of one of these; must be mediated by, or manifest in, one of these; or must be that totality of everything, the entire universe of item 1. If indeed consciousness and the observer do play a fundamental role in the measurement problem of quantum theory, as we discovered with the help of Bell's theorem, then only one of these things can be that fundamental interface between mind and body. Only materials such as these can build a bridge between the physical world and the world of conscious being.

Note that although we have included "everything" and have included individual fundamental particles in this list of ours, we have omitted the idea that the interface with consciousness might be mediated by some kind of atom, some organic molecule, or by say a crystal or other composite structure. For any such larger structure to be the link between consciousness and physical reality, the forces that give rise to their structure (the atomic or molecular forces) would have to be the real link between mind and matter, because such structures have no independent identity apart from those forces. Absent those forces, such objects are just a collection of particles that *we* have chosen to separate out for consideration because of their significance to us, not because of their own innate existence as self-validating entities.

Similarly, we have omitted "complexity" as an origin for the connection between mind and brain. Complexity is not a physical thing and so cannot be the point of contact between physical processes and "mental" consciousness processing. This eliminates identifying consciousness with the "program," to use computer language, of the brain. There is nothing about a computer program, the execution of a program by the computer, or even the computer itself that would justify our treating it as an entity in and of itself, except as those processes of components may be knitted together by some electromagnetic or other physical interconnection. The program per se is an object only to our mind (or brain). In all other respects, it is a collection of separate processes, objects, or components. It exists not as some single, distinct physical thing but as many physical objects, just like the words on this page, which together function as a unit in our brains' processes, but that in no other way have their own independent existence as distinct single entities.

QUO VADIS?

Total Universe

Is consciousness associated with everything (with the totality of the universe)? If consciousness were associated with the total universe, then the conscious stream of any one person should entail everything that goes on in the universe, and in equal amounts. Not only should I experience the writing of these words but equally and indiscriminably experience your reading of these words. And I should be viewing each sight with the eyes of each person and animal on earth. I should simultaneously experience the motions of the planets, the explosion of quasars, and the transformation of each quark in the universe as being myself. This does not happen, and this is not the nature of consciousness. Some may argue that there is a God-Consciousness that spans the universe just like this. But don't rush me. We are a hundred pages away from being ready to think such thoughts. So far, we are talking only about our own human consciousness.

Elementary Particle

Is consciousness tied to an electron, a quark? The fact that all physical processes can ultimately be reduced to single particle-particle interactions or, in some cases, interactions among a very limited collection of particles suggests that consciousness might be associated with such particle-particle interactions: with particle collisions, or the field interactions of such particles. Whether consciousness plays any role in elementary-particle interactions, and whether consciousness is elicited in association with such processes, we cannot say, but it is clear that our own consciousness, the consciousness associated with brain functioning, cannot be tied to single, isolated particle events, and for exactly the same reason we excluded any connection between consciousness and the universe as a whole. What we experience in our consciousness is not compatible with the idea that consciousness resides in a single particle or arises from single-particle interactions. Associating consciousness with a particle or with interaction events of a particle would mean that each such particle would live its conscious existence totally isolated from all other particles or events. Spatial and temporal relationships could not figure in the coupling. Such entities do not have a sufficient range of states to encompass all the vast patterns that fill our conscious experience. Only if we find some field that spans the brain and extends over the data-processing events of the brain will we have an acceptable candidate with which consciousness could be identified.

We would also find any investment of consciousness connectivity in a single item to be inadequate to account for the fact that our conscious experience involves events that take place in various parts of the brain. We would be unable to account for the fact that both the visual perception of the occipital lobes and the auditory perception of the parietal lobes of the brain are a part of one consciousness. I both listen to and watch an orchestra play a symphony. Whatever the source of consciousness is, it must span that space.

Space-Time

What about space and time? Are they the physical interface with consciousness? Just as the connection between consciousness and physical processes cannot be satisfactorily explained by anything that lacks spatial extension, so too it also fails if we try to make the connection with space itself or with time. For if we propose space itself as the connection, we lose the ability to explain the lack of immediate conscious contact between the minds of separate individuals. Indeed to connect consciousness with space, it should span space—all space—and to connect with time, it should span all time. Neither of these conditions has anything in common with the vast, detailed stream of local events that feed the information we experience. I do not mean here that space and time will not be vital to the mechanism linking consciousness to physical nature. Indeed, we should expect that they will play a role. But they will be employed there as parameters to describe interrelationships of a more complex property of the physical world that can support the vast variety of states we experience as consciousness. Space and time, without the actors of particles and fields, simply have nothing left. So we must look elsewhere.

FORCE FIELDS

We now come to a class of physical entities that has more potential for matching the complexity of consciousness as well as for mapping the fields of conscious states onto those of brain states; and yet, we will see that these, too, fail to provide the physical basis for consciousness. Here we ask, "Is consciousness some manifestation or some attribute of a force field?" There are four forces in nature as usually numbered: gravitation, electromagnetism, the strong nuclear force, and the weak nuclear force.[2]

These forces,[3] together with only three of the elementary particles (electrons and the up and the down quarks), are sufficient to account for almost everything we see or encounter in nature. It would seem that somewhere in this list we should be able to discover a mechanism for linking mind to brain. Let us look at each one in turn.

Gravitation

It is rather obvious that gravitation cannot explain the mind-brain link. The claims of astrology notwithstanding, the planets, sun, and moon are not the creatures of ordinary conscious experience. The force of the moon on my body is about 300 dynes,[4] and the force of the planet Jupiter on me is about 2 dynes, but the gravitational force of this book, which has far more to do with my present state of consciousness, is nearly a thousand times less, and the printed words on this page exert a gravitational force ten thousand times smaller still. Turning to the brain, we find that the gravitational effects due to any of these are still vastly larger than the gravitational forces involved in any brain functions processing the information that you are reading.

But my consciousness does not swim with visions of planets and moons. It does not see friends and books as gravitational blobs. My mind contains the information that is being processed by the neurons of my brain and yours, despite the fact that the gravitational effects of this brain activity or of the words on this page are nil. All this means one thing: Consciousness is not mediated by the gravitational force field. Gravitation does not provide any handles that could pull the brain and consciousness together.

But maybe there is some new kind of gravity. Maybe there is a biogravitational field of some sort, as the physicist Jack Sarfatti has suggested.[5] The problem with ideas such as this is that such a field, acting in any way like a gravitational field, would have produced observable effects in experiments that have been done repeatedly. No such field exists. Any such field would have been seen, were the interaction big enough to have any role in the brain's functioning, as an interaction between the body of the physicist and his gravitational test equipment used in past experiments. We can do better than invent mythical fields to explain consciousness.

Strong and Weak Nuclear Forces

Exactly these same arguments can be used to show that neither the strong nuclear force, with all its gluons pulling at quarks in the atom, nor the weak nuclear forces mediated by the exchange of the $W^{\pm}$ and Z^{0} particles could have an effect of any relevance to the mind-brain question. These forces exist. They play an important role within the nucleus of the atom and in the particle experiments of physicists, but they do not reach out beyond the nucleus enough even to play a role in the physics of electronic shell configurations of the atom itself, much less to play a role in the neurophysiological processes of our lives. They do not mediate in the processing of visual, auditory, or other sensory information. They do not participate in eliciting afferent responses of the central nervous system. They are not the answer to the mind-brain problem.

Electromagnetic Force

That leaves us only one choice among the forces, the electromagnetic force. To the extent that they have considered the question, this is where most people have placed the locus of the mind-brain interaction. Those who would place the link in brain waves are dealing with aspects of electromagnetic interactions. Those who envision consciousness as an aspect of the neurological activity of the brain, again, are identifying consciousness with electromagnetic forces that play out their role in the chemical, ionic, and electrical activities of the neurons.

It would seem that the flow of electric currents resulting from the presence of an electric field would be a very reasonable way for the various parts of the brain to be interconnected. However, the currents involved in the functioning of the brain do not flow along the length of the dendrites and axons of the brain's nervous tissues. The various regions of the brain are not electrically interconnected. The currents that flow are exceedingly local, involving only the diffusion of ions in the immediate vicinity of a small portion of a nerve cell (specifically, sections of the cell membrane along the axons and dendrites). Thus the regions that are interconnected via electric currents are extremely small. Nevertheless, the presence of these local currents produces an electromagnetic field, the source of the "brain waves," that do involve the whole brain in a continually pulsing field.

But we encounter difficulties ascribing consciousness to this electromagnetic disturbance. When we go to sleep, our consciousness—except for the intermittent occurrence of dreams—goes away. But the brain waves, though different in pattern, are still very much there, still a robust pattern of steady rhythms. If consciousness depended on brain waves, we should expect a change in the character of our conscious experience when the body is asleep, but we would not expect consciousness to disappear as it generally does.

Furthermore, it would seem that if consciousness resided in the electromagnetic fields associated with the brain's activity, then consciousness could easily be modified by subjecting the individual to electromagnetic radiation. But we are bathed in electromagnetic radiation all the time without any effect on our conscious experience. Radio, television, and a host of other transmissions pass through us continually, but we still have to buy a radio to hear "All Things Considered" or a TV set to see that "I Love Lucy" is being telecast yet again. Were consciousness tied to electromagnetic radiation, we would know it—experience it directly—as soon as the broadcast began, whether we were asleep or awake. Indeed, there would be no such thing as sleep. No matter what we did, the transmitter would pulse into our consciousness; it would *be* our consciousness. When we stand in the vicinity of a TV transmitter, we receive radiation that is more than strong enough to produce effects comparable to the field strengths generated by brain waves. And yet it has no effect on our conscious state. Because of this, it is reasonable to conclude that consciousness is not associated with electromagnetic radiation generated by the brain's activity.

There is also another reason to exclude the electromagnetic force as the link between the brain and consciousness. Electromagnetism covers too much territory. It embraces too much. The forces holding molecules together are electromagnetic in origin, as are those holding the electrons in their "orbits" about the atoms. So are the osmotic forces at the boundary of neurons that govern neural impulses, the forces governing chemical reactions, those that account for the tensile strengths of tissues, and even those that drive the blood through the veins and arteries. In short, electromagnetic forces do everything regarding brain functions, and that is why we have to rule them out. If we were to identify electromagnetic forces as the bridge between the physical realm and the realm of consciousness, we would have no basis for separating those special processes in the brain that handle the flow of data to our conscious experience from the torrent of chemical, biological, and mechanical processes—all tightly locked together as interdependent processes that surge continually through every fiber of our being. Why are we conscious of the flashing lights that send delicate ripples of ionic displacement currents in our brains, involving only 70 thousandths of a volt and yet unconscious of all the chemical processes that keep our brains alive, that involve chemical reactions at potentials of several volts, and that, moreover, consume most of the brain's energy?

Even if we could answer this question, we would still face the puzzle of the unconscious activity of the brain. Most of the brain's activity that filters, sorts, identifies, and makes sense of the river of information coming into the brain through the senses—the drudgery calculations and processing of information in the brain—is done totally without our control and without our conscious awareness. Even in the case of learning a skill in which consciousness plays a part, its contribution is rather limited. As Julian Jaynes points out,[6] "In the learning of skills, consciousness is indeed like a helpless spectator, having little to do." Jaynes demonstrates this by suggesting a simple experiment in which one tries to toss a coin from each hand to the other hand. He says you can learn to do this in a dozen trials. As you do this, he suggests, "Ask, are you conscious of everything you do?" He says, "I think you will find that learning is much better described as 'organic' rather than conscious." The brain carries on a vast unheard communication within; it churns out data, it generates unseen, hidden ids and libidos of a Freudian mind. It does its billions of necessary tasks quietly and remotely from our conscious self. It is not enough to say that we are conscious only of the things we must control, such as walking when we first learn; that only says what we see is a characteristic of human consciousness. It does not identify a physical process present in one case and absent in another, nor would the chemical processes involved in synthesizing the materials needed to form new synaptic connections among the nerve fibers of the brain account for this difference in conscious activity. These chemical processes again pale into insignificance compared with the chemical processes involved in merely burning sugars in our brain to do nothing but keep the brain warm in the winter.

Somehow, a tiny part of the brain's activity that is doing almost nothing else but registering the information produced by the brain's vast analytic machinery is tied together by some unique process, giving us our conscious connection to the brain. Somehow vast electrochemical operations of synthesis, transport, signaling, and processing are all filtered out, fortunately, sparing us an otherwise pervasive din. We do not experience those things, and our consciousness cannot be tied to the brain by means of electromagnetic processes. Electromagnetic processes just cannot provide that one-to-one key we have to find.

Quantum Mechanical Connection

It might seem that we have already eliminated just about everything to which we might hope to tie consciousness. If it were not for the list drawn up with the aid of the Schrödinger equation, we just might conclude that no alternative remained. But in fact, there is one option left, and we are led to echo Holmes: "Whatever remains, however improbable, must be the truth." However improbable, we are left with the wave function—the state vector of quantum mechanics itself—as somehow the access route from the physical world into the mind.

And it is improbable. Quantum mechanics is the physics of the atom. It is the arena of the incredibly minute. Only in the most exceptional situations, such as in the case of superconductivity, do we ever find quantum mechanical effects entering the macroscopic world, and we certainly cannot appeal to superconductivity as the mechanism of mind in the brain. If it were the interface we seek, we could make its fierce magnetic properties evident simply by placing a magnet near the head. As a result of the Meissner effect, even a small magnet would disrupt our consciousness. We would be able to feel consciously the diamagnetic properties of the superconductive currents induced in the brain.

To find the quantum mechanical link, we will have to look elsewhere. We will have to look into the machinery of the key component of the brain computer; we will have to look into its basic switching element, the synapse. There we will find how quantum mechanical tunneling plays a subtle but essential role in the triggering of these switches. There, in those minute switches, at the minuscule intersynaptic cleft—that is where the quantitative link between mind and brain is to be found. That is where the first test came and where the fit between an incipient idea stirring in my mind first met the physical evidence that, as we will see, makes this concept so immediately apparent. It came so smoothly together that I knew this must be the secret doorway into the mind.

And what a beautiful answer! There at the synaptic cleft, quantum mechanics brings together the world of physical phenomena and that of the nonphysical mind. We see quantum mechanics as the mechanism of mind; just as in the measurement problem, we found that the consciousness of the observer was the conclusion Bell's theorem forced on physics. We have seen that quantum mechanics and the measurement problem lead us to the observer, to consciousness, and now we see that an independent investigation of consciousness returns us to the quantum world. This is that completeness we talked about earlier. This is the mark that great truths stamp upon the visions of prophets, the sign that discovery is leading us along a path already mapped out for us. And all this brings us back to the question of what the state vector in quantum mechanics is about—to the question of ultimate meanings and realities woven into the equations of physics and into the fabric of our lives. Our thoughts are the ticks of the clock that beats the rhythm of time. Each flickering spark of life spreads the moments into corridors of space. Mind is the forge of creation, and the echoes are the universe.

◈

How much you write in a diary turns out, years later, only to hint at the life that time has rinsed into a sea of forgotten moments. I read laundry lists and wish I were hearing the words we spoke. I read of where we walked, and I want to remember how I looked into her eyes. I want to read the words there and remember her touch. But it is gone, lost in adolescent jabber. Still, I read the words, looking, and sometimes something strikes me from across the years. A filament tightens across the span of time, pulling at the words as I write them now, as though it had all been one act:

> Monday, January 7, 1952: 7:30 A.M.—First chaotic day of the new year. School opened, of course. Economics first period—talk of upcoming test. American History—same, but spiced with Bull Connor scandal. Third period lunch. I had the usual: milk 7 cents, sandwich 15 cents, two cakes, 5 cents each. Fourth period, Algebra,—34/34 on a test. Latin: 43/50. 29/38; English, nothing as usual; Debating—we did less. I walked with Merilyn between classes and entered into petty yet beautiful conversation. I came home and called Merilyn. The weather was pretty but cool.

How much of my time and how much of my life revolved around her everyday.

> Tuesday, January 8th: I talked with Merilyn between classes and after school. I walked her to Homewood for the club meeting at Dunn's Drug store. I walked her home and I walked back home.
>
> Wednesday, January 9th: Got in trouble for throwing peanuts and peas at lunch . . . . I went home after school, called Merilyn, and went to play tennis with her—I beat her 6–1. I walked her home; sat on the steps with her for a while and then I walked home. At home I called her, after finding the word *enjambment*, and then read some of the book *The Forms of Poetry*. Jeanne Connally said something 'good' about me to Merilyn today at the tennis courts. Merilyn promised to tell me what it was.

> Thursday, January 10th: I walked around with Merilyn between classes. I went home and called Merilyn. We decided to play tennis. I started walking to meet Merilyn. Met my mother on the way, who told me to go home because it was too cold. I walked on. At the courts we pinged a few minutes, but it was too cold so we went to her house. Ceil Keeley came along too. I left at 5:30. Called her when I got home, but she was eating. Called her again. After that I started on my Economics theme on "the Korean Peace Talks." Ending at 3:45 A.M. with 2500 words, 14 pages . . . .
>
> Friday, January 11th: I read the first act of her [Merilyn's] play. I didn't like the language too much, but her idea is good, if she can finish it.

And so it goes—algebra, Latin, English, study hall; Mr. Goodwin grading test papers; jolly debate class; Whitson with an idea to canoe to Mobile in the summer; and walks with Merilyn, always Merilyn. Walking with her between classes; walking her to Homewood for her Les Amies Club meeting; going home with her; phoning her to talk for an hour—going back again some nights. Reading the poems she had written and, sometimes, something I had written. How much of my time and how much of my life . . . even then I did not know.

Saturdays was usually my time for the movies. For some reason Merilyn did not go with me. On January 12, I went to see "Jiggs and Maggie Out West," which failed to win critical acclaim. "Terrible," I wrote. "The show was about animals reincarnated into people!"

Occasionally, I find myself having had thoughts about the previous involvement: "I have been thinking of Maitland today. I get lost wondering about her." And then suddenly I am back again, talking about Merilyn: "I called Merilyn this morning. She talked to me about an idea she had for a party she was planning for Saturday, January 26, with her friends . . . [it's an idea] she wants the two of us to pull off at the party." The party was to be a "Gangster Party!" I wrote, "The girl is wonderful."

Sunday, January 13th: I got up at 10:00. Off to church at 11:00. "Of course I couldn't make it, but it was better than staying at home. I was walking along Oxmoor Road toward Dawson Baptist when Gloria Brower drove up in her Buick, stopped, and I got in. We were driving along when Lumas Langley started following us." As I remember, we spent the hour wildly tearing up and down the streets, Gloria and I, with her boyfriend Lumas jealously following behind. "I came home, called Merilyn," and then went to the Frat meeting. "Worthy Master (Eugene B.) wasn't there; I got back a little after 3:00. I called Merilyn again and we played tennis. I walked her home. I left at 5:45 and ran home in 15 minutes . . . . Merilyn is lovely."

There are some hints about our conversations, at least about a few disputes: "Monday, January 14th: I went to the library to find out which has the most extensive vocabulary, French as MAZ claims, or English as I claim." I didn't record an answer as such, just a note that seventy million people speak French and two hundred seventy million speak English. Later, "I called Merilyn. We talked for one hour forty minutes, having an argument about whether you form a major chord by counting 1, 3, 5 (Merilyn) or 1, 5, 8 (me)." Of course, I settled the matter by looking it up in a physics text which gave a formula!"

On Tuesday, "She insulted me . . . referring to my poetry as 'junk'! At the time it didn't bother me, but the more I thought about it, the madder I became." She was obviously a much better critic than I had given her credit for. My stuff was rather oblique, to say the least. She felt that poetry should speak directly and clearly to the heart. I, on the other hand, felt that it should be avant-garde, with meanings to be imagined and words chosen with abandon. So created, the results were of less than uniform quality.

On Wednesday after school, "I called Merilyn and we went to play tennis. She wore shorts and she looked beautiful. Ceil came down and played me. This made Merilyn jealous. It seems that Ceil had called MAZ up and asked if we were going to play. MAZ said she would call back, but she didn't. Ceil came chasing along anyway. I walked MAZ home. On my way home I fell down running . . . skinned my knees, elbow and hip." Hard to believe that once that was me—my running, my knees.

On Thursday, "We presented my homeroom teacher, Mr. Holliman, with a pen and pencil set because he is leaving. I went to see Mr. Shelton about a permit to take a make-up exam for the Economics test I missed on Monday. I wasn't on the absentee list nor could I find my excuse, but it was finally settled, anyway. I made 72/100 on my Algebra test. Merilyn had a thirty-minute meeting after school, so I went home and put up my books. Came back and walked Merilyn home. I stayed at her house till 5:25. After I found out my Algebra grade, I was miserable the rest of the day."

Friday was a bit more adventurous—another furtive effort to obtain my driver's license: "After school, MAZ and I swiped the car to take my driver's test. But when we got to the motor vehicle test station I was told that they already had too many people waiting ahead of me so we took the car back to the Medical Center." No terror, no machinations; a smooth operation—a failed mission, but at least I did not panic. Afterwards, we went to her house. "We played tennis. I walked home, called Tommy to get a ride tonight. Called Merilyn back. Went over to Merilyn's 7:45 to 11:20, (2+). I told her that she was beautiful."

Saturday morning my "Mother broke two of my records because I didn't stop playing them." Apparently, my mother had wanted to drive to Atlanta, but my father wouldn't come up with the gas money. I had entirely forgotten about the records. I mean these were classical records. What mother could possibly object to her son playing classical records? Looking back on it, I can see now what I did not see then. There were other irritations involved! "I called Merilyn at 10:00 A.M. She was leaving on a Les Amies rush party. We were to play tennis at two or three o'-clock, but she didn't get back till 4:30. She called me when she got back. Nancy Kimbrough called also. Couldn't go anywhere with Merilyn so we talked on the phone for an hour and a half. We agreed to play tennis tomorrow (Sunday). I went to the show, alone. Thought for the day—Merilyn is lovely, and I love her."

> Friday, January 25, 1952: After school, called Merilyn and we decided I would come over at 7:30 P.M. Called Tommy Whitson. At seven, I left for Merilyn's. We talked a bit about what we were going to write (for the poetry contest), and we just talked. She asked me what I thought of her poems, (2+). She wrote on a piece of paper "*Ich Liebe Dich*" and wondered if I knew its meaning. She all but said she loved me and she asked me after a time, "What are you thinking?" I said, "That I love you!" After this she told me several things she wouldn't otherwise. She told me that she had dreamed someone had taken her to Mississippi to force her to marry him and she started screaming, "No, I don't want to marry you. I love Harris!" She said that sometimes she dreams about something happening and several days later it happens. Then she put on my sweater like a Shroud, lay down on the floor with her hands crossed and said to me, "How do you like me in Black?"

In my diary I set this off in brackets: "[Remember the quote, 'How would you like me in Black?'] She said it to me about my black sweater. I said to her again, 'I love you.' "

I do not now remember why I emphasized that statement of hers by putting it in brackets. Did I know this foretold something of the future? I do not remember. I remember though its foreboding implications, literally of death. Her father had been buried in Jackson, Mississippi. It is obvious that the man in her dream was her father, calling for her, wanting to take her away to Mississippi. The disease that would take her on her final journey to Jackson, unknown to anyone, was just about to begin. "Remember the quote, 'How would you like me in Black?' " The thread pulls at my mind across the distance of time as if a stitch, as I write this now, had finally been pulled, and the fabric closed.

Saturday the 26th was the day of the party:

> I got up at 8:45 A.M. Soon after Merilyn called me. She was supposed to wake me up before eight. Then I called Tommy Lindsey about a ride to the gangster party. He said he would call me back. Then I called Merilyn. She said we would play tennis when she got back from her singing lesson. I called back at two, but by then it was raining. I called again at 5:30. Tommy L. called me at 6:30. He said he would be by at seven. I called MAZ again. I wore my red knit tie, a white shirt, blue jeans, my overcoat, one red and one green sock, daddy's hat, and two knives. The party was at Barbara Rid-dlebaugh's house. Barbara and Pat Turner kept acting 'horsey'! I gave Merilyn my

> mother's high school class ring. Both of us were in somewhat of a quiet mood. MAZ's dress was split up the side practically to her hip.
>
> Then came our skit. Merilyn began by singing 'Saint Louis Blues.' I came in, swiped her pearl necklace and slipped back outside. Merilyn screamed. I came running in, now playing the part of her boyfriend, searching everybody for the necklace—which, of course, was hanging conspicuously out of my pocket. Merilyn sees the necklace. She comes at me with a knife and I push her on the floor and shoot her. Then I picked her up and carried her out.

I remember that when we practiced the stunt, it had seemed difficult to lift her, but at the party I tossed her onto my shoulder as easily as if she were a kitten. I remember that so well.

All the fun, and all the life in her. Images of her singing, us playing, and of me telling her, "I love you!" . . . memories of a dream dreamed that someone had taken her to Mississippi to force her to marry him and she started screaming, "No, I don't want to marry you. I love Harris!". . . Sometimes she dreams about something happening and later it happens. Then she put on my sweater like a shroud, lay down on the floor with her hands crossed, and said to me, "How do you like me in Black?" How would you like me in black?

We played in those days. We played then . . . in those days.

NUMBERS

We left off talking about how consciousness could not be tied directly to any of the usual constructs of the physical world, like space, time, mass (as such), or particles, or to any of the fundamental forces. Of course, there are connections to each of these things (each enters as material and subject matter for our conscious experience), but these are not the things responsible for consciousness. They are not the interface. We suggested that the only thing left would be for consciousness to be tied to some quantum mechanical process going on in the brain—whatever that might be, or indeed, you may say, whatever that may mean! The point is, however, we had at least found a place to start looking for an answer to the puzzle of consciousness. But if we are going to find out just what consciousness is and how it works, we must be able to show that our quantum mechanical ideas lead to some meaningful, testable, predictable, falsifiable results. We are going to need numbers. We will need somehow to "measure" consciousness. We will have to do just what it seems our first postulate about consciousness says is impossible. We will have to measure, as it were, the nonphysical.

If the postulates are correct—if indeed consciousness cannot be measured and is therefore nonphysical—it would seem that we have placed consciousness, and the whole study of the mind, beyond the reach of science with our very first step. Getting past this obstacle is perhaps the most important step that we will take in our efforts to bring into being a science of consciousness. We must be able to specify numbers that will constitute a starting point for gaining knowledge about what consciousness is. These quantities will serve as the basis for testing our hypotheses about consciousness. They will let us develop theories and understanding about how physical phenomena relate to the mind.

In *Einsteins's Space and Van Gogh's Sky*, Lawrence LeShan and Henry Margenau try to cope with the seeming ineffable nature of consciousness. They look into the domain of consciousness for some scientific clue by which it may be measured, and they find nothing:[7]

> We cannot, as we have indicated before, quantify the observable in the domain of consciousness. There are no rules of correspondence possible that would enable us to quantify our feelings. We can make statements of the relative intensity of feelings, but we cannot go beyond this. I can say, "I feel angrier at him today than I did yesterday." We cannot, however, make meaningful statements such as, "I felt three and one half times angrier than I did yesterday."

Yet LeShan and Margenau expect too much of measurement in science. Of course we can quantify "anger." We can not only say that we feel angrier today than we did yesterday, but we can say that we were angrier then than the day before and can even note that we may get angrier still by tomorrow morning! Four levels in this scale—two bits of information that can be used to mark the measure of anger on this scale. It is a relative scale, surely, but didn't Einstein teach us that even in the measure of space and time, things are relative? LeShan and Margenau speak as though measuring somehow means that numbers take the place of the reality of the thing measured. That is not what measuring and quantification are about. Quantification simply reflects the fact that for any particular basic characteristic that goes into making up our reality, that characteristic can take on two, three, or an infinity of discriminatable levels of existence.

Position in space can be marked off in an infinity of possible values for a particular object. Saying that the position is 5, of course, would be entirely meaningless without somehow saying what that 5 was measured relative to and saying what the units of that measurement are. The spin of an electron, no matter how we measure it, can have one of only two possible values, not an infinite number. In that spin space, there are only two points.[8] In the "space" of the conscious experience of anger, there are several values that we can experience, and we can even have some idea of a unit that reflects how much that anger changes in a day. But measuring the position of an atom in space does not make space just a number any more than saying "I am twice as angry today" would make anger a number. It remains the thing that it is. It is one of the many coordinates of our consciousness, a constellation of coordinates that form the framework of our identity and being.

But how can we speak of any kind of numbers associated with consciousness and still deny that we can measure consciousness? The answer is that we do not obtain these quantities in the same way we measure physical quantities. Every physical measurement involves a very particular process, a serial sequence of relationships—interactions that connect our conscious experience of the measurement to the event being measured.

To be less abstract, let me give an example. Let us say that we are measuring the time at which an atomic decay event occurs. A collection of such measurements can be used to calculate the half-life for uranium, say. Now we never see the atomic decay itself. Instead, the decay event produces ionization inside a Geiger tube or counter. The wires into the Geiger tube permit this ionization charge to flow as an electric current, which in turn deflects the needle in an ammeter, produces an audible click in a speaker, or, if we wish, trips a timer to display the time of the event directly. We get into the act only at the end of everything. We merely note, consciously, the result. Our conscious participation in this process is almost incidental.

But when it comes to the question of the state of our own consciousness, quantification takes place in a fundamentally different way. Because there is no device to measure consciousness in another individual, what we know about consciousness comes about by associating our own conscious experience with other data that are simultaneously experienced. These other data can either be data about something taking place in the outside world or something else in our mentation. What we do is make a correlation between a physical effect on the brain that produces one conscious experience and a second conscious experience that registers the magnitude and type of the physical action on the brain, or on some other associated condition of the consciousness.

We take on the role of the "counter device" from within our own consciousness. We are forced to make the final "measurement" comparison for ourselves inside our own heads. To obtain a physical measurement, all we have to do is stand at the end of the line. But if we are going to "measure" consciousness—that is, obtain a number associated with consciousness occurrence—we must get into the act directly. We have to complete a *comparison loop*, a loop that can include an external sequence of events, or a loop that can take place entirely internally, involving brain processes to complete the loop. The topology of such quantification procedures is fundamentally different. The loop can also be of a type where the brain is stimulated by a visually observed pattern, say a checkerboard pattern, while the brain and consciousness participate in analyzing some feature of that pattern introspectively. The data obtained this way can be just as good and just as reliable as any physical measurement, but they are not public. Only you can experience this kind of measurement on yourself.

Only the ice cube itself knows if it feels pain when it melts; only you know your own inner consciousness.

I live in an old house out in the woods. Every winter, field mice find their way into the sanctuary of a shed, safe from the large barred owls that live here as well. It is something of little consequence as long as they remain in the remote dark corners "out there." But sometimes by spring their numbers become a problem, something that can no longer be ignored. Once again, out come the little steel boxes, the "humane" traps with which we, the mice and I, have in the past played a game of who can get back home first.

But today, this morning it was all acted out more tragically. The whole gang had been caught at once and held incarcerated in that tiny space. I took the trap down the drive, across the highway, hoping that the road would act as a barrier against their quick return, and there I opened the trap. Two ran out immediately, and a third one that had died overnight had to be shaken free from the trap as its mechanism snapped with each shake. When I got back to the house, I discovered a fourth mouse hanging motionless from the trap, caught by its hind legs in the trap's mechanism. As I attempted to unlatch the trap mechanism, a fifth mouse poked its head out to see if the coast was clear and then, terrified at the struggle going on darted back inside, trembling. Again I tugged at the trap's lid, released the mechanism, and the fourth mouse ran from the trap, uninjured and quick. But the fifth little mouse was gasping for breath. In his fear, he went into shock, and in a few minutes he died. He feared for his life. He feared for his life, and he died.

His life was a life fully as precious and real to him as ours is to us. Terror seized him and extinguished his being, and his consciousness—a consciousness as vital as any. This creature, for all the irritation it may have brought as it scampered through closed closets or up some crevice in a wall, was as precious in the sight of God as you or I. He was not a machine that breaks and fails. He felt. He had conscious being and feared for the loss of it. So intense was his feeling, so large the size of his consciousness, that the terror of the moment took his life away. His was not a small, simple mind. His was consciousness, full and real and fragile and precious.

It is a simple story and a small tragedy—like all tragedies. But it has a lesson. It has a lesson that even scientists should be able to learn.

SOME MEASURES OF MIND

Let us look into our own conscious experience in order to find numbers and measures there. These will be quantities that will enable us to learn the nature of mind and let us quantify consciousness as it occurs naturally in our own inner reality. These will be the quantities, the numbers, with which we will check out our ideas about consciousness as we build a theory of what mind is.

First and most important is the fact that consciousness is an *onset* phenomenon. By this I mean that consciousness springs into being when certain conditions of brain functioning are met (such as when we awaken). Consciousness does not come into being only when the brain begins to think. Even when we are asleep and not dreaming, our brain is usually very active. Thoughts—that is to say, thinking and data processing—are present, and yet consciousness is absent or, at most, is maintained at an extremely low level. If awakened from a sound sleep and asked what we were just thinking, we may indeed be able to relate that our brains were active with thought, but this data processing was not associated with any conscious awareness. These thoughts existed merely as ripples of neural activity ready to rush to the surface of consciousness from an ever active brain when awakened. Consciousness happens when some new process comes into play as our awakening brain becomes somewhat more active.

From neurological studies, we know that the brain, the seat of consciousness, is not inactive when consciousness is absent, even during deep, nondreaming sleep. In fact, from brain wave data and from direct electrical probes placed in brains, we know that the level of activity of the brain changes only by a factor of about 2 between the awakened state and the sleep state. But something very dramatic happens when we awaken. Everything changes. The colors of the sky, the sounds of breezes, the odor of the fresh air of the woods—these did not exist the moment before. Consciousness came into being and gave sentience to existence.

In that moment marking the transition from unconscious existence of the body to conscious existence, as the level of brain activity changed from an idling engine of thought to a vehicle that has lurched into motion, something new has come into being—something that gives rise to this freshly awakened consciousness. Whatever it is that has happened in the brain, consciousness has made a discrete change. It has turned on.

What has happened in the brain is that it has changed in its level of activity. And there must be an associable or correlate number, some quantity that tells us just what level of brain activity had to be reached in order to bring about the onset of consciousness. The consciousness came into being in that moment of increased brain activity. Consciousness is an onset phenomenon and, as such, can be associated with numbers characterizing the physical correlates of the occurrence of consciousness. This fact will serve to dramatically narrow our search for the mechanism of consciousness.

By the way, before we leave this business of the transition from sleep to consciousness, it is worth noting that one of the correlates of brain activity during the onset of consciousness concerns the way dream-consciousness occurs. In dreams, though the state is called sleep, even though the brain is not "awake," we still experience a form of consciousness. It is, of course, a different kind of consciousness, both in content and in intensity, but it is still basically the same phenomenon. As such, it can provide us with many clues about the brain's activity that gives rise to consciousness. As we will see later, dreams occur when only certain limited portions of the brain reach the levels of activity adequate to give rise to the onset phenomenon of consciousness, while the rest of the brain remains at a level of activity that is too low to support that state.

This change in activity between sleep and consciousness seems to be controlled by a structure in the brain called the reticular formation. It should not be too surprising that some master circuit in the brain governs the brain's wake-up messages to its many parts. But the fact that this structure controls wakefulness should not lead one to confuse its function with the occurrence of consciousness. It may trigger functions that give rise to consciousness, but a light bulb does not eliminate the need for a wall switch. The reticular formation serves as the switch, not as the light bulb.

This onset characteristic of consciousness is significant in that it points to a special class of physical phenomena that must occur in the conscious brain. When the atmosphere, filled with water vapor (a gas), is cooled below the dew point, condensation occurs. Droplets of water appear and in aggregates form clouds. A very small change in temperature can thus bring into being, seemingly from nothing, droplets of water, clouds in the sky, and even a rain storm. So condensation, too, is an example of an onset phenomenon.

The condensation of water vapor into liquid water is but one of the many changes of state that can occur in matter. The change from water to ice and the melting of iron, copper, or gold into liquid metals are other examples of what physicists call a phase change. The difference between carbon as a lump of coal and carbon as a diamond is another example in which a very small change in one physical quantity, here the temperature, results in a great change in the appearance of the object, fundamentally changing the thing's nature.

Similarly, a small increase in the brain's level of activity gives rise to something new. In condensation, the form of water in which water exists changes from individual molecules to the vast collections of molecules that constitute liquid drops of water. In like manner, some potential of the brain—something that *may* have occurred before in isolation, as scattered monads of consciousness—now condenses into a whole ocean of consciousness.

The onset nature of consciousness, then, leads us to look for some onset process, something like a phase change in the brain's functioning that results from a modest change in the level of activity of the brain. Whatever we may ultimately find that process to be, one thing is singularly significant about this onset characteristic of consciousness: For the first time in the history of philosophy or of science, we are dealing quantitatively with the question of the nature of consciousness. Recognizing that an associated measurable quantity characterizes the brain's functioning at the onset of consciousness gives us a quantitative handle on this question of what mind is. We are on the threshold of a new science that will disclose the innermost secrets of consciousness.

But onset is only the beginning. When we begin to think scientifically about consciousness, other quantities jump to our attention. What is the physical extent of consciousness? What is its temporal extent? What is its capacity?

In philosophy, any question of the physical extent of consciousness would seem foolish, and in metaphysical or in any mystical-religious context, a question of its extent would probably lead to dogmatic assertions that mind is everywhere or that it fills all space. But in our context, the question has a precise meaning: we wish to know the extent of the physical process that interfaces consciousness to the physical world. Since one's consciousness contains data generated in various regions distributed throughout the brain, it is reasonable to assume that the physical extent of consciousness is approximately commensurate with the physical dimensions of the brain.

How can it be that consciousness can be characterized as having an extent of 10 to 15 centimeters? If consciousness has an extent, why have philosophers said that the question of the spatial extent of consciousness is meaningless? And why would the mystic say it must extend everywhere? The answer to the first question is that obviously consciousness is not length; it is itself as fundamental as space or time or matter. The associable quantity characterizes the interface that, so to speak, makes consciousness stick on a piece of matter—the brain. And if we are to understand what in the brain constitutes that "glue," we must look for traces of that glue spread throughout the brain. That glue must extend as far as the brain extends. There is no single cell, no pineal gland, that contains consciousness, so there must be something else, and we must use our knowledge of physiology and of physics to show how this something else works.

As for the mystic, we will have to wait a while to see the other side of what consciousness is—to see its other extensive characteristics. But just as we saw in Chapter 8 that matter itself has nonlocal characteristics, so too we will see that mind has its own nonlocal nature. We will find that, in a way, mind is everywhere.

THE TICK OF THE CLOCK

How fast does time fly? How long is the tick of the consciousness clock? What determines whether the tick of the clock will seem long or short? Think now, what is the duration of your experience? What is its temporal length? What determines the speed with which time passes, or the duration of the moment now?

It is surely difficult to convey what we mean by an associable quantity for the measure of consciousness length when consciousness is something beyond space itself. How much more difficult, then, to speak of the clock that measures the pace of consciousness time. But this is another basic quantity we will need if we are to understand what consciousness actually is. It is this time unit that measures the briefest chunk of conscious sentience making up the stream of consciousness we experience. Knowing what this is moves us closer to grasping the interface between things physical and things mental. It also moves us one step closer to knowing the nature of all reality.

Watch a clock tick the seconds away and imagine. Consider what our conscious experience would be if our consciousness could not respond with the swiftness with which the neurons of the brain can fire. Imagine that this process, whatever it might be, could not respond to a complete change in our conscious stream more rapidly than once a day! Our neurons would still function as before; our brains would still respond the same; and at present, we do not know of a function our consciousness so controls that life could not go on without its intervention. So possibly the life of the body would go on as before. A million images would flicker through our brain, but only the blur of these images would register on our conscious existence. We would "see" the time on the watch indicating when we should go to work, have lunch, and end the day, but we would not consciously experience any single event separately from the rest. The inertia of that slow interfacing process would not let the images change faster than once a day. Everything would be a blur. The period of a day would be our moment.

Or if that characteristic time were an hour, we would see flashes of blurred events flicker in our lives, experiencing some average effect of light and sensation, and the day would pass in a few "moments" of light and darkness. But that is not what we see. Whatever that basic unit of time for consciousness is, it is much shorter than this. As we watch a clock tick the seconds away, we can experience each display of the seconds or each position of the hand pointing the seconds. Whatever the shortest unit of consciousness time, it is surely less than one second—indeed, less than a tenth of a second.

Now imagine that the unit of time, the shortest interval as measured by a physical clock that can be distinctly experienced consciously, were a quadrillionth (1/1,000,000,000,000,000) of a second. What would our conscious experience be like? Such a time interval is so short that nothing would change in the brain for an enormous number of cycles of our conscious experience. Hundreds of cycles would pass before even one of the trillions of synapses in the brain would fire. It would seem that nothing would change, not even the slightest flickering in our visual field. If consciousness controlled nothing of our brain's functions, we would be unable to change anything we might say, or do, or write about this condition. Rather, we would experience that infinite sameness as the mark of our existence. But again, that is not what we experience.

If the basic interval of conscious experience corresponded to about a millionth of a second, that would be just enough time, on average, for one optic nerve fiber to change its state. (Functioning of any one fiber takes much longer, so we are now speaking as though the visual field we experience depended on discrete states of the optic nerves.) Our conscious experience would then follow each element of our visual experience, changing one pixel at a time. But again, that is not what we experience. Instead, our conscious experience is clearly pushed to the limit watching a digital stopwatch flicker away. Watch one that can count to a hundredth of a second. You will find your consciousness pushed beyond its limit to experience, as discrete states, the display of hundredths of a second. About the best you can do is notice every third or fourth number that blinks past, a flickering of about 1/25 of a second as the digits dance away. It seems that the shortest interval of time that exists as a separate element of consciousness has a duration, measured in terms of physical clocks, of about 1/25 second. Let me write this down so that we can come back to it later. We write, for the minimum consciousness time interval δt,

$$\delta t = 0.04 \text{ s}$$

remembering that we are not exactly sure how accurate this value is. The exact value may be as much as 1/10 second or as little as 1/100 second. Within reason, however, the value given is about right.

In all of this, some may be bothered that what we are really determining is not the temporal length of consciousness but the speed at which neurons can operate. For example, it takes about 1/8 second for a neural impulse to transit the brain. When you drive down the highway at 60 miles per hour, you are really about 11 feet in front of where you think you are just because of the delay in the nervous system. The neural net of the brain itself functions with a characteristic time of 1/10 to 1/100 second. So what do we mean by this characteristic time for consciousness? Isn't the time we have noted just the characteristic time of the brain?

This is one of those times when the answer is a definite yes . . . and no! Yes, the time we are talking about is tied to one of the characteristic times with which the brain itself functions, but no, the characteristic time of concern here is that of consciousness itself. To put it differently, the brain and the consciousness of the brain have obviously evolved as a whole, as an entity. This very fact is a clue that consciousness must play an active role in the successful functioning of the brain. But because the brain's functioning is so intertwined with the way consciousness works, it is difficult for us even to recognize the separate existence of consciousness.

When we speak of consciousness—of its characteristics, of its contents—it seems we are speaking of the brain. This problem mirrors the problem of distinguishing brain functioning from the external world, recognizing that we do not see the external world directly but rather images of the world that are created in the brain. Indeed, this is that hard path to enlightment with which we began. We do not see the outside world, but instead we see the "inside" of our brain! And then we realize that we do not see images created by our brain but that we see instead our consciousness. We discover that what we see is ourselves, our own consciousness. The brain is there and the outside world is there, but consciousness, not these other things, is our existence that we know directly. All of these elements fold so perfectly into one another that we almost miss the reality, miss the hierarchy, and miss the finding of enlightment.

Because enlightment is such an elusive discovery, so much more so is the effort to separate out the characteristics of one phenomenon from the domain of the other. It is difficult to realize that seeing colors as ocular and as occipital is wholly different from seeing color as light frequency in the external world and that, in turn, is still distinct from the red and the orange of conscious experience. But those distinctions exist, and they help us discern the basis of the associable quantity for the temporal characteristic—the rate—of our consciousness experience.

INFORMATION CAPACITY AND DATA RATES

For reasons that I hope will become clear, I will emphasize the visual content of consciousness. For most of us, I believe, it is the most intense aspect of the conscious stream, and it is the most easily quantified. It represents the preponderant contributor to the information capacity and data rates of the carrier phenomenon of the consciousness.

But what do we mean by these ideas? First, to speak scientifically about the question, we must be able to quantify information. With the extensive use of computers of all sorts, the idea of measuring information is perhaps fairly well known, but let us discuss this point at least for clarity. In information theory, we use the *bit* as the unit of information. In many ways, 1 bit of information is like the measure of 1 quart or 1 liter. The quart measures the capacity of a bottle to contain a certain amount, whereas a measure such as 1 pound or 1 kilogram tells us how much we have. Like the quart, the bit tells us about the ability of something to contain something. It is a measure of the ability to contain information. Thus, when we speak of information theory, we do not actually concern ourselves with the study of information, but rather with the capacity to have or to carry information. It is like speaking of the information that a blank page has the capacity to carry. If it is the capacity of, say, this page, a page in a conventionally formatted book, to contain information, then we already know the number of lines of type and the number of spaces on each line. The only question is how much information can be put in each space. If the spaces were filled with only A's and B's, a line would look like A A B A B A A B A B B B A B B A A B A A A B A A B, etc. With such a restriction, one space can be only A or B, and the actual character answers a question that has only two possible answers. "Is the space filled with an A or with a B?" Thus there is 1 bit of information, the answer to the yes-or-no question—one of two possible symbols for each space in the line.

Let us next consider how many bits are represented by symbols that can have more than two values. Let us say that the symbols are a, b, A, and B. How many yes-or-no questions do you have to ask to find out which symbol has been selected for a given space? If you ask, "Is it a capital letter?" The answer will let you decide between a, b or A, B. Then if you ask, "Is the letter a?" (or A, depending on the answer to the first question), the answer will enable you to pin down the symbol. Thus it takes 2 yes-or-no questions to be certain that you will find out which symbol has been selected. Two questions, 2 bits. For 8 possible symbols, it takes 3 yes-or-no questions ($2 \times 2 \times 2 = 8$). For a possible 16 symbols (say, the lower-case letters a, b, c, . . . , m, n, o, and the empty space), each symbol can convey 4 bits of information ($2 \times 2 \times 2 \times 2 = 16$). Each space on this page conveys over 6 bits (26 lower-case and 26 upper-case letters plus the symbols . , ; : ! ? ' " [ ] * @ alone give 64 symbols for the 6 bits—that is, $2^6 = 64$). And, of course, there are many other symbols that may appear. If there were 100 possible symbols altogether, that would mean each space would convey 6.64 bits of information.[9] Yes, we count fractional yes-no questions as well! On this page there are 41 lines and (about) 82 spaces per line, or a possible 3,362 symbols. If each symbol represents 6.64 bits, then each page contains 22,324 bits of information—even the flyleaf page!

We can get much more information onto a page. A printed photograph in a book consists of very small black and white dots, perhaps 300 to the inch. Each square inch of the picture would therefore contain 90,000 of these "pixels," each conveying 1 bit of information. A picture measuring 4 inches by 5 inches would contain 1,800,000 bits of information. If we add color by printing the yellow, cyan, and magenta dots standard for color printing, we will have three times as much information capacity—that is, 5,400,000 bits of information. But it takes that much information density in an image for the image to approach the capacity of just our visual conscious experience at any moment.

The image on a television set comes even closer to what we visually experience. It seems to hand us, ready-made, a complete reality that eats up hours of our time, living its surrogate existence. Without effort or preparation on our part, it fills us with a conscious stream that almost exceeds our mind's capacity. Sixty times each second, an electronic beam paints the screen with 256 scans, 300 pixels in each line: 64 discernible levels of intensity for each of the three colors on each pixel of the image—red, blue, and green. The interlaced lines sweep out a complete picture of 2.8 million bits of information 30 times each second. Information is painted on the screen at a rate of 83 million bits per second—equivalent to 4000 pages of print each second. Television seduces the mind by filling its capacity. What we see on this TV screen is, in many ways, just what happens in the mind.

It takes information at this high intensity to adequately fill our consciousness. But it is not just a data stream that flows into our consciousness. Just as with the TV set, there is a "screen." And just as with the TV screen, there are two different characteristics to this visual part of our conscious stream. First, there is an instantaneous image with its information content. This is our own information capacity, or, as we will call it, the *consciousness field capacity*. This is analogous to the image displayed at any moment on the TV screen. It is the information painted across the mind at any moment. Second, there is the *consciousness channel capacity*, which is given by the rate at which information flows into the consciousness. This measures the data rate of the channel—that is, of the consciousness.

These two terms are going to be important to us. We will get values for each of these quantities a little later. As we will see, the values will be similar in magnitude to the numbers we just quoted for TV images and for the TV channel capacity. The TV analogy just gives us a sense of what we should be looking for and an idea of the values of these two quantities that will be important measures of consciousness. The numbers for the consciousness field capacity (the mind's TV screen, at a given instant) and for the consciousness channel capacity (the whole flow) will turn out to be very large numbers. It takes a lot of data to fill the mind.

The Consciousness Field Capacity

There are many tasks where the brain's data-processing capacity is quite awesome. Simple tests can demonstrate some of the brain's vast power for processing data. The eyes feed data to the brain through the two optic nerves, each of which has 1 million nerve fibers. Each of these 2 million nerve fibers can send dozens, even hundreds of pulses of information to the brain each second. There, in the brain, those data must be turned into meaning—a fleeing cat, a fluttering wing, a flickering shadow, or a fearsome beast. The brain, in an instant, must sort 100 million bits of information to know, to react. The brain carries out pattern recognition tasks that cannot be duplicated by any computer today.

In order to tell just how much information the brain can handle quickly, let's look at the random-dot stereogram experiments that Bela Julesz did at Bell Laboratories.[10] These random-dot stereograms are in many ways just like any stereogram pictures. When one views them through a stereoscopic viewer, one sees a three-dimensional image. What is special is that unlike ordinary 3-D pictures, if you look at the individual pictures, all you see is an array of random dots, an array of up to 1 million black and white spots, seemingly arranged entirely at random. There is no image at all to be seen in either one of the pair of pictures. But when viewed through the stereoscopic viewer, corresponding areas in each picture have identical or nearly identical patterns of dots. To find the image embedded in the random-dot pictures, the brain has to sort out all the various possible clusters of corresponding dot patterns. When the brain finds the dots from each image that match up, it is able to reconstruct the 3-D image, and we perceive the dots as a part of the overall mental picture.

Now because we are dealing with a very specific task, we can calculate just how much computer work would be required to match what the brain does when it solves one of Julesz's picture puzzles. Making sure that just a single dot matches up with a particular position in the second picture by using a computer program would require at least 100 yes-or-no decisions. At least that many positions around that dot have to be searched out and compared to the dot pattern in the second image. After all, one cannot just compare two dots to tell whether they match up—there are dots everywhere. Instead, a cluster of dots, perhaps 10 across and 10 down, must be compared, and that means 100 yes-or-no comparisons just to check one position; then the next position over and the next position over, until the 10-by-10 pattern of dots matches up with a 10-by-10 pattern in the second picture. This has to be done again and again until the entire picture has been deciphered. And yet it takes the brain only a couple of seconds to do the job. Somewhere in the brain, these two pictures that haven't any recognizable features at all are synthesized into a cyclopean image in the mind. The 3-D image sweeps into view with just the briefest delay after the eyes focus on the stereoscopic pictures. To achieve only this requires processing at least 100 million bits of information in only 1 or 2 seconds.

The case of the cyclopean eye is merely a dramatic example of what is really a common occurrence. Anytime one looks at a photograph, the brain goes to work efficiently carrying out hundreds of millions of tests and calculations that it completes in a tenth of a second. Almost any high-quality photograph has 100 million bits of information that must be sorted out by the brain before we see its image as something distinct from any other photograph. Yet the whole thing happens in a moment's glance.

The stereograms of Julesz make it clear that we must think in terms of millions of bits per second or even hundreds of millions of bits per second for the information capacity. Even at a single instant of time, the mind contains some field of information with a data capacity of tens of thousands to millions of bits of information.

But exactly how much information are we talking about? Just how can we quantify this information without depending on pure speculation or depending entirely on physical measures? In our efforts to do this, we will find physical measures, particularly neural capacities—such as the capacity of the optic nerve to carry information to the brain—to be an important guide. But for the numbers to be valid, we must determine the quantities by using the techniques we have already considered for obtaining associable quantities.

Suppose you are looking at a completely blank, white wall extending to the edge of your peripheral vision. What is the information capacity of that blank space? It contains no information other than that it is white rather than black, so does it contain only one bit of information? No. Even though there is nothing to fix on, it has extent and capacity to hold shapes and forms, and you experience that capacity. Just as the blank flyleaf in this book has a capacity to hold its 22,324 bits, our consciousness has such a capacity as well. There is a right side and a left, a top and a bottom. And the right side can be further divided mentally. That capacity has nothing to do with your eyes' acuity or the nerves' capacity. It undoubtedly has something to do with the brain's functioning, but it has its own existence at the same time. To get any good measures of the field capacity, we will have to look at specific test patterns.

Determining the field capacity for visual conscious experience is a lot like taking an eye examination. You need a chart to determine resolution. But where the eye chart uses symbols or letters to determine whether you can "see" to a given level of acuity ("to see" meaning to respond correctly to a question about some eye chart symbol's identity), the challenge now is to determine the detail with which you consciously experience elements of the patterns. A test pattern of the type used to test your vision will not do. The question is not whether you can see to a given level of resolution, but how much information altogether you can experience at one time. For this, we need a test pattern that is filled with simple elements—a pattern of colored dots, for example. By looking at such patterns with finer and finer detail, you can determine the limit of your consciousness to contain distinct elements in various parts of your visual field. The task is not one of remembering or being attentive to each dot, but merely noticing that, yes, that blue dot is a distinct part of my experience, and the red dot, yes, I can still see it as a distinct element even though it is half way into my peripheral vision.

I have carried out such a test on myself. I find that my visual field contains about 100,000 distinct elements, pixels, and that each pixel has a capacity of about 10 bits of information. All together, that is about 1-million-bit capacity for my visual field.

This value should be corrected to take account of the fact that this test merely tests for the visual information content for a flat surface. Since we experience images in three dimensions, this number should be increased. The best way to take account of this 3-D factor is perhaps to note that this illusion of three dimensions simply depends on two flat images that the brain reconstructs as a 3-D image. So we may estimate that the actual 3-D, consciously experienced visual field has about 2 million bits of information.

The Consciousness Channel Capacity

Now let us look at the information rate. As we know, we can see images at the rate of up to 15 to 20 per second, which is why theater films run at 24 frames per second in order to create the illusion of continuous motion. Taking the figure of 20 images per second, and multiplying this by the 2 million bits of information in a single 3-D visual field, we get results of from 30 to 40 million bits per second.

There is another way to estimate this number, and it leads to a similar value for the visual component of the consciousness field capacity. We can get this number simply by noting that there are 2 million optic nerve fibers going from our two eyes to the brain. Now each one of these nerve fibers has a capacity to transmit information to the brain that is coded in terms of pulse trains—that is, a series of successive pulses. From what we know about the way these fibers work, we can calculate the rate at which each can transmit information to the brain. The pulse rate can go from zero up to nearly 1000 pulses per second. This value of a thousand pulses per second is just about at the saturation level for the nerve fiber and could not be sustained. A nominal value is more like 200 pulses per second. We can think of this train of 200 pulses as being about 15 to 20 groups of pulses, each group consisting of about 10 pulses—3.3 bits of information. All 2 million optic nerve fibers therefore can send 15 to 20 images to the brain each second, each with about 7 million bits of information. That gives a result for the data rate from the eyes to the brain of some 100 to about 130 million bits per second.

These estimates of course include only the visual part of our consciousness. They must be increased to take into account our auditory, tactile, and motor experiences, not to mention such things as our verbal thoughts, which contribute further to our conscious experience. But judging from the fact that the optic nerve fibers constitute two-thirds of our sensory input, we probably need to increase our estimates only by about 50%. This gives us, for our first method of estimating the consciousness channel capacity, a result of from 45 to 60 million bits per second. The second method, where we simply looked at how much data the nerves could deliver to the brain, gives us from 150 million bits per second to just about 200 million bits per second.

So somewhere between 45 and 200 million bits of information per second spread out before us as our conscious experience—our consciousness channel capacity. We will favor the result we directly measured over the estimate based on the number of nerve fibers, although the two results are not too different considering how young this science of consciousness is. We will take a nominal value of

50 million bits per second for our estimate of the consciousness channel capacity, and we will take a nominal value of 2 million bits for the instantaneous consciousness field capacity.

There is an interesting point to be made in passing about this quantification. It may help to appreciate what we are really measuring and what we are *not* concerned with. As I have said, we are not trying to copy eye chart measurements. We are not attempting to determine the resolution of the perceptual field,[11] but rather the resolution of our conscious experience as something generated by processes going on in the brain. But when you try to obtain a measure of the resolution in your own peripheral visual areas, you may notice a peculiar difficulty. There is an effect much like trying to discriminate the central visual acuity without your glasses. If you need glasses to see well and remove them when you attempt to read an eye chart, you will notice a fuzziness in your vision. This fact is much more significant for us than you might assume. The fuzziness is there because your conscious visual field has the same data capacity with or without your glasses. There is a place for the extra data in your mind, even though your eyes cannot provide your brain with all the data needed to achieve maximum resolution. Note that when your glasses are removed, reducing the resolution of the data going to the brain, the visual field is not reduced. You do not now see data points, as it were—pixels that are just as fine as though you had on your glasses. Your vision is not poor because things seem too small to see. Instead, the consciousness field is still just as great, but your eyes can only supply fuzzy data.

Now in your peripheral vision, you see the same kind of effect as you would see in your central visual field with your glasses off. Your peripheral visual perception has a greater resolution than the data resolution being supplied by the eyes. The brain attempts to compensate for this by filling in the details, and this sometimes leads to distortions. The result can be an optical illusion such as the well-known Poggendorff Illusion.

And so, this is your reality. We have found a measure of what you experience consciously. This is a quantification of the central essence of your being. This is Descartes's "*Cogito ergo sum*" measured. This is mind in numbers. This is placing a scale beside your conscious mind to read off its measure. But it is not to reduce what this thing is to numbers. Mind is still mind. The fabric is still there unaltered. All we have done is count the stitches and give birth to a new science.

Now that we have some numbers to measure out our path, let us find what consciousness actually is and how quantum mechanics gets into the brain. We have made our way to the gates of the Emerald City. Now we must venture into the city to seek the Great Oz.

12

The Red Shoes

"We will certainly try hard to do as you say," he replied.
"But how shall we bury you?"
"However you please," said Socrates, "if you can catch me and I do not get away from you."

—Plato, *Phaedo*

We have traveled a strange journey to find who we are and what it is all about. It is a journey that begins anywhere and searches for its source. We have looked for reality in the motion of the spheres in the midnight sky and journeyed to the center of the atom, searching for the reason for it all. We have looked inside a grain of sand to find its hidden secrets and have discovered . . . it is filled with stars. Still we search on.

We have cut cleanly to the heart of matter, peeling back each layer of reality layer by layer, believing it was all there waiting to be seen. We have uncovered the pieces of ponderable matter that make the things of which we and everything we know are made—the molecules of our bodies within and things of the world beyond. We have opened those to count their atoms. Like gazing into a peach sky that turns to a star-filled night, we have probed deeper into those atoms to find their point-like nuclei in a vast interior emptiness. Still deeper, we have found the protons and neutrons within the nucleus surrounded by a plethora of elementary particles boiling out of the vacuum itself. Hidden deeper still, three orders of quarks and leptons, and beneath these, who knows what? But have we found reality in these things, or have we only clutched at shadows?

We have probed, too, the mysteries of relativistic time and space, and we have looked on the queer land of quantum things. There at the gate, in an Emerald City, we have found something, someone there that somehow rules this new land, this quantum world. There we found the observer. We have found this consciousness standing there, looking back out at us like an actor on the stage of reality, strangely playing the role of a writer writing the script we play.

We have discovered that the observer is a negotiable instrument of reality, and we have touched our own nature. We have found there that *I, myself, am the world; enlightenment lies within.*

To understand the world, we must understand the quantum picture of that world. To understand quantum mechanics, we must know what consciousness is. We have found that in quantum mechanics, in its formalism and in the experiments of the physicists' laboratories, consciousness arises as something real in its own right. Remarkably, when we turned the problem around, when we probed what consciousness is about, we discovered that its nature must be understood in terms of the nonlocal properties of quantum mechanics playing a role, somehow, in the brain. Two mysteries have come together in the greatest puzzle ever to challenge physics—the measurement problem—and in the mystery that has defied the ages—the question of the source of our own consciousness. Both of these find their answers in each other. The two are intimately linked. Each is the answer that the other asks.

It is a task that has defied the ages, but in the last chapter we discovered some of the keys. We defined the problem. We said that we must find how consciousness interfaces with brain processes, and we found a way to make our study of consciousness quantitative. More than this, we have found that science has advanced to where we have the tools to turn what has always been nothing but philosophical speculation into a quantitative, mathematical science. We have found a way to fuse enlightenment and science.

MIND/BRAIN

Where is the Mind in the Brain? Where is the spark of life, the fire of consciousness, to be found? And what does it mean to try to find a quantum mechanical connection in the brain? What we must look for are the answers to two questions. First, what would a "quantum mechanical process" look like? That is, what do we mean by this term? Second, where should this connection be located? Where inside the brain do we look?

Everything we do in our ordinary experiences has the stamp of classical physics. Everything looks like billiard ball mechanics. We get in our car, turn the key, start the engine, turn the steering wheel, and the car goes where the parts all move it to go. If we are to solve this quantum riddle of what consciousness is, we must find something that transcends this domino world. We must find something that has the stamp of quantum magic about it. It must be something that shows that special paradoxical connectedness that Bell's theorem reveals to us—something that seems to go beyond the bands of space and that can knit together the events of our mind scattered throughout the brain.

What clearly separates quantum processes from classical processes is the way objects move. In our everyday world, an object can go from one location to another only if it moves smoothly step by step, point by point, along a path connecting the two locations. Ordinary things move on definite, specific paths. In quantum theory, we discover that objects move according to the ebb and flow of probability waves. It is possible in quantum mechanics for an object to go from point A to point B even though the two points are separated by a barrier that the particle cannot pass through nor even exist within. Quantum mechanically, an object can go from the inside of a bottle to the outside without removing the top, or breaking the bottle; without punching a hole in it, or squeezing past the cork. The phenomenon is known as quantum mechanical tunneling. Although it is virtually impossible for this to occur in our everyday world, tunneling goes on "everyday" in the atomic world. The radioactive decay of radium and uranium is due to the fact that a part of the nucleus, an alpha particle, can suddenly pop out of the "bottle" formed by the nuclear forces that hold the nucleus together. It pops out even though it does not have the energy to get out! It tunnels out without going through a hole. It is inside one moment and outside the next—without ever going through the barrier, without the barrier ever being removed. It behaves in a nonlocal way.

If we can find something that looks like this going on in the brain, then we will have something quantum-like to tie consciousness to. This might show us how events throughout the brain could interact directly. This might show how some special quantum process occurs, a process that by its very nature involves state vector collapse, where an object changes its state from being in a collection of potential places to being in some new and unexpected single location. By means of this process, it will be possible for the seeds of observation, the monads of conscious being, to spread into all the parts of the brain's machinery.

But where do we look for this quantum process in the brain? Where should we expect to find such a tunneling mechanism? If we find it, it will be significant only if it plays a crucial part in handling the brain's data-processing functions, for that is what we are conscious of. Everything in the brain, of course, plays a greater or lesser part in thinking. But there are two kinds of parts that are basic: First, there are the neural transfer lines of the brain's nerve cells—thousands of spidery dendrites, usually accompanied by one long axon arm, that by their reaching out to other neurons determine the routing of information throughout the brain—and second, there are the synapses—switches in the brain that do the actual information processing. These button-like synapses are the tip ends of the neuron's dendrites that actually form the points of contact interconnecting the brain's 10 billion neurons. It is at the synapses that decisions and thinking in the brain take place, bit by bit. This is where the thoughts begin that move, that motivate, that alter our behavior, and that finally fill our mind. The nerves are much like wires in a computer, necessary to get the information to the work station, but it is the gates and switches, the transistors of the brain, that actually transform data into new creations of thought. It is here at these synapses, at these connections that altogether number about 23.5 trillion, where information as neural impulses passes, or fails to pass, from nerve to nerve, from neuron to neuron, that thought originates. Here is where we must find the contact between brain and mind, for only if it is here can we understand how the data of our senses, processed and refined in the brain, become our stream of consciousness. Thus quantum tunneling must lie at the center of these data switches in the brain. This must be where the spark of life begins to flicker into flame.

THE ENGINES OF THE MIND

The membranes of nerve cells are bathed in a solution containing a variety of salts. Nerve cells maintain an electric field across their cell membranes, even in the very fine branchings of axons and dendrites. This electric field is maintained by certain chemical processes that are referred to collectively as *ionic pumps*. Nerve and neural impulses are very local disturbances in these ionic concentrations that, once started propagating across the neuron, usually run from one end of the cell, out along its branches, to all the other extremities of the cell. The currents that carry these impulses are carried by negatively and positively charged ions that diffuse through special pores or ionic gates in the neural membranes. Before an impulse passes a point along one of a nerve cell's thread-like axons or dendrites, an electric field exists across the cell membrane. This field is caused by the ionic pumps that maintain an excess of positive ions on the outside of the cell membrane. Now when an impulse propagates into the vicinity of any particular ionic gate, the resulting change in the local electric field exerts a force on the molecules forming the gate, which in turn causes these molecules to undergo a conformational change. That is, the molecules change shape. They open up the gate, permitting more of the positive ions to pass through. This discharges the electric charge on the membrane and thus moves the electric impulse farther along the nerve. That, basically, is how an impulse propagates along a nerve fiber.

Once started, this impulse continues propagating in a kind of domino effect. Given the initiation of the impulse, its approach always leads to the conformational change in the ion gates, always leads to the diffusion of ions through the gate molecule, and always continues propagating farther along. Its detailed functioning goes on at the molecular level, so that it must be understood in atomic terms (that is, in terms of the physics of quantum mechanics). Nevertheless, nothing happens because of these quantum mechanical events that might in some special way control the outcome. Nothing takes place here at the atomic level that might depend on or uniquely control information going from the brain to the mind; the impulse just propagates merrily along. Nothing happens that gives rise to any special selection of potential states. We will not have a linear combination of possible states in which maybe the neuron propagates and maybe it doesn't. We will not have any situation in which quantum mechanical uncertainties come into play so that we must talk about the "observer" of quantum mechanics, or in which we must call on state vector collapse or any of the mysteries of Bell's theorem. The neuron just sends along its impulse much as a wire conducts electricity; and in the case of the neuron, that conduction does not involve the whole brain, but just a microscopic region in the immediate vicinity of the neural impulse.

The Synapse: Site of Quantum Choice?

At the synapse, however, things are quite different. The synapse is the junction between neurons, the site at which the impulse from one neuron influences the functioning of the next neuron. In some synapses, such as the neuromuscular and large neuronal synapses, the arrival of the presynaptic impulse always causes the synapse to fire. When the impulse reaches the synapse, it causes the release of certain neurotransmitter chemicals that alter the permeability of the membrane in the postsynaptic region. In the case of the neuromuscular synapses, these chemicals cause muscle contractions. In other synapses, these chemicals can cause the postsynaptic neuron to fire, or they may inhibit a neural impulse from propagating in that neuron. In the case of a few synapses, called ephapses, firing takes place without the release of neurotransmitters.

In all the synapses that release neurotransmitter chemicals, the chemicals are contained in extremely small packets called vesicles. In the case of the large synapses, hundreds of vesicles may release their chemical contents when the synapse fires. In smaller synapses, synapses as small as 1 micrometer (40 millionths of an inch) in diameter, firing is no longer a certainty. The behavior of the synapse—and thus of the neurons as a whole—depends on factors occurring at a smaller scale, factors that may involve quantum mechanical effects and that can bring state vector collapse into play. This means that the decision, if you will allow me to use the word for a moment, to fire or not to fire, constitutes two potential states that the synapse can go into. This opens up the possibility for the brain's behavior to be affected by quantum mechanical processes and opens the door for all the observer characteristics of the quantum world to enter as well.

It probably would be adequate at this point merely to point to the fact that whatever takes place in these synapses may involve quantum uncertainties welling up from events at the atomic level. But if there were no order to this process, if this were random noise in the synapse, then we would still not have a tie-in between quantum processes and the information handling of the brain. We would only be talking about quantum noise randomly affecting things in the brain. At best, it could be no more than a headache in an otherwise mindless brain. If such were the case, noise would be all we would be conscious of, not the orderly patterns of sight and music and feelings we know to make up conscious existence. What we need, what must be present, is some mechanism whereby the functioning of the synapse itself depends on some interplay between the presynaptic and postsynaptic neurons, a quantum choice as to whether or not any particular neuron will fire next. Only in that way can the observer properties of quantum mechanics play a role that corresponds to what we experience as consciousness.

But how could quantum processes play any significant role in the way synapses handle information? Although synapses are incredibly small, quantum effects are vastly smaller. The Heisenberg uncertainty principle that tells us when to expect quantum effects to be important shows that even individual atoms cannot make "quantum jumps" across the cleft that separates one neuron from the next at a synapse. And even if an atom could make such a quantum jump across the synaptic cleft, what good would that do?

Even electrons, the smallest particles we could imagine playing a role in causing the synapse to fire, ordinarily can jump a gap only a few atoms wide. When an impulse arrives at a synaptic cleft, the energy across the cleft is 0.07 electron-volt. If an electron with that much energy tries to penetrate a barrier that has a barrier height of, say, 0.12 electron-volt, we would expect it to make it through by tunneling only if the barrier were only about 7 atoms thick (that's 7 angstroms thick, where 1 angstrom is 10^{-10} meter). That is too thin to be of any significance. The synaptic cleft is much thicker than that.

But the electron is constantly in motion. Over and over again, it tries to penetrate the barrier. In the millisecond that it ordinarily takes a synapse to fire, this electron could make a hundred billion attempts to tunnel through the synaptic membrane. And when we work out the details more carefully, we find that the electron has just about a 50% chance to penetrate a barrier 180 angstroms thick. Thus it turns out that with the electric field across the synapse, an electron can tunnel through a barrier 180 angstroms thick, which is the thickness of the synaptic cleft. Electron tunneling can occur at electrically polarized synapses.

The fascinating fact is that when we look at the structure of a synapse, we find that the synaptic cleft *is* just about 180 angstroms thick, just barely thin enough so that electron tunneling can play a role in synaptic functioning. But the tunneling of a few electrons would not be sufficient to cause the synapse to fire. If the tunneling of a few electrons is to govern what happens at the synapse, if they are to trigger the switches in that vast computer, then the current of these electrons must be greatly amplified. One way this could occur would be for the tiny energy of these tunneling electrons to alter special molecular gates at the synaptic cleft. These are the gates that control release of the chemical transmitters in the vesicles and act on the postsynaptic neurons. As the electron tunnels through the barrier, it carries 70 millivolts of energy—just enough energy, it turns out, to induce conformational changes in molecules. This then opens up the molecular gates and "fires" the synapse. Looking more closly at the detailed structure of the synapse, we find that it is designed with a very special architecture; it is tailor-made to exploit electron tunneling across the cleft. First of all, there are patches of material that face each other across these clefts. The patches on the postsynaptic side (that is, in the next neuron), which are called postsynaptic elements, serve as the source of the tunneling electrons; the patches on the presynaptic side are known as the dense projections of Gray. These dense projections are macromolecules. They act as the acceptor sites for the tunneling electrons and also provide the gate mechanism where vesicles attach themselves and through which the contents—neurotransmitter chemicals—of these vesicles are released. Figure 12.1 provides a schematic of the synapse and its components. Briefly, when an impulse arrives at the synapse, the cleft becomes electrically polarized. When this happens, vesicles carrying neurotransmitter chemicals attach to the vesicle gate molecules. At the same time, ions flow through calcium gates in the presynaptic membrane to hook up with these vesicle gate molecules.

The result is that the cleft itself becomes strongly polarized between the postsynaptic elements and the gate molecules. This large electric field causes the electrons to tunnel across to the macromolecules that make up the vesicle gates on the presynaptic side of the cleft. The attractive force between these electrons and the calcium ions that are now attached to these molecules results in a distortion, a conformational change in shape of the macromolecules that make up these gates. As these gate molecules flex, they open a channel through to the cleft and onto the interneuronal medium. Simultaneously, the flexing of the gate molecules tears open the vesicle itself, letting its contents flow into the cleft. These released neurotransmitter chemicals then diffuse into the synaptic cleft and onto the postsynaptic membrane. They interact with the postsynaptic membrane, causing the ion gates there to open and thereby allowing the excess positive ions to flow through. This starts a new neural impulse in the next neuron. Meanwhile, the excess energy and electric charge in the vesicle gate molecules begin to dissipate. The electron that had tunneled through migrates over to the calcium ion, which causes it to detach from the vesicle gate. After the release of the neurotransmitter chemicals into the cleft, the vesicle gate closes, as does the vesicle itself. Neuro-

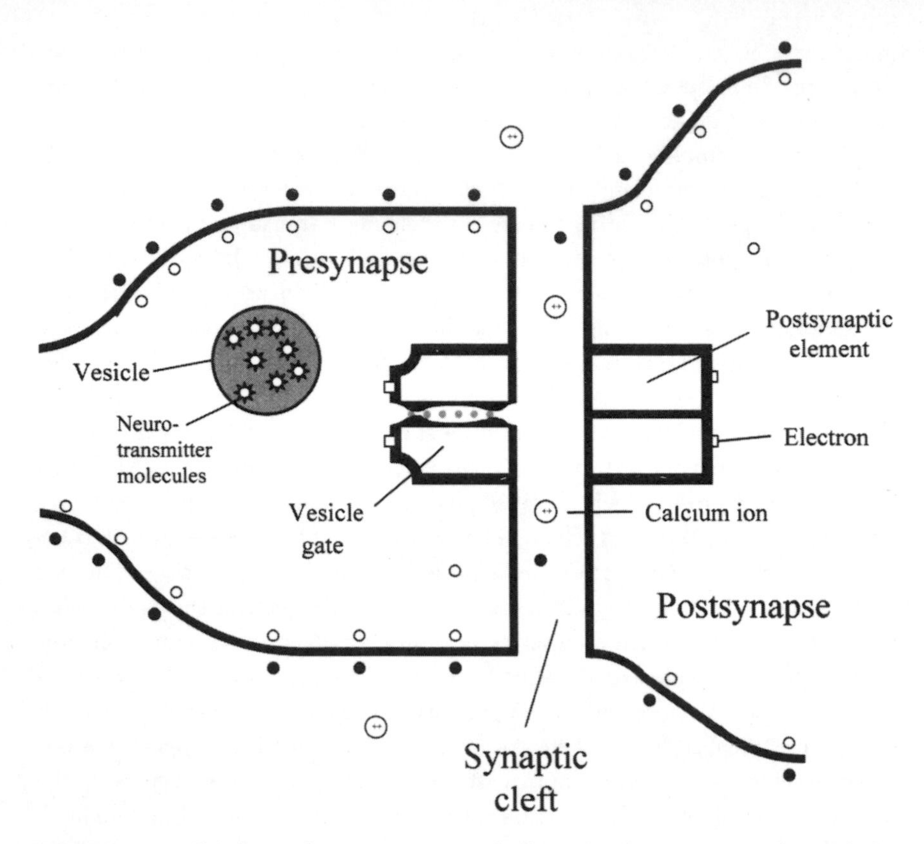

FIGURE 12.1 This figure shows an exaggerated schematic of a synapse together with the postsynaptic structure to help explain how everything functions in the synapse. The black dots represent positive ions of potassium and sodium that are usually found in excess on the outside of the neuron's membrane. The slightly larger circles with two plus signs represent doubly charged calcium ions. The open circles represent negative ions of chlorine. In addition electrons are represented by the squares. These are shown attached to the macromolecules that make up the dark projection, or vesicle gates, and the postsynaptic elements. The large circle in the presynapse represents a vesicle containing neurotransmitter chemicals. The molecules of the neurotransmitter chemicals are represented by small octaspiculated circles.

transmitter molecules begin again to build up in the vesicle. The neurotransmitter molecules on the postsynaptic membrane diffuse away as the impulse in the postsynaptic region propagates on into the neuron. Calcium ions diffuse into the synaptic cleft again as the ionic pumps reestablish the excess of positive ions outside the presynaptic membrane so that the synapse will be ready to fire when another impulse comes along.

All this complex activity actually works out very smoothly. Each step is necessary, and the whole thing functions to provide the great flexibility that chemical transmitters give to the brain. But more than this, this synaptic structure seems to be just capable of letting the fluctuations of quantum uncertainties—the dispersion of quantum mechanics, with the attendant need for observer effects—come into play in the brain. The synapse—the whole exotic machine—is poised ready to fire in the flickering passage of a thought.

In the foregoing discussion, I used numbers that were chosen somewhat arbitrarily. They reflect what goes on in these synapses well enough to make a good case for quantum mechanical tunneling being involved, but of course, these numbers do not offer a complete description of how things work. The details of this quantum tunneling mechanism in the brain's switches, however, has been worked out thoroughly[1] with actual measured numbers and observed details of synaptic structures.

This quantum tunneling account of how synapses work not only shows us how quantum mechanics gets into the brain but it also explains more about the morphology, functioning, and behavior of synapses, and more of the measured data, than any other explanation that has been proposed so far. Thus the details of this mechanism are important. Just as it was necessary to examine the details of Bell's theorem to show that consciousness is relevant to the physics of quantum mechanics, the neurophysiology of synaptic structures is important if we are going to understand where the mind really hides in the brain!

THE MECHANISM OF SYNAPTIC FUNCTIONING

Currently, in research on synaptic functioning, the "calcium hypothesis" is used to explain synaptic firing. This hypothesis says that calcium ion diffusion causes vesicle release, just as sodium and potassium ion diffusion causes neural impulse propagation. As we have seen, however, there is good reason to believe that electron tunneling plays a basic role in synaptic functioning. How can both the calcium hypothesis and the electron tunneling mechanism be necessary? Why should the calcium hypothesis need any revision? The answer is that the calcium hypothesis by itself doesn't tell us enough about how vesicles work. Energy is needed to open the vesicle gates and to open up the vesicles themselves. The tunneling electrons provide this energy. It turns out that they provide exactly the right amount of energy. On the other hand, we need the calcium ions to get the vesicle gate molecules ready to accept the tunneling electrons. The tunneling electrons start out with no place to go. They don't carry enough energy to jump to a sodium, potassium, or calcium ion directly. In order for them to jump to the macromolecule in the vesicle gate, the electron that is already there must first be neutralized. The calcium ion does that job when it attaches to the gate molecule. Then the postsynaptic electron jumps across.

Also, each vesicle gate actually consists not of two macromolecules, as they appear in Figure 12.1, but rather of some six to nine of these big molecules. That is, it takes up to nine electrons to open the gate. This plays an important role in gate functioning. The energies involved for a single electron are they taken alone, the effect might be swamped by thermal noise as the electrons got jostled about by the normal heat motion of the atoms. But the fact that nine electrons have to jump across at almost the same time (something that is not likely to happen by thermal effects alone) ensures against the synapse firing at just any time. Only when the calcium atoms first set things up, and then only if the several molecules open up at the same time, does the synapse fire.

The overall theory explains all kinds of data about how synapses work. It explains in great detail how spontaneous quantal release works, how temperature affects things, the differences between synapses in mammals and amphibians, osmotic pressure effects, ephaptic transmission, details of the morphology of synapses and ephapses; and all the effects found associated with the calcium hypothesis. It also explains the details of something called synaptic delay. All this gives us confidence in the idea that quantum mechanical tunneling plays a vital role in synaptic transmission—in the basic switches of the brain's mechanism. The theory is also supported by experimental results related to spontaneous quantal release. Two researchers named Katz and Milhedi,[2] who are well known for studies of the synapse, gathered the experimental data shown in Figure 12.2. It shows three histograms that plot how long it usually takes for a synapse to release the contents of a vesicle after an impulse has arrived—how long the synapse usually pauses before it fires. The histograms are given for three temperatures: 17.5° C, 7° C, and 2.5° C. Figure 12.2 also plots the curves predicted by the tunneling theory. There is no fudging here; everything is worked out from physical principles, and all the numbers that might have to be fitted to make the theory work have already been fitted using entirely different experiments. As you can see, the theoretical curves fit the experimental data that Katz and Milhedi measured. These just could not have matched so well if the synapse did not work the way the theory says it does.

◈

An electron, from nowhere, pops through a proteolipid membrane, and a microscopic packet of chemical soup spills into a crevice between two neurons. A neural impulse lurches forward, into the labyrinthine recesses of the brain's tangled net of fibers, quickened with each electrochemical impulse that moves it along. A maze of chemical reactions, pneumatic bellows, and hydraulic pipes and pumps assembled into one splendid creature there, then, and her tremulous voice speaks into my mind across the stellar stretches of time. Mentally, I reach out grasping for reality, but the layers slip away as though I had clutched at shadows all along. Where is she?

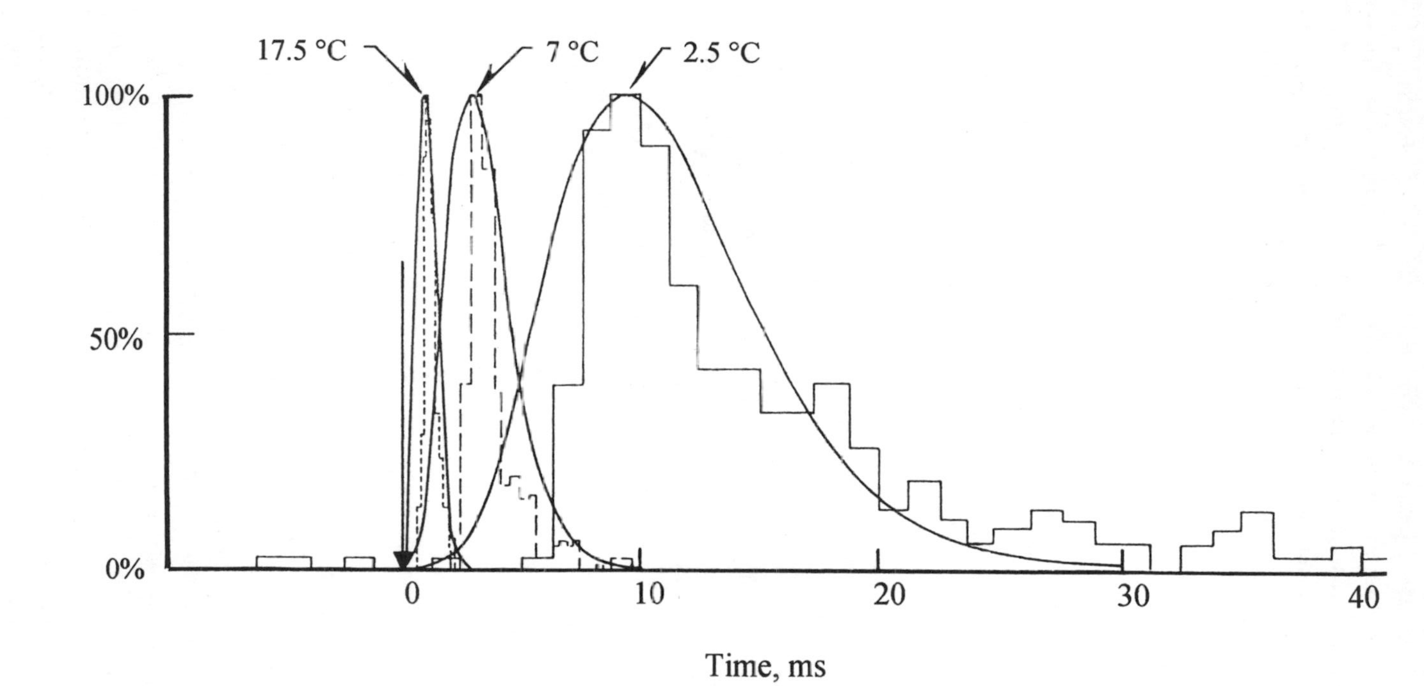

FIGURE 12.2 A comparison between the experimental data for the way synapses delay before releasing a vesicle's contents after being electrically stimulated (as shown in the three histograms) and the predictions of the electron tunneling theory of synaptic functioning. The results for experiments at three different temperatures are shown. The plots show the relative chance for a vesicle's contents to be released as a function of the time delay in milliseconds. For something biological, the fit between this "first principles" theory and experiment is extraordinary.

> Tuesday . . . Miss Aires was our substitute homeroom teacher. We had a problem to work in Algebra. I didn't have my Latin homework. I worked on poems in Study Hall and Debating class . . . I walked Merilyn to Homewood for her Sorority meeting. I called Merilyn at 7:10. [She] told me that her mother noticed my ring (my mother's Ensley High School graduation gold ring that I had given her) on her finger tonight, but she didn't say anything.

This may have initiated some of her mother's apprehension that would later so affect our relationship.

> Sarah J. Cook has been after me to take Merilyn to a 'Mardi Gras' dance Friday. MAZ said she'd like to go, so I asked Violet Bailey if we could ride with her and her date.

Wednesday turned out to be uneventful. On Thursday, I called Whitson about double dating the next day and then I called MAZ.

> Friday, February 1, 1952: I was late to school this morning so I went straight to Mr. Walker's office for an excuse for my absence yesterday. In Debating class we saw two movies about marriage—from the book *Marriage for Moderns*. After school, I called Merilyn at 3:30 and talked till 4:30. Tommy picked me up at 7:20. He was with Betty Barnet. We went by to get Merilyn and then to the Alabama where we saw the movie "Sailors Beware" (2). Then we went to the Pigtrail Inn and got a drink (2). Then we went out and parked (3). Merilyn said over and over that she loves me and I told her that I love her, too. Tommy took Betty home, and while he was with her inside, we talked in the car about many things. I told her how I loved her. She said, "So long as you can say, 'I love you, Merilyn,' that is all that matters." We talked about religion—she said that she believed that "anyone who believes in God, goes to church fairly often, and practices his religion will be saved."

Her father had been Catholic, and she considered herself to be half-Catholic, half-Protestant.

On Saturday we, my parents and I, left for Atlanta to visit my brother, who lived there with his wife and first son while working at Lockheed and attending Emory University.

> We got to Jimmie's at five. Merilyn told me last night to listen to the "Hangout" tonight because she was going to dedicate a song. I tried to pick up WBRC 960 with *that* Arvin from 10:30 till midnight. But I couldn't get it. I finally gave up and went to bed sick: sore throat, neck muscles sore, headache. I found out the next day that Merilyn had dedicated a song to us: " 'Because of You' for Harris and Merilyn." It was played a little after eleven CST.

Somewhere now far and deep in space, somewhere just past Arcturus, the faintest signal from Earth carries her message to me still as if it were happening again. "Because of You," but now, only a whisper.

> Sunday, February 3, 1952: We drove out to see where Jimmie works. The plant where the planes are built is six stories high, three blocks long, and one and a half blocks wide. There are no windows, and the entrances are by tunnels. They will soon be building B–47's. We went through Marietta and Smyrna going back . . . got back home at 7:30. I called Merilyn at eight . . . she told me that she had missed me. In so many things we are so much alike . . . .
>
> Monday, February 4, 1952: School. Economics . . . Algebra . . . Latin . . . . And of course we cut up in Debating class as usual.
>
> After school I called Merilyn and we decided we would fly a kite. We flew it on [where else] Kite Hill [a knoll on Red Mountain behind her house].

I have only the most fleeting memories now of that day. It was, as it should have been, a moment of carefree joy. Just the two of us and a kite that refused to go upward. We ran and ran trying to make it do something, all the while I remembered a box kite I had seen when I was about six or seven. With no one running or struggling at all, it had magically floated almost straight up in a gentle wind. Ours fluttered, spun, and crashed to earth, but it was wonderful.

On Tuesday, the diary tells me that I stayed home sick. Nothing much happened except that Merilyn called me at five. I went to school on Wednesday. I sat with Merilyn during Auditorium period.

> After school, I called Merilyn. We played tennis—I won 3 to 2. I walked Merilyn home and stayed till slightly after six, (3). Merilyn said 'It's a long time,' referring to how long before we could marry. I said, 'Time is a figment of the imagination,' [not knowing I was quoting words of consolation that Einstein had once spoken on hearing of the death of an old friend.] She said, 'I guess you're right.'

Nothing happened. Nothing happened in any of this, and that is what is important. We peered into our future at an illusion of passing time, and we each tried to touch a moment of time yet to come.

31 GRAMS OF MIND

We have found a way to bring together the idea that consciousness stems from a quantum mechanical process and the requirement that consciousness be connected to the basic hardware involved in the logic-processing functions of the brain, the synapses. Moreover, we have shown that this insight into the way the brain must function in order for consciousness to be explained in terms of quantum mechanics has yielded a new understanding of the structure and functioning of synapses. This is the kind of confirmation that builds a conviction that we are on the right track. But it is now time to see how well these ideas fit the phenomenology of consciousness itself. We must ask, "If consciousness arose from quantum mechanical tunneling of electrons at individual synapses in the brain, what would be the nature of our conscious experience?" We should ask, "What spatial extent should we expect to characterize consciousness according to the picture of its mechanism as so far developed?" What we have so far is merely consciousness tied to individual synaptic firings. There is, in fact, no spatial extension at all. There is nothing here that would give rise to a process integrating synaptic functioning into a single holistic entity. We have not shown why a synapse that fires in one part of our brain should be integrated into a conscious experience involving some other synapse in our brain, nor have we shown why a synapse that fires in *your* head should not contribute to my conscious experience just as strongly as any synapse firing in *my* head.[3]

So far, we have tied quantum mechanics to the individual synapse, but we have done nothing to show what integrates synaptic firings into a single quantum mechanical conscious existence. The mere existence of the neural network, without further justification, no more accounts for this interconnectedness than it does for consciousness in the first place. We must find how synapses scattered throughout the brain are nevertheless bound together into one functioning quantum mechanical whole. The obvious way to get some kind of interaction among the synapses is to try to extend the idea of quantum mechanical tunneling to the intersynaptic distances. But all one needs to do is look at the physical equations governing the tunneling process to discover that a direct tunneling from one synapse to even its closest neighbor is hopeless. Even the electrons cannot make jumps of micrometers, much less span the dimensions of the brain. So, we need to find some way that distant synapses can interact via a quantum mechanism. But whatever we do, we must be sure that the mechanism we postulate preserves the electron tunneling going on at individual synaptic clefts; otherwise, the mechanism will no longer involve data-processing events at synapses. At least we are not searching for a phenomenon that may or may not exist. We already know that quantum tunneling is there. We already know its features and much of its mechanism. All we need to discover is a way that electrons at synapses can make the long jumps from one synapse to the next. We need to look more closely at the path between synapses that might in some way facilitate the movement of electrons between synapses.

The space between synapses is filled with a conductive ionic medium. But these ions are too massive to be able to conduct the charge between synapses by any quantum mechanical process. Ordinary conductive transfer of charge is not possible, because the direct electric circuit that passes between distant synapses through the fluid surrounding the neurons is buffered by the ionic pumps that maintain a steady ionic concentration there. There is, however, another way the electrons can get from one synapse to another. It is called hopping conduction.

In hopping conduction, an electron can pass from one molecule to the next, just as you might walk on stepping stones across a pond. Hopping conduction is found primarily in organic materials that otherwise would not support conduction. In metals, the outer one or two electrons of each atom are essentially free to move through the metal. This freedom of the outer electrons accounts for the ease with which electrons can flow as an electric current, and it is one of the reasons why good conductors such as gold, copper, silver, and aluminum are so malleable. But organic compounds usually do not have free electrons. Although in some of these materials the electrons can move rather freely along the length of the molecule, when the electron reaches the end of the molecule, the only way for it to move farther is to hop across the gap to the next molecule in the same way that synaptic electrons jump across the cleft. Thus conduction in these materials proceeds via a succession of quantum mechanical tunneling events—hops from molecule to molecule through the organic conductor.

If we look at the medium between the synapses again, we find large organic molecules that may serve as stepping stones for the synaptic electrons. There are, in fact, many kinds of molecules present. But in order to be specific, I have assumed that only one kind of molecular stepping stone is involved. These molecules are called soluble ribonucleic acid (RNA) molecules. RNA molecules are the so-called messenger molecules. They are the intermediaries in the control of cell chemical activities. They "take orders" from the cell's DNA and then control the production of all the cell's proteins.

There are several reasons to choose soluble RNA molecules as the "propagator" molecules, or stepping stones, for the synaptic electrons. First, these molecules are about the same size as, and somewhat similar in structure to, the molecules that the synaptic electrons start from on the synaptic interface. Second, the structure of RNA molecules is such that they should provide many close-lying energy levels, making it easy for electrons to find a site to land on, no matter what the electron's initial energy happened to be. Third, the arrangement of bases (that is, the amino acids) forming a particular RNA molecule will result in slight changes in the energy levels of these molecules that can act as a kind of information coding, possibly serving as part of a molecular memory mechanism in the brain. To suggest this, of course, is to get ahead of our story, but it is a consideration, a possibility we should keep in mind. Fourth, the RNA molecules that carry copies of segments of the DNA already are designed to carry a form of memory, the genetic code. They may furnish the key to certain remarkable abilities of the brain's and body's functioning and provide an easily understood path for the evolution of memory and consciousness. Fifth (and this must be the real reason), it works. I can write down the equations and use this RNA hypothesis, together with experimental data, to get numbers that agree with the numbers describing the mind that we got in the last chapter. It fits hand in glove. But let me show you how.

HOW ELECTRONS FOLLOW THE YELLOW BRICK ROAD

Once again, let us follow a signal propagating along the pathways of a nerve cell in the brain. Let us watch as this electrochemical impulse races into the tiny neural tributaries, propagates down into the dendrites and into a synapse, and there comes to a halt at the synaptic cleft. At this point, the interior of the synapse becomes electrically neutral. Opposite the cleft lies the next link in the brain's labyrinth, the molecules of Gray's dark matter. These molecules are the source of the energetic electrons, because they lie in a region that is still at an elevated electrical potential. These are the circumstances that can lead to the normal functioning of the synapse. A few electrons begin to tunnel across the cleft to the group of molecules forming the vesicle gate. As the gate opens, an attached vesicle is also opened; this results in its contents being released into the gap between the synapse and the next neuron. The released chemicals affect the permeability of the membrane of that neuron, and, as a result, the postsynaptic neuron fires. If a vesicle is already on the gate, and if the energy states of the gate molecules will accept the electrons, all will proceed along the prescribed stochastic course.

But what if there is no vesicle at the gate? This would alter the available energy levels to which the electron could jump. Or what if the various molecules have a fine structure that acts like a code, so that only when the molecules on both sides match perfectly can they fire.[4] Or perhaps the energy levels at the available sites in these gate molecules depend on the recent history of the synapse. A recent sequence of pulses arriving at the synapse may condition the gate molecules to "read" the encoded molecules in the postsynaptic neuron. For whatever reason, we may suppose that the electron that would otherwise cause the synapse to fire cannot make the jump. However, suppose that there is a neighboring synapse that has gate molecules meeting exactly the firing conditions for this electron but has no electrons available locally. If the electron can reach that neighboring synapse by hopping conduction, then that neighboring synapse will fire. This would then give us the quantum mechanical connection between neighboring synapses that we have been looking for.

For this hypothesis to work, quite a number of conditions must be satisfied. First of all, there must be a sufficient number of molecules to serve as "stepping stones" for the electron to successively tunnel to the neighboring synapse. Second, the tunneling must take place during the brief interval of time available. The electron must begin and end its journey during the time interval in which both the sending and the receiving synapses have become activated by the arrival of neural impulses—the time interval during which these two synapses will stand poised, ready, and waiting to fire before reestablishing their resting electric polarizations. That gives it, at most, only some 3 milliseconds. There is even less time––0.3 millisecond—if the electron is to make its jump to a synapse during its latency period (the normal delay time for a synapse to fire).

In that brief time, the electron will have to tunnel hundreds, thousands, even hundreds of thousands of times to arrive at the distant synapse just in time to complete its mission to fire that distant synapse.

Now there is only about 31 grams (1.1 ounces) of soluble RNA in the average brain. The typical molecular weight for these molecules is some 25,000 daltons.[5] This gives about 7.45×10^{20} soluble RNA molecules floating about in the brain. The average distance between them will be about 100 angstroms. If we use the same energies and potential barrier numbers as we used earlier (70 millivolts for the energy of an electron and a barrier height of 118 millivolts),[6] we find that each hop of the electron will take 8.4×10^{-12} seconds. In calculating this value, we had to consider a lot of factors, such as molecular size and shape and even the random way long-chain molecules coil around. But the result shows that an electron could manage to travel as far as 10 centimeters—the distance across the brain—in just 0.084 millisecond. This is just about one-third of the time it takes the average synapse to fire.

Therefore, there is enough time for an electron to travel not only to the nearest neighboring synapse, but even to the remote synapses in the brain. It is possible, therefore, for this quantum mechanical interaction to join events taking place in the brain's switches into one grand, unified process. This holistic process brings the mysteries of quantum mechanical uncertainty into play in the functioning of the brain. It links "thought"—that little fire of the fleeting, jumping electrons—and the specter of the "observer" of quantum mechanics into one orchestration of mental phenomena. Without this connection, the brain might, for all the world, be nothing but a big billiard table of senseless chaos. But this contact between synapses and electrons, reaching across the space of the brain, turns on the light of consciousness.

We have much more to do, however. We have to explain the characteristics of consciousness and show how the associable quantities come to have their values. We have found a means by which quantum mechanics can be at work in the brain, but let us look at just how this happens. We will see, as we proceed, why this must be the way consciousness works.

We have described how two synapses somewhere in the brain interact by means of a quantum mechanical tunneling process. We now must look at the conditions required to keep this process running so as to give rise to a continuous state of consciousness. We must have more than just a momentary bridge between two synapses. We must have an ongoing process like consciousness itself. But what is necessary to keep such a process going in the brain? If the conditions essential to this process only happen occasionally, we would expect the corresponding conscious experience to be sporadic. The brain must meet some set of requirements for this ongoing process to happen, and whatever those conditions are, they must be sustained.

ARE WE IN KANSAS YET?

First of all, there will only be a certain number of molecules on a given synapse that can donate electrons to effect the firing of some other synapse. From photomicrographs of synapses, I have estimated that about 20% of the area of the synaptic cleft has donor molecules. If these areas consist of typical proteolipid molecules, each can be taken to be cylindrical in shape and about 20 angstroms (about 80 billionths of an inch) in diameter. For a synapse 1 micron in diameter, that gives about 200,000 molecules and a corresponding number of electrons (one from each molecule) that are available as we search the regions of the brain for a suitable acceptor site. This gives us 200,000 electrons starting off from a given active synapse and hopping off on the 745 billion billion available "stepping stone" molecules searching for another synapse to fire. If this happens, and if it continues time after time as an ongoing process, then consciousness can occur as a result of this quantum mechanical interaction.

But we need something more for it all to work. We need a sufficient number of "resting places" where the electrons can finally go. We need acceptor sites on the neighboring synapses to complete the quantum mechanical interaction of the synapses. To show how this works, we need to write a few equations, but they are quite simple. All we are going to do is find the chance that a suitable receptor site will exist at the same time as the first synapse is active. *We multiply*

1. The number of synapses that are active (about to fire) at the same time as our donor synapse is active *by*
2. The chance that any particular donor electrons will be on one of the 200,000 donor molecules in another active synapse, *and multiply that by*
3. The number of donor electrons traveling out from the donor synapse, *and finally multiply that by*
4. The number of hops each electron can make during the time that the donor synapse is active.

Then this will give us what we can call the coupling factor Q for the donor synapse (see Appendix I). It is the number of opportunities to cause another synapse to fire. If we now multiply this coupling factor by the chance that this single electron will then cause that synapse to fire, and if the resulting number that we get is 1 or larger, then all the conditions necessary to sustain our electron-tunneling process in the brain will have been met. Electrons starting off from one synapse can make it to another, distant synapse and then to another and another. The whole thing becomes self-sustaining—an ongoing process. This then can tell us what conditions must be satisfied in the brain for consciousness to be present. Our Red Shoes will have taken us home to Kansas!

The details of this calculation are given in Appendix I. However, it is important to emphasize that what we have here is not smoke and mirrors—so many words describing what might be—as has so often been the outcome in efforts to understand the nature of consciousness. We have very restrictive conditions on whether or not those Red Shoes will get us to Iowa. If the numbers do not work, we fall out of the sky!

For the whole mechanism to become self-sustaining, the brain's synapses have to satisfy *this* firing condition on the minimum synaptic firing rate:

$$f_{min} = M\tau/nNt^2$$

where M is the number of molecules in the 31 grams of RNA, τ is the time for the electron to make one quantum jump, n is the number of such electrons that travel out from one synapse, N is how many synapses are in the brain, and t is how long a synapse usually pauses before firing.[7] That is the equation I said we would get. It is the equation that sets a severe limit on the occurrence of consciousness in a human or similar brain. All the numbers that go into this equation are easy to get. The equation is either satisfied or it is not. It is either right or it is wrong. The numbers that show that this works are given in Appendix I. The condition is met.

But there is more. Consciousness must occur for values above this minimum firing rate, and it must be absent—that is, sleep must set in—when the firing rate falls below this number. This gives two more conditions. Again, when the numbers are plugged in, they show that consciousness does occur above this limit and sleep ensues below the limit. The quantum mind is our consciousness.

But there is even more that checks out. Remember that coupling factor Q we figured out a while ago? It gives us the number of synapses interconnected at any instant. All we need to do is divide that by the time, t (the time it takes for the coupling to occur), and we get the number of synapses that contribute information to our conscious experience each second. And if we then multiply that number by how many bits of information one synapse handles at a time, we get the consciousness data rate—the channel capacity that carries our stream of consciousness! It is all right there in the theory already. If we use the letter i to represent the information that one synapse sends, then the consciousness channel capacity, C, is iQ/t. If we write all these factors out, we have

$$C = in^2tNf/M\tau$$

Again, the numbers for each item are given in Appendix I. The purpose here is to make clear that these are definite attributes of what we consciously experience that can be treated quantitatively by the quantum mind approach. If we put their values in, we get for the consciousness data rate (channel capacity) $C = 47.5$ million bits per second. That is a pretty good result. Remember that the information channel capacity of consciousness was one of the associable quantities we said had to be explained if we were to understand consciousness. In the last chapter, we came up with measured values for the consciousness channel capacity. We took the nominal value to be about 50 million bits per second. (Our various ways of estimating this quantity gave values ranging from 45 million to 200 million bits per second.) Our calculated value of just under 50 million bits per second fills the bill just about perfectly. The result arises out of the very simple assumptions we made at the beginning of this chapter about how consciousness should work.

Look at all we get. We understand now how the "observer" of quantum mechanics is also really a quantum system himself. We have seen how synapses work in much more detail, why the areas of Gray's dark matter sit poised opposite the vesicle gates, and even how the ephapses (the electrical synapses that are really just like other synapses but have narrow clefts to permit a higher current of tunneling electrons) work. We have found that hopping electrons can connect distant synapses, and we have discovered that a very stringent constraint that *had* to be met in order for this hopping to happen fast enough *is* in fact met in the brain. We have also found that the consciousness onset requirement is a characteristic of this theory, perfectly mirroring the fact that both consciousness and sleep characterize our brain's functioning. We computed the critical level of brain activity required for consciousness to begin, and we found that the result exactly matched experimental data. And now, we have found that the information-handling capacity of the quantum mechanical mechanism in the brain matches almost exactly the estimate we made in the last chapter for the information channel capacity of consciousness.

I believe this represents a significant advance in our understanding of a subject that heretofore not only had not been quantified but had been thought impossible to treat scientifically. We have laid the cornerstone for the science of mind; for a new, more comprehensive psychology; and for a knowledge of what we are that reaches to the foundations of physics. We will find, anon, that this discovery touches the stars in the night sky and reaches to the beginning of time.

In our search for reality, the quest has turned to our own nature. As our understanding has broadened, we have come to realize that we reside somewhere within the brain and in the consciousness of that brain. Looking more closely, we have discovered that we are only that single electron that reaches across the space of our brain. The Great Oz is but one electron.

THE OTHER SHOE

In 1873, James Clerk Maxwell set down his famous equations that for the first time fully embraced all electric and magnetic phenomena. The simplicity and symmetry of these expressions not only convinced Maxwell of their validity and completeness but also suggested one further discovery that lay hidden in these four equations. With just a couple of manipulations, Maxwell derived two new equations describing waves that propagate through space as ripples of electric and magnetic energy. A check on the numbers revealed that, yes, these new equations gave the correct description for a wave moving at just the right speed. Maxwell realized that for the first time the nature of light itself was explained by his formulas.

This is how it usually is in science. Long before Maxwell laid down his formulas for electromagnetism, there had been many suggestions and speculations—hypotheses about the many aspects of how electric charges and magnets behave—had been tested and combined with other hypotheses. As this process developed, refining into a more and more powerful theoretical understanding of the experimental facts, a point was reached where one could finally grasp the nature of light. It was only when the scientific work had been completed—after the formulas were on the paper and the numbers checked—that one could then turn one's attention to describing all the details and all the fine points that are a part of the complex nature of light as revealed by Maxwell's equations.

We have taken a path somewhat like this. We made and tested hypotheses about electron tunneling in synapses and long-range tunneling over "stepping stone" molecules. We put in numbers and checked everything out to see if the equations we had derived would fit the numbers that describe the things we know about consciousness.

And now we come to the last step. We come to the question that we have already answered with all the science we have done—answered, but not stated. What is it that consciousness really is?

THE ELECTRON PICTURE

We have talked already about the electrons that use quantum mechanical tunneling to cause synapses to fire. We have also talked about the long-range tunneling of electrons by means of hopping conduction on RNA molecules that deliver the extra bit of energy that tops off the energy supply on a vesicle gate at a distant synapse to cause it to fire. We derived equations (see Appendix I) that gave us the conditions that have to be met for this long-range tunneling to occur and to be self-sustaining. *This long-range electron tunneling that connects synaptic firings throughout the brain into a self-sustaining pattern is consciousness.*

But our equations tell us there is an interesting complication. Once the level of brain activity reaches the conditions such that one active synapse can bring about the firing of a distant synapse, it would seem that this electron-mediated intersynaptic influence would fill the brain. But we do not get our best fit with the data if we assume that these interactions increase in such an open-ended way. The equations tell us that it is as though there were just one, or just a few, of these electrons that run through the sequence of firing synapses in the brain: a self-sustaining series of long-range intersynaptic influences that threads from synapse to synapse. It is as though other processes of the brain restrained the spread of consciousness, limiting it to just "one thread of thought."[8]

But that is not the whole picture. Beginning with the initial synapse, there is not only this one electron that leads to the firing of the distant synapse; rather, there are many electrons from that synapse that spread out to many synapses. This collection of synapses that these electrons reach are also a part of the consciousness, though there is more to the story, as we will see in later chapters.

Running through all this sequence of collections of potential synaptic firings, there is this thread-electron (one or, more likely, a few that are coordinated so as to mutually sustain the process) that brings about the firing of one synapse, and then the next and then the next. The actual firing, the actual synapse that is selected—that is the thing that is the input information.

Obviously, we cannot take these two things (the collection of synapses that could fire and the one that does fire) apart from each other. They both go together to form the one entity that is the mind.

One more thing: The one electron that will eventually bring about the firing of the distant synapse cannot be distinguished from any of the other electrons that spread out from the source synapse. This is because of the indistinguishability property of elementary particles that we learned before to be an aspect of quantum mechanics. In a sense, this electron that does the distant firing is, at the same time, all of the other electrons. Moreover, it cannot even be fully distinguished from one of the receiving synapses own electrons that could have fired the synapse. It is as if it were one and the same electron that goes out from the initial synapse and continues from synapse to synapse. But even as that electron fires the distant synapse that is poised and ready to fire, that synapse itself has sent out its own set of electrons that are searching for their own distant synapse to cause to fire. Our electron that brings about the firing of this synapse cannot be entirely distinguished from any of those electrons that were sent out by this synapse that just fired. Hence that electron, in a certain sense that stems from the electron's indistinguishability, keeps on going—on to the next synapse and on to the next after that.

THE STATE VECTOR PICTURE

Incredibly, consciousness is that one electron. But consciousness is more. Or should we say that it is less? This electron and this collection of synapses that this electron and all of the electrons reach are all part of what consciousness is, but not as the electrons per se, and not as the synapses per se. Consciousness is not this set of objects or electrons, as such, at all. These exist to bring into being the potential interactions—the quantum potentialities—that are the state vectors. Consciousness is the collection of potentialities that develop as these electrons and these structures of the brain interact. Consciousness is the bringing about of this ongoing state vector of possibilities that runs through the brain. It is all of these branching and interlaced collections of quantum potentialities weaving together the possibilities. And beyond consciousness, something more. By creating the possibilities that we experience as consciousness and by selecting—by willing—which synapse will fire, mind brings into reality each moment's thoughts, experiences, and actions.

Years later, long after Maxwell had shown us how to picture the electric and magnetic waves that intertwine into the process of light, long after Maxwell himself had passed to whatever might come beyond, we learned more about the nature of light. Years later, after Planck, after Einstein, after Bohr, Schrödinger, and Dirac—particularly Dirac—we began to see just how much more there was to learn about light than could ever have been imagined in Maxwell's day. It will be that way with consciousness as well. There is much more to what consciousness is than we have discovered with our present notions and tools. But now, at least, we have a fledgling science on which to build.

13

To Sleep, Perchance to Dream

HAL 9000: [Facing a shutdown] "Will I dream?"
Dr. Chandra: "Of course you will dream. All intelligent creatures dream. Nobody knows why."
HAL 9000: [Facing death] "Dr. Chandra, Will I dream?"
Dr. Chandra: "I don't know."

—Arthur C. Clarke, *2010*

The rays of ultraviolet light that can be so damaging to tissue also activate the production of a chemical, a dark brown pigment that transforms the color of the skin. This pigment is called melanin. In small amounts, melanin gives hair a blond color. In larger amounts, it produces colors ranging from red to brown to black. In addition, it is present in the skin of all races in varying amounts. Melanin itself is not a single chemical but an amorphous mixture of organic macromolecules, probably mixtures of polymers in the form of very tiny granules. It is well suited to its job of absorbing a broad range of electromagnetic radiation and the energy of excitation in nearby molecules. The function of melanin in the skin is to absorb the ultraviolet radiation from sunlight, thus providing protection from sunburn and reducing the hazards of cancer that can be caused by the sun's rays. The melanin in hair also absorbs light to reduce the skin's exposure to the harmful effects of the sun's ultraviolet radiation, particularly on the otherwise exposed top of the head. Inside the eye, melanin occurs in the epithelial cells of the retina. It absorbs whatever light gets through the sensitive layers of the retina and thus prevents scattered light inside the eye from blurring our vision.

Melanin also helps protect the iris of the eye. Its presence can produce brown eyes. In small amounts, it contributes a yellowish tinge that combines with the blue light scattered by the cells of the iris to produce green eyes. The presence and effects of melanin in all these areas of the body are well understood.

But there is a mystery. Melanin is found in one other place in the body. Curiously, it is present in the cortex of the brain. It is melanin that gives Inspector Poirot's "little gray cells" their distinctive color. The next question should be obvious. Why on earth does the brain, protected beneath half an inch of skin and bone, need a suntan? Why is there a layer of tissue wrapping around the brain that is filled with melanin? Why does the brain's skin need a "radiation" absorber? Any radiation that could get through the skull could hardly be absorbed by a bit of melanin. The answer lies elsewhere. The answer has to do with another mystery, the mystery of sleep, and even a mystery beyond that, a mystery of time that meters the tick of the consciousness clock. And something else is curious: We will see how the brain's "suntan" is tied to the shape of our TV sets.

WHY DO WE SLEEP?

Obvious answers to this question immediately leap to mind. We sleep because darkness inhibited the day time activities of the aboriginal man. We sleep because the body is tired; we sleep to restore the chemical balance our body needs to function efficiently.

Because sleep does occur, the body certainly makes use of it to knit up the raveled sleeve of care, to rest the machine, to restore the chemical balance in the body. But the heart does not sleep. It slows down. It takes advantage of sleep. But it does not stop and rest for some eight hours. The brain also slows down, but it does not cease its function. It would seem that the pressures of evolution would have wiped away sleep as the most hazardous of all possible behaviors. Surely the body should rest at night, and the brain should think at a pace consistent with what we find in the sleeping brain. But how could the pressures of evolution permit the existence of such a hazard as the complete lapse of attentiveness that turns the body into nothing more than an awaiting meal for any wild beast? Why do we need sleep so desperately? It must be that sleep is as vital as life itself; it must be vital to the very existence of consciousness.

The answer to the puzzle of sleep lies in the fact that the mechanism of consciousness that we described in the last chapter cannot go on indefinitely without periods of restoration. Consciousness has to pause. Every time another synapse enters into the quantum mechanical consciousness mechanism that couples distant synapses, some 200,000 electrons begin their hopping conduction into the brain's sea of soluble RNA. A few of these electrons will play a part in firing synapses immediately. The brief presence of these electrons on the soluble RNA disturbs nothing. They disappear from the original synapse and reappear in the synapse they fire. In between, they behave in that mysterious quantum mechanical way: Although they need the stepping stones to make the journey, they are never in the space between.

But many of the 200,000 electrons that start out never finish the journey. Their presence helps to smooth out the transition from one intersynaptic event to the next, smoothing out the consciousness experience so that the briefest consciousness interval is greater than the time required for a synapse to fire. But if this outpouring from excited synapses were permitted to continue uninterrupted, if there were nothing to drain off these excess electrons, consciousness would become one grand blur until the glut of electrons choked off consciousness altogether.

And this is where melanin comes into play. To prevent an excessive blurring of the consciousness interval, melanin is present in the outer layer of the brain. There it absorbs some of these electrons—some of this excess energy. Of course, just the right amount of melanin must be present there. Too much, and the melanin damps out the consciousness itself. Too little, and consciousness blurs.

If the quantum mechanical interconnection of synapses were limited to the average synaptic delay time during firing, our minimum conscious experience would be as brief as 0.3 millisecond—clearly shorter than we ordinarily experience it to be. Allowing for the full range of synaptic firings would give as much as about 1 to 3 milliseconds for the minimum consciousness time interval. But there is only a small amount of melanin present. The cortex is only a pale gray. It absorbs only about 10% of the radiation that falls on it, and so we can estimate that perhaps only about 10% of these electrons will be absorbed during the interval of 1 to 3 milliseconds. For the melanin to damp out most of the activity from any synapse, it requires 10, 20, or 30 times as much time. As a result, the characteristic time interval for consciousness is not 1 millisecond, but more like 30 or even 40 milliseconds. During this time, 80% to 85% of the excess electrons are absorbed. This, then, agrees with the time interval of about 0.04 second that we previously found for the minimum time of consciousness—the length of the tick of the consciousness clock. This is another point of agreement between theory and the way we actually experience our own conscious existence. Everything fits. Everything is tied to the way everything else works. The machinery that runs our consciousness is a finely tuned engine. The strange presence of a "tan" on the brain, the minimum time interval we consciously experience, and the quantum mechanical nature of consciousness are all woven together into this silken fabric of reality.

But something else can also happen to the hopping electrons. They can stop on one of the stepping stones. And when one of the electrons ceases hopping and stops on a given RNA molecule, that stepping stone is no longer in the game. It is no longer available to help maintain consciousness. If this goes on long enough, the RNA stepping stones become used up and consciousness stops. Sleep comes and continues until this excess energy clogging up the engine of consciousness is bled away by chemical or other means that can restore these radicals (the excited RNA molecules) to their normal "ground" state. This process takes hours—much too much time for it to occur while the brain is conscious and continuing to use up the supply of unexcited RNA molecules.

During the conscious state, we have 23.5 trillion synapses, each of which fires an average of once every 67 seconds, with about 200,000 electrons being released to the RNA each time a synapse fires. At this rate, the 745 billion billion soluble RNA stepping stones would be occupied—used up—in only 3 hours.

The presence of the melanin significantly reduces this problem. The fact that we can stay awake 16 to 20 hours without difficulty reflects the fact that the melanin damps out between 80% and 85% of this electron activity. But the remaining 15% to 20% continues to use up the available material necessary to maintain the synaptic intercommunication of quantum mechanical consciousness. As more and more of the molecules are used up, the brain reaches a condition in which consciousness can no longer be maintained. The brain must sleep. Sleep has to begin so that these excited RNA molecules can make their natural transitions back into their unexcited states. By morning they will have to support consciousness again. If the brain doesn't sleep, the supply of "stepping stone" RNA disappears and unconsciousness eventually results. In addition, the brain must also quiet down. It has to reduce the rate of its synaptic firings. This means that two things must happen during sleep: Consciousness has to stop, and the level of synaptic firing must diminish to permit restoration of the pool of soluble RNA back to the "ground" state.

With this rather simple idea in mind, it is straightforward to show in detail how sleep works. The rate at which we lose the RNA (that is, the rate at which an electron gets stuck on one of the molecules, removing it from the game) depends on the rate at which synaptic firing is pumping out electrons into this sea of 7.45×10^{20} RNA molecules and also on the rate at which the RNA is recovered. This recovery rate for the RNA (the number of RNA molecules that return to being normal, per unit of time) depends only on the number of excited-state RNA molecules in the brain at any particular time.

Thus, beginning with the level of available RNA at, let us say, 90% of the total, and with a level of brain activity at twice the minimum needed to just maintain consciousness, the supply slowly drops. As we saw in the last chapter, the minimum synaptic firing rate that just maintains consciousness depends on how much RNA is available. If we have the total supply available to serve as stepping stones, then we calculate the minimum firing rate to be about 0.015 per second (once every 67 seconds on average). If during our waking hours we maintain a level of activity twice this (0.03 per second, or once every 33 seconds for the synaptic firing rate), then consciousness can be maintained until we have "used up" half of the RNA, which will require 16 or so hours. After 16 hours, when the amount of available RNA begins to drop below the 50% level, a synaptic activity of 0.03 firing per second will no longer support consciousness. Sleep comes, and in order to restore the available RNA, the brain drops back to a level of activity of about 0.01 firings per second, or about once every 100 seconds for the average synapse. This continues for 6 to 8 hours, after which time most of the supply of RNA has become available to support consciousness again. Thus the presence of melanin and the natural transitions from the excited to the unexcited state of the RNA build up the supply of these molecules over the next 8 hours or so of sleep, restoring the RNA supply to its initial level. At this point, the cycle can repeat.

Many people, especially "night people," run on a different cycle. These people do not start the day at an activity level like the 0.03 firings per second for the average synapse. Rather, their morning activity level barely rises above the minimum to achieve consciousness. If that describes you, you will spend most of the day operating on half your cylinders. You will be conscious, of course, but with just a small drop in the level of the available RNA, you just might drop back below the level of consciousness—back into sleep. Toward the end of the day, however, night people still have a large reserve of conscious activity because of the available RNA. If you are one of them, that's when you enter a period of heightened conscious activity. For a few hours you really feel alive, sharp, and bright. But, of course, all that extra nighttime sharpness rapidly uses up the rest of your reserve. At one or two o'clock, finally, you can fall asleep.

There is another scenario. It is also possible to force extended consciousness. Say one begins the day at an average synaptic activity of 0.03 firings per second. When half the available RNA is used up, this level of activity is then the minimum that sustains consciousness. By applying further stimulation—drinking quantities of coffee, perhaps, or experiencing a heightened level of sensory excitement (say, to a level that corresponds to an average synaptic activity of 0.04 firings per second)—one can still maintain consciousness even when the available RNA sinks below the 50% level, to as low as, say, 38%, or lower. This heightened synaptic activity compensates for the drop in available RNA molecules so that consciousness can be sustained, extending consciousness for hours more before sleep becomes necessary.

We see, then, that this picture not only explains what sleep is and how consciousness works but also explains particular details about sleep. We have all experienced the fact that excitement can overcome even severe drowsiness. This model shows why that is so.

Of course, this is not all there is to how sleep works. As a physicist, I have greatly simplified this account of what the brain is. In reality, the brain has many special structures, such as the thalamus and the reticular formation, that monitor and coordinate the brain's levels of activity, controlling the brain so that it will efficiently carry out these requirements of the physics of consciousness. We still have much more to learn, but this quantum mechanical theory of consciousness lets us see beyond the activities of such brain structures so that we can understand the basic rules of consciousness and sleep.

SUCH STUFF AS DREAMS

As we sleep, the brain continues to function; synapses continue to fire, but at a reduced level of activity. Even in sleep, the brain forms thoughts, and the presence of excited RNA molecules continues to affect those thoughts. At times, the thoughts have ramifications that excite portions of the brain to heightened activity. At such times, the level of brain activity may be so great that it exceeds the requirement for the onset of consciousness locally. This is the dream state. Dreams are the thoughts we have in sleep that bubble up into periods of consciousness, spanning lesser portions of the brain. These unspent thoughts of the waking mind await the quiet of sleep to act out their suspended visions and yet unspoken words.

Just such an effect seems to have been discovered in the dream experiments conducted by Otto Poetzl, a Viennese physician acquainted with Freudian theory. In 1917 he described experiments he had done to discover the connection between dream imagery and the images of our conscious experience. Using a tachistoscope (a device used to flash slides briefly on a screen), Poetzl showed subjects unfamiliar slides and asked them to describe in detail all that they could remember about each picture. The subjects were then asked to watch the imagery in their dreams following the slide show for analysis the next day. Poetzl noticed in these descriptions a tendency for the dream to contain information that was present in the slide images the subjects had seen but that they had not mentioned in their descriptions of the slides on the previous day. The dreams seemed to skip what had already been described, as though these thoughts had been completed, whereas the dreams contained the thoughts left suspended and incomplete in the mind.

The "Poetzl phenomenon" has been studied many times over the years, with varying results. Many investigations have yielded results like those of Poetzl, including some studies by David Foulkes, author of the popular book *The Psychology of Sleep*. But these results are still only suggestive. A number of careful experiments designed to look for Poetzl's phenomenon have failed to find evidence of it.

With only limited and somewhat shaky data to support it, Poetzl's hypothesis about the content of dreams remains speculative. But if we entertain Poetzl's idea about the content of dreams for a moment, we can see a possible tie-in with the function of sleep. To see the connection, we must make a hypothesis that may have some merit, though it goes beyond any of the requirements of the theory of consciousness itself. What if these excitations on the soluble RNA actually represent some form of encoded information? What if nature has taken advantage of this aspect of the consciousness machinery to design a vast memory resource, a data storage capability far greater than anything the neurons alone could provide? Then, during sleep, as the unused information stored on these molecules bleeds away, some of this excess energy could produce mental activity, mental images—dreams about those unused pieces of information from the previous day's activities left over as uncompleted thoughts. We get a picture of a process that sounds much like what Poetzl described about dreaming. And we begin to see the purpose of the dreams themselves.

This hypothesis of molecular information storage is not new. The roots of the idea appear in the works of Heinz von Forester and in the memory transfer experiments carried out by Babich, Fried, Horowitz, Zelman, and others in the 1960s. But these hypotheses have always carried with them such a heavy burden of proof that even positive results in experimental tests have been ignored. The problem with this hypothesis is that conventional wisdom among neurophysiologists has always held that information in the brain is stored by the growth and formation of new neural connections. How could any information be stored on individual molecules? How could any such information ever affect the activity of the brain's neurons?

But if we entertain the idea for just a moment, we find that we already have a way to encode and read the information stored on these molecules. A sequence of pulses at a synapse, rather than a single impulse as often supposed, may actually be necessary to fire the synapse. After all, with more than 2000 synapses on an average neuron for each of the 10 billion neurons, and an average of only 7 firing each time the neuron fires, we are already talking about several hundred impulses for each one that fires a particular synapse. Trains of impulses may in fact be required to cause a synapse to fire. Such a train of impulses may be necessary to raise the level of excitation energy in the molecules in the proteolipids at the synapse. Such sequences, if properly spaced, could key in to exactly excite a sequence of possible energy levels in the chain of amino acids forming the protein portion of these molecules. Similarly, such excitations may wind up stored temporarily on the soluble RNA molecules until they can, as we have already said, be de-excited or perhaps play a role in permanently fixing memory traces by modifying the vesicle release mechanism, inserting new proteolipids at the synaptic cleft or through the growth of new neural processes.

Looking further, we find that the length of the soluble RNA and the length of the proteolipid structures at the synapse are such as to have just enough amino groups to hold about 100 bits of information. This is just enough to encode the addresses of two synapses (about 44 bits would be needed to address any one out of all the possible synapses), together with a few bits left over to indicate what the receiving synapse should do with the information. The point is that this is how computers are typically designed. For example, the instruction may say, "Add the information at location A to the information at B," or it may say, "Take the information at A and store it at B." Variations exist in computers, of course, but still this is the basic pattern to be found in computer architecture. Is it not conceivable that nature has been able to design the brain to achieve a similar level of sophistication?

I have stressed this point because the brain is still far more powerful as an information-handling device than we can begin to envision in terms of our present neurophysiological models or theories of its functioning. Even the boldest of the current theories about brain functioning falls far short of being able to show how the brain could even match present-day computers. The gap between the capability of brain models and the performance of the human brain cries out for some new and better understanding. How does the brain handle all of its incoming information? Perhaps the idea that some of the memory resides on these "stepping stone" molecules will help answer this question. Perhaps.

To have spoken at such length about sleeping and dreaming has been something of a digression. But as we should expect, understanding consciousness leads to our understanding many things that previously existed only in the murky crevices of science. Consciousness has become a glowing light casting its brightness into the night world of sleep and dreaming.

THE MEDIUM IS THE MESSAGE

Now let's go back to something we have looked at before, back to the consciousness data rates and to the consciousness field information capacity—things so esoteric and yet so close to what you and I are.

In the last chapter, we went into some detail to explain just how the quantum mechanical interconnection of synapses works and to obtain a specific relationship giving the minimum synaptic firing rate at which consciousness begins. We even went so far as to hazard writing a few simple equations to make these concepts exact. We needed it right there so that no one could charge that what we are doing here is vague handwaving. We also needed to be able to refer to those ideas whenever the need arose. Our derivation of an equation giving us the minimum synaptic firing rate for the onset of consciousness meant that we had to figure out just how many of the synapses were interacting with each other at any particular instant—and this resulted in an equation for the quantum mechanical consciousness information data rate, *C*, as shown in Appendix I. This derived value proved to be essentially the same as the 50 million bits per second that we obtained experimentally in Chapter 11.

Now we are ready to see where the individual consciousness images come from. As you will recall, we found that the presence of melanin in the brain gives us a minimum consciousness time interval that agreed with what we had obtained experientially: about 0.04 second. Now as data flow through the consciousness at a rate of nearly 50 million bits per second, all the data received in an interval of time equal to the length of the minimum consciousness time interval will be perceived as part of one image. We referred to this in Chapter 11 as the consciousness field information capacity, represented by the letter F. As pointed out in Appendix I, the product of the consciousness data rate, C, and the time interval gives us the value of F. The result is 1.9 million bits in a single consciousness image. That's our mind's TV picture! This theoretical value for the consciousness field information capacity is almost the same value as the 2 million bits we obtained in Chapter 11 introspectively.

And this, by the way, is why the TV set is the size it is. The TV set must provide enough picture information at any instant to supply this 2 million bits of information, and it must cycle these pictures 30 times a second to keep pace with the rate at which we consciously experience images.

There is one more fact that we need to mention, and this has to do with dream consciousness. During sleep, despite the fact that mental activity continues, the brain as a whole does not reach the level of activity needed for full consciousness. However, local regions of the brain that become more active may satisfy the conditions for consciousness onset for brief periods, resulting in a dream state—that is, dream consciousness. The level of consciousness attained, of course, is nearly always much below that experienced during waking consciousness, perhaps with a dream consciousness data rate, C_D, of about 10^5 bits per second. Such a low value for C_D may tell us a lot about where the content of our dreams comes from.

HOME TO KANSAS

Let us look at what we have accomplished. We set up the standards whereby we can turn the philosophical problem of consciousness into a scientific study. We determined what is to be "measured" or quantified to form the basis of our science. We proposed a theory—that consciousness is the result of a quantum mechanical process that involves the information-handling functions of the brain—and we tested this theory, first against the neurophysiology, finding that it accounts for the thickness of the synaptic cleft, and then against the morphology of the dark projections of Gray, the similarities in morphology between the ephapse and synapse, the gap thickness and electrical properties of the ephapse, and other esoterica that has been published in excruciating detail in the scientific literature.

Then we tested the theory against the new data, the experimental-experiential data of consciousness itself. We found that the theory could be used to derive an accurate value for the synaptic firing rate at the onset of consciousness, the consciousness data rate, the minimum consciousness time interval, the consciousness field capacity, and the characteristic extent of consciousness (as extending throughout the brain); characteristics of sleep, dreaming, and perhaps account for the Poetzl phenomenon. It looks pretty good. Considering that no one has ever been able to explain any of this before, it is really quite good—like wandering through fields of ripe golden wheat, back home in Kansas.

Little happened in the next weeks as the weeks played into the spring months. Little happened but that a youthful love matured, and yet remained youthful each day. A youthful love matured into its own minor crises. Gathering emotions turned our views of each other and of this new world we had found together into colored flashes of light like sequins glittering across a ballroom floor, dancing, waltzing, spinning us around and around into our future.

On Friday, February 8, the school's debating class spent the day in a city-wide debate on the issue "Resolved: All American citizens should be subject to essential service in time of war." The long shadow of the Second World War still lay on every political and social concern. I was not supposed to debate, but the absence of another student put me suddenly in the middle of the contest. As my diary records it,

> The first debate was unjudged, so I was all right. The second was judged, so while the first two spoke, I wrote my speech. I think I did better than I ever have. In the third debate the 'negative' team was horrid; their points were ridiculous. We should have won. After the debate, I called Merilyn. She couldn't play tennis because there were brush fires near her home. I called Tommy and we decided to go see "When Worlds Collide" at the Ritz with dates. I called MAZ. Tommy picked me up at 7:30. After the show . . . we parked under a beautiful moon (3). Merilyn again told me that she loved me. I said, 'I want to marry you . . . someday.' Merilyn said, 'In a million years.' I said, 'no.' She said, 'Well it will seem like it.'

This, of course, moved our relationship to a new plateau—with new rules. I was never much good at clearing the hurdles of societal circumstances that the finest of women occasionally place in the path of the men they would domesticate. My ineptitude had an early beginning and has flourished over time. Saturday evening I went to Merilyn's home again. A poem had begun to form in my mind as I walked to her house. The first thing I did when I arrived was to jot down some notes about the poem. It was an excellent beginning. Then, for no reason that I now can fathom, I told Merilyn the terrible secret that I had been keeping from her: I told her "about Maitland." I was not entirely adroit in my approach to this confession; I told her everything. I could have held back a few things, but I was young, I think I told you that, and I was naïve. If I have not told you that, it might help to keep that in mind. I was naïve. A clean breast of the entire "sordid" affair—well, as sordid as it could have been under the limited circumstances of my age and the almost constant observation and hostile supervision from her mother, who hated my guts!

I told Merilyn what I had felt about "the other woman." I told her what had happened and when it had happened and how it had happened and where. Where—that is always the last and always the worst. Where inflicts physical reality; it makes it all real. OK, so what if I was a kid? It is a relative thing. In our young lives, learning that the one we love has—strayed—was just about as bad as things could get.

But it had all ended. I told her it had all ended. The whole ugly unfaithful affair was over. And I told her how it had ended and when it had ended and where it had ended.

She said she had one thing to ask, "Did you ask her the same thing you asked me last night?" I said nothing. What could I say; I was dead.

"That silence must mean you did." I remember her words, all right. It may have been some two hundred years ago, or seem that it was, but those words have not been forgotten, not lost in time, nor dimmed by the passage of a hundred thousand hours.

But Merilyn was sweet, and she was forgiving, and she was something true. She did not scold me, or curse me. She could not have thought of such. Instead, she then told me about a dream she had had.

"I dreamed about the two of us last night and about my mother's [class] ring [that she had given me]. It was ten years in the future, and we were married. You came in the room and told me you had lost my ring."

The symbolism of the ring is at least in part obvious. She was concerned about whether I would still love her in the future—whether she, like her ring, would be lost. And of course the ring also symbolized her sexuality—the love she wanted to give but that she was afraid would be love lost. We talked on and on that night. But the dream had meaning, and it had come at a meaningful time, juxtaposed as it had been against my terrible revelations about the other woman.

We argued, but it wasn't about the other woman. We argued the evening away about religion, about Harry Truman, about the Vatican, about what good Baptists should be, and about Red China. In all the arguing, she won all the important points. She let me have the one about Red China. She cleverly let me have my way on this, that for all that, was for us an irrelevant cause.

When the arguing was over, she gave me something of a reward for having lost so well. She said, "Since you have told me about Maitland, I respect you more, and if possible, I love you more." I had lost. Surely, I knew that that evening I had taken a course that had no likely outcome but a tactical defeat. So I had lost. But I had lost to someone who knew how to win so sportswomanly.

The evening was not yet over. We still had time to make our relationship just a few moments closer and a time to remember, and so we did.

I went to Sunday school and church on February 10. Later I called Merilyn several times.

> Tommy Whitson came over about 7:30. He had his car, so we went over to Merilyn's. All of us then went to the Pig Trail Inn, where we got an Orange Crush apiece. We rode around on Shades Mountain while Tommy told ghost stories. Tommy suggested that we picnic next Sunday . . . the four of us.

But it was not to be. There were clouds on the horizon.

> Tuesday, February 12, 1952: I walked Merilyn to Homewood for her sorority meeting. When I got home I saw our new TV set. An RCA Victor 21-inch cabinet model—the Meredith!

Important things are, after all, important things.

The next day after school, I walked Merilyn home and stayed several hours. She was very loving. Later on I had an argument with my mother about getting something for Merilyn for Valentine's Day. I do not know what prompted this problem, but it was a disagreeable situation for me. And the weather, normally so good even in February, had become violent.

> There was lightning and strong wind. There was a tornado which hit the county and several nearby counties. Twenty-seven were injured here.
>
> On Thursday after lunch period, Merilyn made mention of someone telling her about my fight with Jimmy C. It spoiled my whole day.

Strange how these things from high school affect us. For years I couldn't think about this individual without having some of the most un-Christian ideas about how I hoped his whole life had been spent cleaning toilet bowls, or something else appropriate. But he is dead now—cancer, I was told. I learned of this just a year ago, at a school reunion. My reaction was sudden and instinctive. "Great! Couldn't happen to a nicer guy." Thirty minutes later I was embarrassed at my behavior. I should not have felt that way, and certainly I should not have carried *him* around with me for so long.

> After school Merilyn and I went to play tennis. I won, and she wore black shorts. Later I walked her home, and stayed (3), till six—and I walked home.

It had been a Valentine's Day.

Friday was my birthday. I barely made it to school on time.

> In Economics class, Mrs. Jones got mad at someone for disagreeing with her. She was writing a slip to send him to the office when I said, 'He's right,' and she sent me with him. Then two more were kicked out. Two of the group went back to try to reason with her, while I went on to the office.

I was allowed to cool my heels for two hours waiting in the office for the vice principal. Later, we had an ice cream party in Latin class. That evening, "Merilyn gave me for my birthday a small portable chess set."

Saturday I went into town to Loveman's department store. A girl's sorority from Ensley High was having an initiation on the corner in front of the store. The girls were lipsticked all over their faces. One was doing acrobatic stunts.

"Tommy called at twelve noon to say that if the weather did not get better, plans for the picnic tomorrow were off. Merilyn showed signs of a cold." Was this the first sign of what was to come? Was this the beginning of what her stepfather would later call "the trouble"? Did it all begin with just a virus months before anyone thought anything but that she had a cold?

On Sunday:

> I awoke at nine A.M. and started getting dressed for church. Merilyn and I had planned to meet at Trinity Methodist. She called at 9:45 to tell me that she could not go because of her cold.

I was still coughing myself from my bout with whatever it was that I had had. The weather had also turned too cold for the picnic we had planned with Tommy and Betty.

> I didn't go to church. I called Tommy at 1:15; he suggested that we go out target shooting, but I couldn't get a gun. I called Merilyn and talked for an hour.

The next day at school, "Whitson told me that Saturday night he and Betty almost left for Mississippi to get married." Later on, that is just what they did do. "I called Merilyn after school and talked to her for over an hour." But a person's feelings are complicated and mixed—always mixed. "I have been feeling in a philosophic mood today, thinking about the past and about Maitland. I wrote a letter to her." Still, it was just a momentary lapse, a moment of nostalgia: "Bed at 11:40 P.M. I love you Merilyn Zehnder."

> Tuesday, February 19, 1952: Substitute teacher in Economics. Substitute for Mr. Goodwin in Study Hall. Merilyn made me somewhat disgusted when she slapped me

> after I said an Apache is an Indian—between 5th and 6th periods. After school I walked with her to Homewood to her L/A meeting. I called her tonight and talked to her for well over an hour. I mailed the letter I wrote yesterday to Maitland.

On Wednesday I wore my brother's old Army clothes to school. I remember now how well I really liked that khaki outfit. But now I understand—it is always much better to wear someone else's old uniform than to have one of your own.

> We had a test in American History which I probably failed. I took an Algebra test and a Latin test. After school, Merilyn and I went to town. We stopped off at Five Points so that she could get a stub from her mother to get her watch out at Bromberg's Jewelers. We went to the library and then left for home (2+). I stayed till six and then walked home. I got a letter from Maitland.

What those letters contained, I do not know. None of them has survived, and my memory has not been able to reach back that far.

> Thursday, February 21: Weather, chilly this morning, warm this afternoon—clear. The B-36 bombers were flying over all day. During Algebra class we heard the air-raid siren being tested, two or three miles away. It is ten horsepower.
>
> Friday, February 22: I gave Mrs. Stephens a poem I had written for the school magazine—'Elegy to a Flea.' The senior class has sold $2,450 of magazines. Merilyn escorted a student from Jacksonville State Teachers College around school.

I remember to this day being just a bit jealous . . . and a bit insecure . . . and just a little proud of this beautiful *woman* of mine. But she was beginning to feel sick. "Merilyn was with the flu," I wrote. The words trivialized the situation.

> We planned to go out with Tommy W. I called Merilyn after school and then went to town to pick up my suit. When I got home, Merilyn's mother had put her to bed. I got mad at Merilyn because of the way she announced she wasn't going, almost as if she didn't care.

She probably didn't care—and how could I have blamed her? There may well have been more to her illness than just flu. She had a fever of 104 that night. I went to the movies in Homewood that evening alone.

On Saturday it rained all day. I called Merilyn at noon, at 3:00 and again at 8:00. "I got a letter from Maitland and I wrote her one. I have been trying to think of a plot for a long poem . . . one to take a year to write." Sunday I went to church. At 1 o'clock I called Merilyn and she asked me over. I took my second book of poems and a short story to her. When I got there her parents were out. Merilyn was still sick and was lying on the sofa.

> She looked so beautiful, her straight, shortcut brunette hair . . . said, 'I love you.' She smiled, quietly; she looked at me a long time and then she said, 'I love you, too.'

INDIVIDUAL IDENTITY, WHERE?

The picture of consciousness as a quantum phenomenon has relevance to another question that has long concerned philosophers. It is the question of who we really are in our own identity. The child is father of the man, but in what is either? What is there that stays the same? In much of religious thought, there is assumed to be a soul that embodies one's true identity. The body may perish, but the soul endures—a Christian theme. Hindus and Buddhists also believe there is some identity that persists, that carries Karma, perpetually, lifetime after lifetime—an identity that lies beyond an association with a specific material body. But there, the personal identity is thought of as an illusion that one must escape to reach a union with the true reality in *nirvana* or some similar ultimate transcendence of the individual.

Science, having grown out of a rejection of philosophical speculation on the nature of such questions as soul, has generally assumed individual identity to be identical with the body mechanism—or, more specifically, with the brain as a special-purpose machine-stimulus-response converter. Gordon Pask, a psychologist and philosopher, pointed out a second conception of identity that has become popular since the advent of computers. His is a view that arises out of the functionalism we discussed previously. He sees the brain as a general-purpose programmable machine and thinks of mind and consciousness as the collection of programs that inhabit this computing machine. Individual identity, rather than being attached to the brain-machine (the hardware), is thought to be attached to the functioning programs (the software of the machine), regardless of which machine is involved.

The concept of individual identity emerging from our quantum theory of consciousness differs from either of these two concepts. Though the similarities to religious ideas are only slight (at least at this point in our story), what we have in the quantum mechanical picture is closer to a conception of a soul-like consciousness inhabiting and animating the machine. (I didn't start out with this as a goal; it is just the idea that seems to work best at present.) The classical machine cannot have consciousness, and it cannot have any identity of its own. It is we, of course, who anthropomorphically imbue the collection of mechanical parts with its machine identity. But there is a transformation that takes place with the onset of consciousness. Something changes when the brain undergoes the transition to this new mode of functioning that lies outside the capabilities of all present computing machines. When this happens, we acquire our identity—an identity that exists in and as that consciousness state. Individual identity resides in the continuity of this quantum mechanical process.

Questions remain, however. Sleep interrupts the continuity of this quantum consciousness. But do we ever awaken again as the same conscious identity, or is a new consciousness created? Aside from the physical matter of my body (which is continually being replaced anyway) and aside from my memory continuity, is there something of me that remains the same from day to day? Or am I today as different from who I was as a conscious entity yesterday as I am different today from another person around me? Must we confront each night with the knowledge that "This night thy soul shall be required of thee?" Do we assume that we remain the same consciousness from day to day because we retain a particular set of memory traces, or because of some continuity of consciousness maintained despite the interruption of sleep? And, if the latter, how?

From what we know at this time, there appear to be only three possibilities. First, it is reasonable to assume that when the brain goes to sleep, a portion of the brain retains a sufficiently higher level of activity where consciousness resides so as to hang on to the same identity. It functions as the rest of the brain sleeps. In such a brain function, we might reasonably assume that none of the usual thought or memory functions from the waking state are maintained, and we might also assume the consciousness data rate to be significantly depressed, resulting in a consciousness more like that of one of the lower animals. This assumption is not entirely unreasonable, since it is known that during sleep there are parts of the brain that go from a low level of activity to a higher level. This answer would account for consciousness continuity, but it leaves us with even more questions. What, for example, is the purpose of such a soulless personal identity? What does it accomplish for us that is not adequately achieved by the continuity of memory traces? Why should such a mechanism of personal identity have been a concern, as it were, of evolution?

A second possibility is to be found in the Heisenberg uncertainty relations. As we saw earlier, because of these relations, the concept of the path of a system ceases to have the kind of meaning it had in classical physics. Thus in quantum mechanics, identical particles are said to be indistinguishable. As a consequence of this principle of indistinguishability, an exact replica cannot be distinguished from the original.

You will remember that in the case of thermodynamics, it is possible to determine exactly how the collisions of atoms and molecules bring about gas pressure on the walls of containers. But to make these thermodynamic calculations, it is necessary to include every possible collision that can happen. We need to know how many times atom A collides with atom B and how many times atom B, coming from the opposite direction, collides with atom A.

But whereas we can color marbles to distinguish red balls from blue balls, this cannot be done with individual atoms. One atom of oxygen-16 (its isotopic number) is just like another atom of oxygen-16. The interchange of atom A for atom B, where both are oxygen-16 atoms, does not give us something new. It gives us exactly the same thing. As a result, when we calculate the force on the wall of the container, we have to count this kind of collision in the gas only once, not twice. The atoms' indistinguishability, therefore, has a real effect. This means our picture of reality is wrong. We imagine these two possibilities to exist when in fact they do not. The two are not only indistinguishable; they are, in fact, identically the same thing.

And this error in our picture of atomic reality may have something to do with the question of our own individual identity. We have seen that our consciousness really boils down to what some electron is doing in our brain, so perhaps the same arrangements of energy states, though separated by the hours of sleep, are really indistinguishable. If the quantum mechanical state of some portion of the brain at the inception of sleep were to be identical to that of a part of the brain at the moment we awaken, then in spite of the temporal displacement, perhaps continuity would still be maintained. This can only be offered as a possible solution to the puzzle of the seeming continuity of individual identity, and even at that, it would seem to have only a slim chance. But it is a possibility that perhaps in the future we can subject to a more rigorous analysis.

The third option is that when we awake, we are indeed someone new. It may be that the *you* who now reads this, perhaps just before retiring, are about to pass into oblivion. It may be that each morning a consciousness is born, lives one day, and dies to eternity—no soul, no greater existence, no further purpose; a legacy only in what *you* pass to tomorrow's inhabitant of your borrowed body.

Maybe there is a fourth possibility. You will have to read further to find out what it is. Perhaps you should keep on reading; postpone that nap. Consider the possibility that oblivion is only a few moments away, but hope that perhaps there is another answer that awaits you in the next chapter.

Consciousness may be associated with all quantum mechanical processes in nature. What is special about our own consciousness is that it is a part of a vast logic machine, which in turn is the brain of a particular kind of physical system, a living organism. Life, thought, and consciousness are three separate things. An organism does not have to be conscious or to be a thinking machine in order to be alive. A brain does not have to have consciousness in order to be capable of thought (that is, data processing or computing). Any of these attributes may exist independently of the others or in conjunction with only one of the others. A nonliving computing machine that is capable of both thought and consciousness is thus a real possibility for the future. And let us hope we know what we are doing before we create it.

Consciousness may also exist somewhere without being a part of either a living body or a data-processing system. Indeed, because everything that exists is ultimately the result of one or more quantum mechanical events, the universe is inhabited by an almost unlimited number of rather discrete, conscious, usually nonthinking entities. These conscious entities determine singly the outcome of each quantum mechanical event, whereas the Schrödinger equation constrains their freedom of action collectively.

For the first time in our history of wondering who and what we are, we have a way to find the answers. We have a way to know the truth, and we have taken a step toward the realization of that hope. We can see that this quantum mind of each of us is part of the fabric of reality.

But there is another part of the mind, something beyond consciousness, something incredible, something that is already within our grasp. It is something that shows why our minds transcend being machines. It is the true wellspring of our identity. It is our link to the infinite.

14

A Matter of Will

. . . a still small voice.
—1 Kings 19:12

With our discovery of consciousness as a fundamental constituent of reality, we have moved out of the twilight of classical mechanics. We have opened a door to a radiance of enlightenment. Now ahead, another door, another wonder waits for us.

A parting kiss, an inscription chiseled in a marble headstone, a laser beam flickering a code from earth to the moon—all are the same. Each is a message, cut into pieces of matter and configured according to some collection of rules, to transfer information to others at other places and other times. The kiss lasts but a moment. The letters carved in stone will last a hundred generations. The laser signal may last an instant or, traveling into space, go on forever. But as different as each of these is, each is a particular way of using a configuration of matter to represent information. This is the classical approach to the representation of information. It is this kind of information expression that we use when we write books, and it is the kind of information representation that computers use in their lightning fast calculations that always yield one deterministic final answer. But there are other kinds of information.

On the basis of what we know about quantum mechanics, we can see that there are two more unique ways in which information can be handled. Specifically, we can represent information in terms of the potential states in quantum mechanics that can come into being as the outcome of material interactions—the potential states of an atom that becomes excited by radiations that strike it for example. The condition of such an atom is given by the transition probabilities that can be calculated from the state vector that represents the excited atom. The state vector description of the atom as given by quantum mechanics, therefore, contains information about the physical system it describes. Thus any quantum mechanical system has this second kind of information.

A third kind of information is also at work in the brain. As we have already seen in quantum mechanics, something eventually has to occur. One potential state becomes the actuality, the observed event. One synapse does fire. Something does happen. It is this selection of which event will occur, out of all that could occur, that is the third type of information at work in the brain. This third type is distinct from the other two, and yet it is so intimately entwined with the state preparation that gives rise to consciousness that it is difficult for us to distinguish and recognize it.

This third type of information describes what *we do*. It is the information measure applied to state vector collapse, the thing that *becomes* from out of the realm of what can be. Here, as part of the operation of the brain, our consciousness observes the brain in terms of what can happen and then, as observation of the quantum mechanical state brings about state selection, brings into being the one state that actually occurs from the range of the possible. From this description, we can immediately identify what this process is. This is the *will* of the conscious mind. It is everything ever meant by the concept of will. *Will* is that third form of information creation. It is state vector collapse as one synapse—out of all that could fire as the result of a remote interaction in the other synapses—does fire. Thus we see that mind actually has two parts: consciousness and will.

There are three processes in the mind and in the brain that involve information processing. The first consists of the classical unconscious brain's computer functions. The second is a quantum mechanical sampling of the first, the collection of potentialities that may follow that gives us our conscious knowing/experiencing of what the brain does. The third is that which goes from what could be to what is. We have opened the next door.

WHERE THERE'S A WILL

It is important for us to be quite clear about what we mean by will. We must establish a sound basis for our claim that the occurrence of state selection in association with consciousness in the brain's functions should be identified with the will, a concept we are borrowing from philosophy. Too many have talked vaguely about will. We must make it clear that we have not merely forced a tired and discredited concept from seventeenth-century philosophy into this story about consciousness. We must show that the facts justify our use of the word *will*.

In the last century, as we have noted, James Clerk Maxwell brought the rigor of mathematical formulation to the imaginative "pith ball" and "ice pail" experiments of Coulomb and Faraday. He found that when he wrote these equations so that they applied to the electromagnetic field for free space, the equations looked like something physicists were already familiar with. They described a wave motion that propagated through space at the speed of light. Maxwell had found a formulation that fit all the requirements for describing light. He had discovered that light is made out of electromagnetic waves.

This is a model for us. Maxwell determined what light is by showing, via mathematics, that electromagnetic phenomena have all the characteristics of light. Now what are the characteristics of will?

There are several. First, the word *will* refers to a state of mind or a capacity of mind. It is in some way associated with conscious experience. The philosophical concept came into existence because we have a direct sense that "free choice" is a human capacity. Second, for will to have any meaning, it must be possible for the mind to affect events—for the mind to control the body. The concept of will is not compatible with the classical conception of physical processes. Classical physics would demand that nature grind out blindly and automatically the consequences of any initial action. Any mind attached to such an automaton would be only a passive observer. Such a mind would not be able to control any aspect of its body's behavior. It would be a captive bird in the brain cage, and there would be nothing to call "will." Thus the concept of will demands that before the mind comes into play, before the mind acts on matter, the physical laws must allow—must specify—a range of potentialities as to what the body could do, and the process that selects from that range of possibilities which possibility will happen must be clearly outside the prescriptions of the physical laws. That is to say, the physical laws must be underconstrained.

Finally, the concept of will requires that when the mind does interact with the brain, the physical brain and body will then do one of those things that physics permits so that the thing the mind willed becomes the state of the brain and the action of the body.

Like Maxwell looking at his new differential equations, we too have seen all this before. We have already seen in quantum mechanics that before a measurement, before observation, the state of the system is described as a collection of possibilities. We have seen that the laws of physics as given by quantum mechanics are underconstrained. But we have also seen that after observation, one state (one of the allowed potentialities) does occur. In the case of the brain, we have described what the quantum mechanical process is and how a range of possibilities arises—namely, in the particular synapses that have the potential to fire. We have shown that consciousness is associated with the process of creating these possibilities and that, therefore, when observation takes place—when one state is selected—

that process occurs in association with our consciousness. Indeed, just as for consciousness, there will be a data rate, a will channel capacity associated with the process. We will presently calculate just what the channel capacity is for this will process.

Finally, when the "observation" happens—when state vector collapse occurs—one synapse, from all those that could have fired, does fire. And the state selected by this synaptic firing, by this process associated with consciousness, specifies just what the brain, and consequently, what the body, will do next. This observation process brings our brain's next thought and our body's next action into being.

This is a perfect description of will. Like Maxwell discovering that light is an electromagnetic wave, we have found that will is quantum mechanical state selection going on in the brain.

This is something doubly remarkable. It is remarkable, certainly, for any scientific investigation to reveal some of the details of this "machinery" of mind that lies quite beyond the realm of ordinary brain or of physical functions, but it is equally remarkable that philosophers have realized that will must be a part of our mind—that the feeling of free will should have implications for the nature of body and the world around us as well. It is unfortunate, however, that confronted with the conclusions of a more limited science, philosophy has come so close to abandoning its finer moments, often only to become a mere echo of dated scientific notions.

Let us turn now to find out just what the numbers are, for doing so is going to reveal even more about our own reality, the nature of will, and the much greater reality that lies beyond. We have only cracked open that door we spoke of earlier. Now the light is about to flash through that opening. There are surprises ahead.

As I consider the mind and the distance of its reach, I have to believe that somewhere out there, something of that beautiful young mind still remains, something of that life still lives.

It is nevertheless a difficult belief to hold. I wonder about her survival; I wonder what is left of her. And then I read my own words from all those years ago. I read what I wrote and I wonder what is left of me.

"She seemed a bit less than cordial . . . ." Merilyn's illness was not over; very possibly it had just begun. She was not very cordial that Sunday afternoon nearly 50 years ago. She would have been old and perhaps tired now, but instead she is 16 and a memory and perhaps nothing else.

I was there as often as I could be there, and when not there at her house, with her at school, or going about town with her on some errand, I was on the phone with her, whiling the hours away.

That gets to parents. Parents eventually begin to apply pressure, and the pressure eventually has its expected effect. The pressure probably did play a role in what happened.

On Monday, March 3, 1952, in English class, we had to write an essay on "What I will be doing in ten years." I wrote, jokingly, that by then I would be on Mars! My diary does not record how my humor was received.

Mama (my grandmother on my mother's side) came up from Brighton that Friday to stay with us. Mama was old and withered away, little more than four feet was left in her hunched, decalcified body. She was frail and so very old and so very loving. The story was that my grandfather had married her because she was the only girl in Bessemer who would sit on the porch with him and chew tobacco. I never asked her about that myself. Mama came to visit, and while she was with us, she would dust and clean and do what she could do to still feel that she was someone who had some use. Eighty years of dusting and cleaning, I guess, and that was her use.

On Saturday, Merilyn and I planned to meet in town in front of the Darling Shop in Five Points. But she had already gone by the time I got there, just five minutes late by my watch. I had to scamper off after her to catch up with her at Mrs. Pumpfry's, where she took her music lessons. I now remember absolutely nothing about Mrs. Pumpfry—where she lived, what her house or studio looked like, absolutely nothing. But my diary tells me that I called Whitson from Pumpfry's, hoping to make some Saturday night plans with him. He wanted me to help him get out of the house so he could go see Betty. We were not able to come up with anything workable. Merilyn and I left Pumpfry's and went to the library. By this time, however, we were having an argument. I was still irritated that she had not waited for me a few more minutes in front of the Darling Shop. She was irritated with my being irritated and, though she didn't say so then, with the fact that her mother was putting pressure on her about seeing so much of me. The subtleties were lost on me. Subtleties are not a strong point of youth.

These little tiffs didn't last long:

> Sunday, March 9, 1952: I love Merilyn. Went to Sunday School, then caught the No. 43 bus to Trinity Methodist with Merilyn. We walked to her house afterwards. Later, I went to the Frat meeting. We planned a hayride/rush party for next Saturday night. I was put on probation for missing so many meetings. I called Merilyn at six. I asked Merilyn to go on the hayride.

On Monday I wrote,

> Merilyn did my Economics homework. We saw segments of the show 'Romeo and Juliet' in Large Auditorium for English. Sarah Emerson's father gave Tommy and me

a ride to Tommy's. Merilyn had to stay at school to practice for Comargo Club. I have had a headache and felt generally bad today. I called Merilyn from Tommy's. I ate supper at Tommy's. He told me that he had told his father this morning that he was going to marry Betty. I called Merilyn when I got home. I wanted to tell her that Tommy and Betty were talking about getting married. I told her about being sick and she told me to 'get to bed.'

So I was sick for the whole following week. I talked to Merilyn on the phone, had a temperature of 102, and got penicillin shots when the doctor came to the house. I missed the Wednesday night Frat meeting, which made the flu more acceptable, and on Friday, March 14, I got a letter from Maitland. She was "sick." My diary says cryptically only that "she knocked herself out." On Sunday I wrote, "My throat is still sore. I called Merilyn later today. I have missed Merilyn a great deal."

I went back to school on Tuesday the 18. My diary was again filled with school-day trivia:

Sixth period English we saw the first half of 'David Copperfield.' Latin class: had a test—we had a supply teacher. On Wednesday the supply teacher was late coming in. I rode a tricycle around on the stage of the auditorium. That afternoon, I played tennis with Merilyn (winning of course), and then I walked her home (3). I left at 6:15 P.M. She acted very much as if she wanted me to leave, and it made me mad all the way home. I walked. When I got home, I called her, and this time I told her how she had been acting toward me. She cried, and she apologized, and she said that she had not meant this.

I guess I said then that I was sorry that what I had said made her cry. I hope I did—I hope at least that is what happened then.

Thursday afternoon,

I called Merilyn about going to play tennis. We did and I won 3–0. We had to stop because I broke the strings out of my racket. Went to her house (3) and stayed till 6:30. I walked home.

Friday, it was raining. Merilyn was in a bad mood. I called her at 7:00 and she asked me over.

We 'listened' to TV. The weather was beautiful this afternoon and tonight and particularly tonight. [Saturday,] I called Merilyn. I called her at 4:00 and again at 7:00. I didn't go over to see her. I walked to Homewood to buy a copy of *Collier's Magazine*. Merilyn called me at 8:15. I fixed up our entries for the high school anthology of poetry. The weather has been cloudy all day with short heavy rains.

On Sunday we met at Trinity. Merilyn was upset. She told me her mother had gotten after her about going just with me. She said, "I love you. I love you, and I want to be with you. I don't want to go out with anybody else. Mother doesn't like me being with you all the time. She wants me to stop seeing you. She wants me to go out with other boys. She's afraid I'll get too involved with you." It was sleeting when I left her. I had walked her home, sharing an umbrella against the freezing rain as we climbed the hill to her house—close together and worried about our future—not realizing the real future held worse tragedies than the ones that worried us then. She had cried when she told me, "She wants me to stop seeing you," and the rain had turned to sleet as I left her. By Monday, the sleet had become snow that covered and quieted the hills of north Alabama.

For the next couple of days, it was more difficult finding opportunities to be with Merilyn. Her stepfather and mother were going with her on the trips we usually took together. But we were still able to be together. A little less often, but we were still able to be together some. Wednesday night and again on Thursday evening, we were together. We had a wonderful time—a wonderful time, and we played, and we loved, and we laughed, and we cried, and we parted.

THE MEASURE OF WILL

You will remember that in Chapter 12 we talked about a quantity we simply called Q. It's a number. It tells us how many other synapses one synapse can influence quantum mechanically. One synapse has the ability to cause to fire any one of the many possible synapses associated by means of the quantum mechanical interaction. You will remember that we calculated the value of Q and we found it to be 200,000. Now the chance that any one of these synapses will fire is 1/200,000 (that is, $1/Q$). Using this probability of $1/Q$ we can easily calculate just how much information[1] is "input" every time one synapse fires by means of this quantum mechanical interaction. The result is 17.6 bits of information. Now this happens at a rate of once every 0.3 milliseconds. Therefore, we can get the information rate by dividing by this time. The result is just under 60,000 bits per second.[2] That is the channel capacity of the will. This will is what we really are. Let me say that again. This will is what we really are. We will soon find this *will information capacity* to be an extremely important quantity, so let us call this number W, just as we have called the consciousness channel capacity C. C is what we experience; W is what we are.

There is also another quantity that we will find to be quite important. It is the amount of will information in one "frame of the film," so to speak, through which we see ultimate reality. We showed in the last chapter that there is a characteristic interval of time over which our consciousness is smeared, a time of about

0.04 second. This is the length of the tick of the consciousness clock. You will remember that in Chapters 11 and 13, in addition to the consciousness data rate, we obtained a quantity, *F*, which we called the information content of the consciousness field at any moment. The consciousness channel capacity, *C*, consists of frames or images with data content, *F*. Correspondingly, for the will data rate, *W*, we calculate a *will field*, which we denote by the letter *G*. We get the value of *G* by multiplying the will data rate by the consciousness time tick of 0.04 second. This gives for *G* a value of nearly 2400 bits. Make a note of these four measures of the mind: *C*, for the consciousness stream; *W*, for the will stream; *F*, for a frame of a conscious image; and *G*, for a moment's will, an act of will, if you like.

Let's summarize the values of these quantities, because we will be referring back to them:

Consciousness stream	*C*	47.5 million bits/second
Will stream	*W*	58.7 thousand bits/second
Consciousness moment	*F*	1.90 million bits
Will moment	*G*	2.35 thousand bits

Of course, the accuracy of our calculations does not warrant the precision implied here. We have already pointed out the limitations in obtaining accurate values for these quantities. Values will vary from individual to individual by as much as perhaps a factor of 2, but listing the exact numbers removes any temptation to stretch them elsewhere when we are using them to explain how some other phenomenon works. It will help us to see how well (if that is how it turns out) we have explained what we observe in the real world.

This will data rate was not something we had intended to figure out at the beginning of the work. Instead, it is something that has been thrust upon us as a result of our efforts to understand consciousness. This means that we have to look at how this aspect of the mind ties in with physical processes in order to understand just what capabilities the will possesses. The will is indeed a part of our mind, but it has other aspects as well. The will is the channel that determines what our next move, choice, and thought will be. It selects the path our mind takes through the images of things the brain scatters before us. The will is our innermost nature, our being that is there even when the things we might see go blank. It may even be there when all else is gone . . . but for that, we must await more understanding. It is perhaps this aspect of the mind that comes to the fore when one is in deep meditation, a state of consciousness designed to remove thoughts and sensory contact with the world.

But there is something more. From what we have already seen in the past chapters, from what we know to be the nature of state selection on observation in the quantum mechanical process of state vector collapse, and from the surprising characteristics proven in the tests of Bell's theorem, we should immediately see just what *W* means and what we have in the will data channel.

Before an event occurs in quantum mechanics, objects have interacted and entered into a set of potentialities: the states of quantum mechanics. After observation, on state vector collapse, everything goes into one state. Everything. No matter where the objects are. This is the incredible discovery that the tests of Bell's theorem prove. When a particular synapse, among the 200,000 possible synapses, fires, everything else—everywhere in the whole universe—that might be tied to that choice by the will channel must also go into that same overall quantum mechanical state! Unlike the consciousness, our will is not restricted to the tiny space of the brain cavity. Everything we touch, see, or experience in any way that becomes caught up in that special aspect of the quantum nature of matter will retain a link back to our mind. And observation of it or our subsequent observation of its consequences can impact on the state that it is in. The will channel is a link that transcends space, and because physicists have found that something called Lorentz invariance always holds,[3] this will must transcend time as well.

We will make these rather imaginative notions about will much more concrete presently. It must be kept in mind, however, that this thing we have quantified is caught up as part of our brain's functioning and is also tied to exactly the central point tested by Bell's theorem. Nothing we have done, nothing physics has proved, takes the events of the physical world out of the domain of quantum mechanics and puts them into the realm of that billiard ball physics of the classical or objectivistic conceptions of reality. It is by means of the quantity *W* that we as observers determine state vector collapse in the things we observe, and conversely, those external events are tied to the states our brain becomes.

What we are saying is that this will channel causes the events in the brain and those in the external world to go hand in hand with what happens in our consciousness. We are saying that our mind can affect matter—even other brains—and that distant matter and minds can have an effect on us. What we have here, what is forced on us by the formalism of quantum mechanics itself, is something that sounds like telepathy or psychokinesis; it is exactly what Einstein feared to be the implications of quantum mechanics. Why is it so elusive? Why does it seem so strange? Because the signal, *W*, is small compared to the noise of our everyday consciousness given by *C*. That is, because the signal-to-noise ratio, *W/C*, is very small. What we have found in our quest for the tangible fabric of reality has carried us past objectivity, beyond mind even, and into the realm of things paranormal!

In all of this, the observer is tied to all observers, and these observers collectively select the reality that occurs. But much more, the observer—the perfect observer—always gets what he or she wants. Indeed, for the perfect observer, there cannot exist even the concept of desiring a state (from among those that quantum mechanics permits to occur) without that state at the same instant becoming reality. We do not explain how these state selections happen—at least not for a few more pages. Rather, this is the nature of observation. What we can point out, however, is that we are not perfect observers. Our desires that are a part of our consciousness are not always a part of the will channel that could bring them into being. We do not always have, as the mystics might put it, a perfect purity of mind and heart.

Instead, our ability to distinguish the will channel, from all the thoughts of our consciousness, as a means to affect outside occurrences is nearly nil. This is the discovery locked up in the numbers for *W*, the will channel, and *C*, our consciousness channel. Both of these channels are a part of the mind. Both stem from the same kind of things, quantum mechanical events going on at the synapses. Our will images are so dim compared to those of our daily conscious existence that they are almost always lost in the torrent of our consciousness stream. You see, the quantity *W* measures the channel capacity that can affect things and events, directly and globally, through space and through time. But the quantity *C*, our consciousness stream of information about the world here and now, is, as far as such efforts are concerned, just so much noise. We may experience our desire to influence an object or contact the mind of another. It may enter our consciousness *C*, but it will have no effect there. Thus *W* is the measure of the signal that can do these things; *C* is the measure of the noise to which we misdirect our efforts; and *W/C* is the signal-to-noise ratio. Its value is small: 0.00124 using the values of *W* and *C* we gave earlier. It says that if one tries over and over to determine the flip of a coin ahead of time, one probably will succeed by means of this perfect channel only once in a thousand tries. All else will be the result of chance. But that one try is not a try. The wish is the completed fact. The word is a perfect act.

15

Quantum Miracles

There were mutterings in some quarters that Walker was smuggling pantheism into physics disguised as quantum psychology, but many younger physicists—especially the acid heads—accepted the Walker solution.

—Robert Anton Wilson, *Schrödinger's Cat*

Modern physics gives us such a complete picture of the world, and it is now so thoroughgoing, that anything violating its basic principles would not simply be miraculous—it would simply not be. And yet nonlocality and will are both part of what mind is all about. What, then, truly are the bounds of mind? Can the quantum mind alter matter? Can the entangled quantum states of consciousness see into other minds and beyond the limits of space and time? Can these things happen despite the fact that they seem so far removed from what we commonly understand to be the nature of the world?

These things seem to fly in the face of the laws of physics. Yet these questions now stand at the frontier of physics thinking. And, indeed, at the atomic and subatomic levels, such things are in fact quite familar oddities. To understand the subatomic world, one may think of electrons, protons, mesons, and neutrons constantly vanishing from one spot and appearing in another. Alpha particle radioactivity, the kind of radioactivity that usually takes place in radioactive materials such as radium and uranium, is due to exactly this kind of happening in the atomic world. There, the alpha particle disappears from inside the nucleus of an atom and then suddenly appears on the outside, carrying away excess energy from the atom in the process. This is the quantum mechanical tunneling we talked about earlier in connection with the firing of synapses in the brain and as involved in consciousness. The particle does not travel from the inside to the outside; it translocates. It disappears from one place and then, in the next moment, appears somewhere else. The laws of physics not only permit this odd translocation of matter, but it is ingrained in the very fabric of these laws. At the very heart of reality, that is how things work. This "translocation" is not the exception; it is the rule. It is actually what quantum mechanics substitutes for what we are accustomed to think of as "path."

These things happen all the time in the atomic world, but for ordinary objects they are so incredibly unlikely to occur that physicists commonly regard their occurrence in the macroscopic world around us to be impossible. The chance for the smallest dust particle that can be seen glinting in a beam of sunlight to translocate by only its own diameter is so small that were the whole earth to be examined, not one such dust particle would be found ever to have made such a jump—at least not if chance quantum jumps alone determined what happens. What at first sight would seem an easy explanation of spooky things in our world turns out to give us little real expectation. An enormous chasm would have to be bridged in order to usher such things into the everyday macroworld.

But that is not the end of the story. To understand what actually goes on, we must first understand how this minuscule world of quantum wonder might affect the world we experience. Then we must understand how the consciousness and will that we have already shown to have a quantum mechanical nature can affect matter by manipulating the indeterminacy that underlies every physical event.

QUANTUM JUMPS

We have mentioned this before, but it bears repeating. It is the matter of what the term *quantum jumps* means. We have just used this concept because in the context of the motions of atomic particles, it is commonly used to depict graphically the implications of quantum mechanics. But as with so much that is said about quantum mechanics, this word picture, quantum jumps, falls short of faithfulness to the physics.

If any two atomic particles collide, they do not bounce away with a single possible outcome; they bounce away with a collection of possible outcomes. This is what we found out earlier in this book when we talked about quantum mechanics and about Schrödinger's equation. All these possibilities exist simultaneously as potentialities, and when a measurement or observation is made on the object, it "jumps" into one of the possible states as an actualization. If we could observe the atomic world, we would see something that indeed appeared to be the object jumping from one place to another discontinuously, but the reality is different. This is why the Schrödinger equation is so important to us here: It lets us see what the reality is behind the appearance of things. The Schrödinger equation lets us see that these jumps are not there until our observation creates them. Hence we observe the object in one location when one particle of light (that lets us "see" the object) arrives to indicate the object's location. Then, after its encounter with the particle of light, the object moves off like Lord Ronald who "flung himself upon his horse and rode madly off in all directions" at once. The object would move off in an infinity of possible paths—all real but existing as potential states. Then the next particle of light illuminates the object, and our observation collapses these possibilities—collapses the state vector into one state, one actualization. We see this as a jump. What we observe, therefore, looks like objects jumping about as though we were looking at things with a syncopated strobe light with everything jumping and dancing wildly about.

If, however, we look at things not on the atomic scale but on the macroscopic scale, most of these jumps average out so that on the large scale we usually see a smooth motion, the classical smooth clock-like motion of planets gliding in their orbits about the sun, or billiard balls rolling deterministically across a pool table. That is what we usually see, usually.

DIVERGENCE

Usually. There are times, however, when these possibilities, these quantum mechanical potential states, diverge. Rather than things looking smoother and smoother on the macroscopic side, the possible states diverge from each other so that macroscopically distinct events become possible. Indeed, if it were not for this, we would not even know about the atomic world of quantum mechanics.

To see how this can happen, let us look at something that would seem at first to be entirely in the macroscopic domain. Let us look at what happens when we throw dice on a craps table. We can follow any cube as it impacts the table. If we know exactly the velocity (speed and direction) of the center of the cube, if we know the angular velocity (rate of rotation and axis of rotation), and if we know exactly when and where the cube first impacts and its orientation angle when it hits the table top, we can calculate exactly the motion of the cube after the collision by using ordinary classical laws of mechanics. It is then possible to follow the bouncing motion of the cube from there on. This sounds like a lot of work, but it is basically quite easy to do. The calculation is tedious, but there is nothing exotic about it.

Now quantum mechanics, specifically Heisenberg's uncertainty principle, puts a limit on the precision with which we can know all this. We will not be able to know both how fast this die is traveling and exactly its position on impact. In principle, we can know either one exactly, but not both. Similarly, we cannot know exactly both the angular velocity of the die and its orientation on impact. If we cannot know all of these exactly, then we cannot know the future motion exactly, or, to put it differently, the future motion will then be represented by a dispersion of states having motions that fall within these uncertainties.

Let us see what effect this has on our following the entire motion of the cube. Any uncertainty in the velocity on impact will lead to an uncertainty in the velocity computed after the impact. Now when we take that inexact velocity and use it to calculate the position of the cube on its next impact, the uncertainty in velocity also leads to a further uncertainty in the future position on impact. Or, if we assumed we knew the velocity more accurately to begin with, we would have a greater position uncertainty, according to the Heisenberg uncertainty relations, and this would lead us to uncertainties in the future velocity of the cube.

The situation is similar for the angular velocity and orientation angle. Uncertainties in the angular velocity lead to uncertainties in the orientation angle, and vice versa. And, of course, uncertainties in the velocity or position will lead to uncertainties in the computed angular velocities and orientation angles on the next bounce. To make matters worse, these uncertainties grow with each bounce of the cube.

According to the Heisenberg uncertainty relations, we can know any of the quantities exactly on the first bounce, but if we attempt to calculate any of these beyond the second bounce, uncertainties exist in all of the quantities. For example, for a typical die, the uncertainty in the orientation angle by the second bounce will be about 10^{-13} degree as a result of quantum mechanics alone. And that uncertainty grows with each bounce, until after some 50 bounces, the orientation of the die is entirely unpredictable. Even though at the beginning of the throw, the object may be considered to have, for all intents and purposes, an exact location, speed, spin, and orientation, by the end of the bounces, that die has no single orientation. It exists as the "linear superposition of states." It exists as though it had all possible orientations at once. Only our observation of the object leads it to take on one out of all its possible orientations and come to rest with one of its six faces up. It is this quantum mechanical behavior—this Heisenberg uncertainty—that is resolved by observation and leaves the die in one place rather than another on the dice table. And this too lies within the reach of what the will of the mind can touch.

THE EHRENFEST THEOREM

What we have just talked about, the growth of quantum mechanical uncertainties from the atomic arena up to a level we can easily interact with, may seem at first to violate an important idea that was proved by Ehrenfest. From the beginning, the founders of quantum mechanics—Heisenberg, Schrödinger, and Bohr—realized that this new theory of matter would have to agree almost precisely with the physics of Isaac Newton when we get to the large-scale arena, the objects of everyday life. Ehrenfest was the physicist who showed how this happens. He showed the conditions under which Schrödinger's equation causes objects in the macroscopic world to obey the laws of classical physics. The usual jumps in the motion of objects at the atomic level (that is, the differences between the states allowed by quantum theory and those of classical physics) get smaller and smaller as objects become larger until they would seem to have vanished from the macroworld altogether.

But as we observed in Chapter 7, if that were the case under all conditions, we could never know anything about quantum mechanics at all. The exceptions to Ehrenfest's theorem concern the behavior of objects subjected to rapidly changing force fields, situations in which objects are suddenly and rapidly accelerated. Under these conditions, the range of possible motions of the object do not narrow to the classical limit but diverge. But this is exactly what we have to have in order to amplify our atomic measurements so that we can see what goes on in that microscopic world, and it is just what happens in the case of the bouncing dice. In both cases, rapidly changing forces increase the dispersion in the states that describe the allowed conditions of the objects.

Thus the Ehrenfest theorem agrees with what we have already discussed about the growth in position and orientation uncertainty that becomes macroscopic with successive bounces of the dice.

LOOKING AT THE DICE

Until a measurement is done, until we look at these dice that have bounced across the table, the correct description of what is there is this superposition of possibilities. It is as though there were a book of pictures showing each possible turn of the dice and each possible placement of the dice on the table, one to a page. This entire book of pictures would then be the correct representation of the reality already out there before we even look at the table. But when we look, one of these possible events occurs. We find one reality with the dice in one and only one of the possible locations, and with each die showing one observer-caused face up. It is as though at the moment we look, one page is snatched at random from the book of all possible events and then becomes real.

But what if we were to assume one small change in this way of looking at the reality that quantum mechanics describes? What if we were to assume that we could look through the book first before pulling out the page that occurs? Well, in that case, within the range of what the physics allows—that is, within the selection of pictures in the book—we could get whatever we wished. As we discussed in detail earlier, we know that state vector collapse—that is, the selection of one of the pages from that "state vector" book—occurs because of observation. Isn't it therefore a reasonable hypothesis to consider that a correlation may exist between the state the observer causes to happen and the state of the observer who selects that state? We already know that quantum mechanical state selection is nonlocal. Everything must be part of the state that observation selects. Even the states of the brain, responding to what is seen on the table when the dice come to rest, must be a part of each picture drawn on the pages of that book of potentialities.

This being the case, isn't it really the *hidden variables* of our consciousness and of our will that do the state selection—that create the events of the next moment we see?

We have already seen how quantum mechanics tells us that objective reality actually exists as a collection of potentialities like pages in a book, and we have seen how tests of Bell's theorem show that these states are selected as whole pages nonlocally, irrespective of spatial relationships. We have seen that the principal interpretation of quantum mechanics says that this state selection process (the page pulled from the book) is caused by observation, which ultimately means the consciousness of the observer. And we have seen that the consciousness is a quantum mechanical process that has associated with it a will channel that connects our consciousness experience to those events in the outside world to bring about state vector collapse. The will selects the state of the brain that we consciously experience, and the global nature of quantum mechanics of necessity links this brain state to the external event that occurs.

This ties everything together, and it says just how the mind can affect things in the outside world. We have seen how everything is connected, but how is it that an event that the will wants actually occurs? The answer to this is rather different from what one would expect. The event does not occur because we punch in the result we want like using a TV remote control. Instead, what the will selects is that which *is*, and that which *is* is that which is the will. These two cause each other reciprocally. In a way, they are really the same thing.

Now this surely sounds like Zen talk, but let us examine what it means. It is a lot like what Isaac Newton said about how forces act and react. Newton said that "for every action there is an equal and opposite reaction." The phenomena that are controlled by one's will work exactly the same way. The state that occurs is the state that is willed, and the state that is willed is the state that occurs. If the consciousness channel were equal to the will channel, then whatever one wished would also be exactly what happened. Every thought would be a perfect act. One could not even conceive of having a wish that would not be fulfilled.

But the will channel that causes the events that occur is only a small part of the mind. The consciousness channel, which is of quantum mechanical origin just like the will, is much greater in magnitude than the will channel. Our problem with understanding how these phenomena occur arises not with figuring out how the mind can cause a particular thing to occur, but in understanding why only a small part of what we consciously wish for is carried into reality.

The answer to this question is to be found in the small magnitude of the will channel as compared to the consciousness channel of the mind. The will channel carries the "signal" that selects events by means of state vector collapse. The consciousness channel is—as far as having any effect on the states of external objects—simply so much "noise." The signal-to-noise ratio of the mind is simply a very small number.

WIGNER'S FRIEND AND THE COUPLING OF MINDS

In an earlier chapter, we talked about the phenomenon of state vector collapse on observation, noting that the simple act of looking at things enters into the physics of what happens. Eugene Wigner added a corollary to that phenomenon. It is called the Wigner's Friend paradox. If Wigner observes a quantum mechanical process—the reading of a meter, the photographing of an atomic particle track in a cloud chamber, or, as we have seen, the rolling of dice—then before the observation, the proper description of any of these was a linear superposition of possibilities. When Wigner looks at the system, it collapses into one of the possible states. But now what about Wigner's friend? What if this friend asks Wigner what happened? For him, the overall system is the photograph of an atomic-particle track plus Wigner observing and then telling him what was seen. For Wigner's friend, the linear superposition of states includes not just all the possible particle tracks that might be in the photograph but also all the descriptions that Wigner might have given in response to his observation of each possible picture.

Are we to assume that Wigner himself is subject to state vector collapse when his friend observes him? No, there is a better way to solve this problem. For now the solution will seem to be just a *post hoc* hypothesis. Later, in Chapter 16, we will explain why this postulate is the only reasonable choice—why it comes out of the physics itself. For now, however, let us merely note that all observers must be treated equally. We cannot always state unequivocally even which observer is the first observer. Einstein's theory of relativity gets in the way. This is because there is no such thing as an absolute frame of reference. There is no single reference point at which the observer is the only one who sees things correctly. In addition, the theory of relativity shows that time runs at different rates in different frames of reference (see Chapter 4). There is no one first observer; there is no absolute observer who is the unique cause of state vector collapse. Every observer must be treated as the observer collapsing the state of the system, and every observer must also be constrained to go into the same final state.

Everyone collapses the state vector. The consciousness-linked hidden variables that cause state vector collapse on observation are nonlocal. They exist everywhere and nowhere, and they are independent of the time at which the particular observer looks. Irrespective of space and time, an observer causes the system to go into the one observed state from the allowed collection of potential states, and these hidden variables are held in common for all observers. If there are only two possible states, one bit of information in the will channel can cause state selection on observation; and if there are two possible observers, or ten possible observers, there will still exist only one bit of information that causes the state selection. But that one bit will be a part of the mind experience of each of those observers. To the extent that they limit themselves to that particular part of their minds, they will have a common conscious or mind experience. The will channel, consisting of those hidden variables of the mind, links us together as though we were one so that if any one of us touches some physical object, some same system, then to that extent we are the same mind. A piece of my mind writing this now is in your mind now and is identical to some small fragment of your mind as you read this. This is so irrespective of our spatial separation. Irrespective of the time on the clock, this must be so. If it were not so, then we would have to violate Lorentz invariance (which says, in essence, that all observers are on a different but equally valid footing), or we would have to fall subject to one or another of the problems about Wigner's friends in which we have to assume that one of the individuals is really unconscious.

The hypothesis we have advanced here, that all observers share a fragment of their mind experience, nonlocally and nontemporally, is forced on us by the physics. The problem of Wigner's friend came up long ago. But this resolution to the Wigner's Friend paradox gives us a much better understanding of how state selection works. We see that although it works by means of state vector collapse on observation, the minds are linked by this observation process. Whereas our control might have appeared to be a kind of shadow hand in which we control quantum mechanical states directly, here a part of our reality is the fact that about 1/10 of 1% of what we are in our mind's being is shared; it is identically the same as the mind-being of all others who exist. This is an incredible realization.

Later, we will see how basic this idea is. We will see how this assumption that physics forces on us is justified by the results it leads to. It is a process in which a collection of alternative realities is poised on the brink of coming into being. It is a vision of potentialities cascading from the depths of our own brain's quantum machinery or spilling from the turbid sea of atomic uncertainties suddenly coming into being through the action of observation, at once individual and universal, unbounded by limits of either space or time.

Friday, March 28, 1952: The day was warm for March, warm and clear. I happened to see Maitland at school. I don't know what she was doing there. She should have been at college. We had no chance to talk. At Activity the class went over plans for the Junior-Senior Prom. Of course, I had asked Merilyn to go with me. I turned in my book report on the *Spoon River Anthology*. At seventh and eighth periods, there was a chance to be with Merilyn. I still have a page scribbled with foolish notes we passed back and forth to each other that day. So much foolishness scribbled on a piece of yellow paper 50 years old.

That evening Merilyn, Violet Bailey, and I went skating at a rink out by the fairgrounds. The only skating I had done before was on pavement in street skates. The experience was different, but Merilyn helped me get started. I was surprised to see her skating not only backwards but also artistically. That night she made up for a lot of her losses on the tennis court. All I remember now is entering the long, low, white, wooden building at dusk; inside, the dim lights and the echoing sound of the skates going round and round. I remember my own inability to gain control until she took my hand and slowly, caringly, showed me how to keep from falling. We went to her house afterwards, and I stayed with her till the early morning—holding her. Saturday evening, Maitland called. She was home from college for a few days. She asked me to come over. I didn't go.

Thursday, April 3, was a holiday. I called Merilyn, Maitland, and Tommy to see what was up. Merilyn had to clean house. Maitland had gone to some luncheon at the Country Club. Tommy couldn't get away until later. I worked on some kind of acoustic motion detection gadget that used old radio parts. Later, Tommy came over. We went by Maitland's to see if she had gotten back; then we went over to Howard College (now Sanford University). We had heard that the physics department was building a cyclotron. We saw it. Modest. Later still, Tommy and I picked up Betty and Merilyn and went to the "leadout," the sorority dance. Afterward, when we took Betty home, they went inside for a while. Merilyn and I stayed in the car. The weather was warm, the company was friendly, and life was beautiful in every way.

On Saturday, Merilyn and I went to town together. We went to the library. There were things I wanted to show her in the science section. In that large, empty building, there were places we could go to be together. When I got home I called Maitland. She had a date and was about to leave, but she asked me to come over at 11:00. "She called me from the show. I started to go tonight and had left the house, when I decided not to. Maitland called me at 11:20." She must have asked why I had not come. I must have told her at least part of the reason, but in reality, I probably did not know myself.

"On Sunday, Merilyn and I went to church. At 3:00, I went to her house to work on my Latin project." Later, I went over to Maitland's without telling anyone, and I got into trouble when I came home. I went over to visit Maitland again on Monday because she was upset at being by herself. I guess I went because it was beginning to be more difficult to visit Merilyn, and after all, Maitland was still a good friend. Even then, there were good memories. Still my diary records, "I wished I hadn't . . . . I got home at 11:30. Worked on History and went to bed at 2:00."

On Wednesday, "Maitland called me from the airport. She left at 8:00." She was feeling lonely. She wanted to be with me. If it hadn't been for Merilyn, I would have wanted her as well, but I couldn't afford the complication. "Merilyn and I played tennis this afternoon, and then we went to her house to work on my Latin project."

The entry for Monday is a bit odd. It simply says "Blank Isn't It." I had thought this had been entered simply because I had not gotten around to filling in the page. But on Friday, April 18, I find "I helped Merilyn with the scenery (for the Les Amies Club Dance) this afternoon and tonight. She had me burn a poem I gave her yesterday about what happened Monday! (3) Got home at 1:30 A.M." So I am left with a puzzle. What happened that Monday? Perhaps it was little more than an argument we had about any of a hundred things that we could find to disagree about. Or perhaps it was a poem that spoke too clearly about some of our more pleasant and private moments together. Perhaps I will remember some day. Perhaps it is gone forever, like so much else of life.

On Thursday, April 17, "Merilyn and I found out that our poems had been accepted for publication in the Anthology of High School Poetry. They took Merilyn's 'Gypsy Violin' which is very good, and my poem 'Man's Golden Pages'—a sorry poem . . . . I sent Maitland a letter today." The next day (Friday again), "I took Merilyn in to see Mrs. Borders this morning so she could see the certificates the publisher has sent. Mrs. Borders said she is going to present them to us at the next assembly program. Everyone in school knows about it." Everybody in school knew about "us" is what I meant. At that next assembly, Mrs. Borders did hand out the certificates. We stood on stage. There we were, writing poems and standing on the stage for all the high school to see. Nothing was left to the imagination!

Saturday, Merilyn was in her bathing suit when I came over. She could look so appealing. In the afternoon, we went for a long walk. We talked about the future together. She asked me to prove my love. I vowed to bring her a rose every day for the rest of her life.

On Monday "I took Merilyn a rose to school . . . . Mama told me about the purple and green lights that were seen in Bessemer"—the aurora borealis, which can only rarely be seen in the South. My grandmother also told me that "written on my great-great-grandmother's tombstone is the epitaph: I was once as you are; you will soon be as I am."

Tuesday, April 22: "I took Merilyn a rose . . . . Merilyn's and my names were in the Valley Echo (the school paper) about our poems." The next day, I noted that "many people remarked about the item in the paper." I was embarrassed by the attention and I worried about us as a "couple"—what it might lead to. I felt like the teachers were saying, "Isn't that cute" and probably adding, "Maybe we shouldn't let them spend too much time together." Still, I continued as usual. "I took Merilyn a rose again. I gave a report in English on Shelley. Got a letter from Maitland. Started thinking about a color organ design. . . mailed a poem, 'Vestavia—Mood Pink,' to the newspaper."

Friday, Merilyn and I went to the Alabama theater to see *Singin' in the Rain* with Debbie Reynolds and Gene Kelly. Merilyn identified with the role Cyd Charisse had in that movie. For all her young innocence, she wanted to play the role of the dancing seductress. And she looked the part. On the other hand, I was by nature so Victorian that even the dance scene in *Singin' in the Rain* seemed salacious. When I see that film now, though, I see something entirely different from what I saw then. I see the technicolor red lipstick and the black silk stockings with seams that had to be kept straight. I see the curly-short hairdos and cotton dresses. Soft cotton dresses. All remembrances of Merilyn.

There are no entries in the diary for the next several days, not until the Junior-Senior Prom nearly a week later: Friday, May 2. We rode to the prom with Frank and Ceil. "Ceil asked MAZ if she had on Falsies. After we left, we went to a movie." Later we went to Merilyn's home. We sat on her back porch for a while. I remember that moment vividly. We sat close together on the glider, in the dark in the stillness. I held her for a while. I remember the touch. I remember how she felt in my arms. Then I left her, and I walked home alone.

16

From Epicycles to Loops

A moment in time but was made
through that moment: for without
the meaning there is no time,
and that moment of time
gave the meaning.

—T. S. Eliot, *Choruses from "The Rock"*

Time is just a parallelogram.

—Jerry Lewis, *Visit to a Small Planet*

Time. Time is the real mystery. We seem easily to accept the space that separates us, space that spreads out above and beyond. We cope well and poorly with its provisions and with its walls, but time defeats us. Time takes the dear things we have and spirits them into the lost past. We hope for futures. We grieve for vanished pasts. We remember. . . only remember . . . and *now* is never grasped. Time is the puzzle.

For all the myths about what science is, about what the scientific method is, about how science works, about myths of hypotheses tested and paradigms supplanted, science really works by solving puzzles—puzzles like that of time. Science is knowledge, of course, but it advances by solving puzzles. The pieces are many, and they are often dimly seen with often only the accuracy of the probing instruments to show the outlines of the jigsaw parts. We find the pieces, fit the shapes, and then we see the picture. That is how we know when a new theory is right. It is right when the picture is right. Sometimes we put two or three pieces together and we think we have a fit—we think we have solved part of the puzzle—but the puzzle is not solved until all the pieces are placed together. The puzzle is not solved until we see the whole pattern emerge. It is that emerging pattern that tells us that we now know more, and that our science is knowledge.

We have put together many of the pieces here in this book. We have fit together pieces that show the quantum mechanical uncertainty and some of the Bell's theorem results; pieces telling facts about consciousness and mind, about neurons and synapses, sleep, and a will within that touches other minds. We have fit together many of the pieces. A few, surely, are dimly lighted, their outlines poorly measured. As we learn more about these pieces of the puzzle, some may have to be moved, some changed. But the fit overall is good, and the pattern overall is true.

I say all this because there are more pieces of the puzzle to be fitted together. We will not be able to fit these next pieces together one by one. Instead, we will have to examine them and see, all at once, how they fit together. Let us look at some of these pieces.

WHAT'S SO SPECIAL ABOUT YOU AND ME?

First of all, there is a glaring, weird question we have so far avoided. What really causes state selection itself? What causes state vector collapse? Why is it that the "observer" causes state vector collapse? What is it that makes the observer an observer? Of course, the consciousness of a human or animal arises out of the quantum mechanical interactions in the brain so that in an observation, we have one quantum mechanical process impinging on another. But in all other cases where we discussed how quantum mechanical systems behave as time passes, we have said that the Schrödinger equation would not yield a state vector collapse for such an interaction, but merely a larger collection of new potential states. Why would anything different happen for the brain? Why would the interactions of atoms lead to more and more possibilities but the quantum events in the brain bring this all to an end in a conscious experience?

According to the Schrödinger equation, successive interactions do not lead to state selection, but just bigger and bigger collections of states. Why would our interaction be different? Why are *we* observation when all the rest of the world's interactions are just interactions? The circumstances would suggest that the human or animal observer completes something—but in what way? It is as though there were something more to that mysterious equation Professor Schrödinger invented. It is as though there were something else hidden from view.

A PLETHORA OF PUZZLES

A second puzzle concerns the dual structure of quantum mechanics itself. For every "answer" to a physical problem, quantum mechanics turns out to have a peculiar quirk. It gives us not one answer but two. One is the usual state vector symbolically represented simply by the Greek letter Ψ—the collection of possible outcomes of any measurement or any physical event. The other solution is the first solution's mirror image. It is called the Hermitian adjoint of Ψ. We represent this Hermitian adjoint by the symbol $\Psi^\dagger$.[1] (For a clarification of the notation, see Appendix II.) The basic difference between Ψ and $\Psi^\dagger$ is that $\Psi^\dagger$ is obtained by changing the signs of all the imaginary terms from plus to minus in the formulas for each of the states that make up $\Psi^\dagger$. Whereas Ψ consists of all the states Ψ_i, $\Psi^\dagger$ consists of all the states ψ_i^*.

Each one of the states ψ_i is a picture of something that could be, a picture from a book of all possible potentialities. Each one of the states ψ_i^* is also a picture from a book of potentialities, but it is the mirror image book, with each page having a mirror image picture of what could be. And the mirror that does the reflecting, rather than sitting in the real world, sits straddling across into an imaginary world, one in which we would be able to see that imaginary store front we talked about earlier.[2] The $\Psi^\dagger$ is the collection of all the ψ_i^*, and each of these is just a negative of the corresponding ψ_i picture from the usual solution to the Schrödinger equation.[3] Now each of these quantities, the Ψ and $\Psi^\dagger$, contains all the potentialities of the physical system, all the quantum mechanical states, all the possible pictures. Physicists are not puzzled about these two expressions; they use them to calculate probabilities. The actual values for the probabilities are obtained by putting the corresponding pictures together, which means that we multiply each of the quantities ψ_i by ψ_i^*. When we do this—when we calculate the product $\psi_i^*\psi_i$—something special happens. We get a real number for the probability. And we get real pictures for what can happen, rather than the murky half-submerged pictures we had before.[4] Now we are dealing with real numbers that can happen in the real world. The result is a number that turns out to be the probability for a real event to happen. We get numbers we can actually measure in experiments.

But just before we measure anything, we have the quantity Ψ, which describes the unobserved world, and we also have the quantity $\Psi^\dagger$, which is a second way of describing the same thing. In addition, there is the Schrödinger equation, which gives us Ψ, and there is also a second equation, the conjugate Schrödinger equation, which gives us $\Psi^\dagger$. We know what we have to do with Ψ and with $\Psi^\dagger$ in order to get numbers that we can compare with things in the outside world that we measure. We know we have to put Ψ and $\Psi^\dagger$ together to get real numbers out. But what is nature's machinery that brings Ψ and $\Psi^\dagger$ together? Something missing from the Schrödinger description of the physical world must carry out that operation in which Ψ is mated with $\Psi^\dagger$. What is that missing part, and why is it hidden?

The mathematics gives us two equations and two complete sets of solutions (see Figure 16.1). Often in physics we find that when an equation gives us extra solutions, these extra solutions tell us something very fundamental about reality. Often, they answer an unasked question. This is how Paul Dirac found that positively charged electrons—positrons—should exist. His equation for the electron turned out to have two complete sets of solutions: one set for the electron that everybody already knew about, and a second set that looked like electrons but with the opposite electric charge. Dirac did not entirely believe what his equation was telling him. To some extent, he missed the chance to make a really bold prediction that positive electrons would be found. Nevertheless, 4 years later, in 1932, Carl Anderson discovered these positrons. Dirac's equation told him more about nature than he had expected to find. Since that time, we have learned to consider very carefully what our equations are trying to tell us about the reality they describe.

So what is it about reality that the existence of both Ψ and $\Psi^\dagger$ indicates? What is it that is completed in nature when $\psi_i^*\psi_i$ happens? Where is the machinery that carries out this coupling operation?

But there are yet more puzzles. There is the puzzle of something that at first seems to have no puzzlement to it at all. This is the puzzle of Hilbert space. There is no puzzle here at all, and yet there is a profound puzzle.

You see, in physics, we deal with mechanics. We, of course, all know what mechanics is. Think intuitively for a moment about what mechanics is, what mechanism is all about. It has to do with objects in space, their motions, and their interactions. In quantum mechanics, we want to learn about these motions, but we do this by working out the problem in a different kind of space called Hilbert space. This is a space that has many more dimensions than the three-dimensional space we are used to; in fact, it can have an infinite number of dimensions.[5] But this is OK. We merely have a different kind of space that implies a more complex kind of mathematical machinery, rather than the one we are used to. It is just part of that more complicated machinery of quantum mechanics that holds these greater secrets of reality.

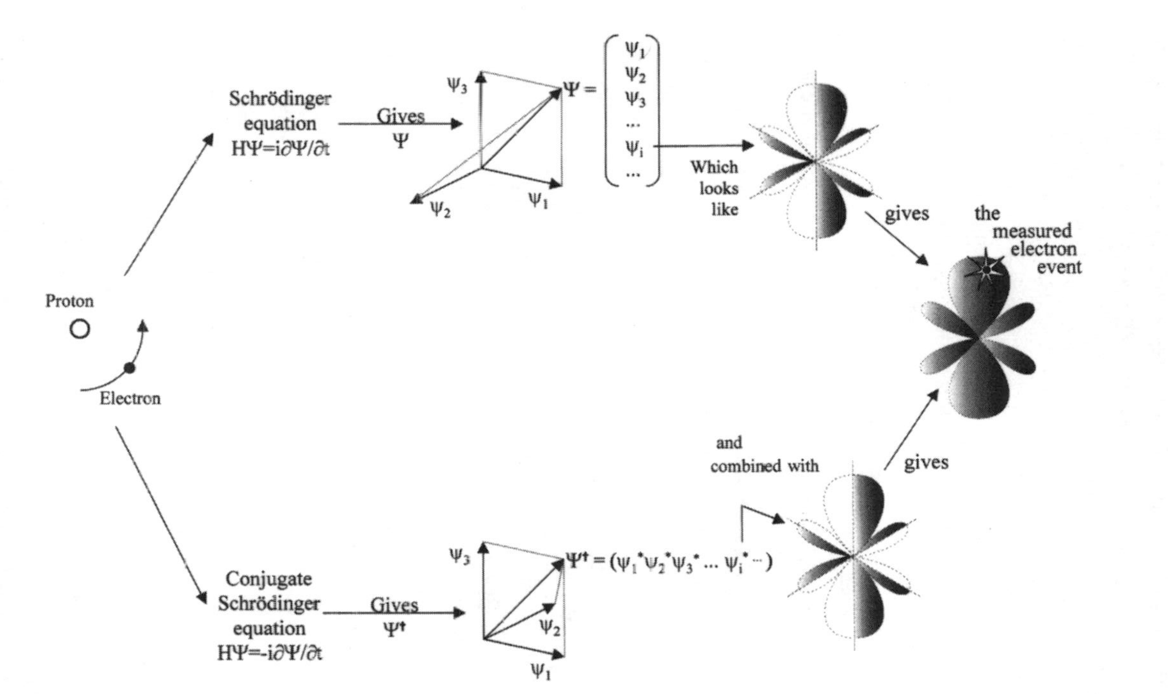

FIGURE 16.1 This figure shows in more detail the steps involved in solving a problem in quantum mechanics and illustrates the relationships between the two forms that the Schrödinger equation and its solution can assume in the dual-space representations. At the left, we see the beginning stages of the problem, where we take the system we are dealing with to be that of the simplest atom, a hydrogen atom, which consists of a proton and an electron attracted to each other by an electric field. We next find an expression for the energy of this system, which includes the energy in the motion of the particles as they change positions and the energy in the electric field. Together, these are given by H. This value of H is then put into the Schrödinger equation or into its conjugate form. These equations give the solutions Ψ or $\Psi^\dagger$, which consist of many possible potentialities; one example is selected as an illustration. Note that neither Ψ nor $\Psi^\dagger$ alone is real, but they become something real when combined.

That would be OK, but that is not what Hilbert space is. Hilbert space doesn't describe the machinery at all. It is space constructed out of those pieces of the probability in quantum mechanics that we have been talking about. It is made out of the $\psi_1, \psi_2, \ldots, \psi_i$. These are the directions in Hilbert space, just as we have length, width, and height in our ordinary, three-dimensional physical space. More than that, $\Psi^\dagger$, which is the Hermitian adjoint to Ψ, forms what is called a second space, the two spaces being referred to as the dual space. As we have already said, we use ψ_1 and $\psi_1{}^*$, ψ_2 and $\psi_2{}^*$, and so on to generate probabilities. Now here's the tough question: Why should probability be described by a space? Is this just an idiosyncrasy of our mathematics? Every time we sit down to do math, do we have to express everything in terms of some space? Or could it be that there is a more fundamental connection between the physical space we live in (the space we know of as our physical reality) and the probability that is the basic idea of quantum mechanics?

And while we are puzzling over what that might mean, let us take a few more steps down this road. We have already talked about measuring information. We talked in the last chapter about the three fundamental kinds of information that we have discovered to be basic to quantum mechanics and also basic to the functional relationships among brain, consciousness, and will. We measure information in terms of bits. We calculate how much information there is from probability expressions, and we use information measures to measure consciousness. Probability, information, consciousness, and space, for that is where we put the things of our consciousness. What is the tie-in among all these things?

Closely related to these questions is something in quantum mechanics called the dispersion measure. It is a simple quantity. It is just the probability minus the square of the probability.[6] Von Neumann, one of the great figures in the development of quantum mechanics, invented this expression in order to give a measure of just how disperse the potential states are at any moment. Once we have observed a quantum system, it has to be in the one state in which we see it. As a consequence, it has no dispersion. It is in one state rather than being in a lot of potentialities. Its dispersion is zero. But before we measured it, it was dispersed over all the potential states. Von Neumann's dispersion measure is a measure of this dispersion, or the "spread-outness," of reality among the possibilities permitted by quantum mechanics.

Von Neumann's dispersion measure is just an arbitrary formula, however. It is not the only formula that could have been chosen. What is interesting, and this is a piece of the puzzle, is the fact that the formula we use to calculate information measures—to figure out how many bits of information there are on this page that you are reading—has the right properties to serve double-duty as a dispersion measure. It gives a value that gets larger as the uncertainty about the state of the physical system grows, and it gives us a value of zero dispersion when finally we know which state the system is in. This suggests that perhaps the information measure plays a role, but just what is its role?

That piece of the puzzle turns out to be easy. The only place to use the information measure is in the Schrödinger equation itself. If we are going to get all this to be a part of physical reality, state vector collapse, information, dispersion, observation, then it all has to be stuffed into that Schrödinger equation that represents reality. But just how would that work?

We get the answer to that by looking again at what causes the measurement problem in the first place. As we saw in Chapter 7, the measurement problem arises from the fact that the Schrödinger equation never gives us state vector collapse. That is because this equation is "linear." There is no term in which the value of one state component acts on another to cause one to grow in probability at the expense of the other states. What the Schrödinger equation needs is another term, a "nonlinear" term, a term just like the information measure! But the term must not affect what we finally measure! We must be able to do something to the Schrödinger equation that will cause state vector collapse, but that otherwise will not change any of the answers.[7]

Well, because the information measure becomes zero when state vector collapse occurs, if we use such a term in the Schrödinger equation, it will disappear when a measurement is made, leaving us to see only that part of this new equation that we already see using the usual Schrödinger equation. In addition, this new information-dispersion term is also a nonlinear quantity. We can put it into the Schrödinger equation and set it up so that it drives the potential states into one observable state and then vanishes. This information term—this hidden-variable term, this consciousness term—will cause state vector collapse and then disappear from the material world when the observer enters the picture. The observer effect enters physical reality to cause state vector collapse, and then it vanishes so that you cannot actually find any outward physical trace of consciousness itself.

We have assembled a small part of the puzzle, but we have not put all the pieces together yet. This term, this new term that we add to Schrödinger's equation, must not disappear at all times. Indeed, it must play an active role in physical interactions until something happens to turn it on, so to speak—to turn it into the dispersion form that then suddenly forces the system to go into the collapsed state. What we would like is for the physical system itself to signal the appearance of the $\psi^*\psi$ probability form, which would then lead to state vector collapse automatically. We can get this to happen in a particular way, a way that involves the special kinds of things that go on in any interaction that looks like the Einstein-

Podolsky-Rosen (EPR) interaction, the kind of set-up used to test Bell's theorem, where we first found that a strict objective reality could not exist. And it also occurs in the kinds of interactions that are basic to the design of our brain.

EASY MATH: HOW 2 AND 2 = 22

Before looking into this, let us first consider how physicists represent interactions between two systems (refer to Appendix II). The reason for us to do this should become apparent quickly. Remember, we would like to see a way in which the physical system itself, rather than the observer as such, could be the origin of the probability terms of the form $\psi^*\psi$. Now let us assume we have two objects that can interact in some way. For simplicity, let us say that the first object can exist in only two possible states—two possible pictures of what we can find on measurement—which we will symbolically represent by θ_1 and θ_2. For example, θ_1 might be the state of an electron spin in the up position and the state of θ_2 an electron spin in the down position. Before an observation, the system can be in both potentialities simultaneously, both pictures that give us the state vector we symbolically represent by Θ. The state vector Θ simply says that the electron manifests both spin up and spin down as potentialities.

Now let us say we are going to let this electron symbolized by the set of pictures Θ interact with another object, maybe a measuring device that detects the spin of the electron. Now that detector will have to have two possible readings: a reading that will let us know that the electron has its spin up and a reading that will let us know its spin is down. The measurement apparatus has to be designed so that it can show us the results of the measurement on the electron. We can symbolically represent these two possible states of the measuring system by m_1 for a meter reading that says the spin is up and m_2 for a reading corresponding to spin down for the electron. The overall state of this measuring device can be represented symbolically by M. Now one of the nice things about quantum mechanics is that after the interaction is over, it is very easy to say symbolically what the state of the overall interaction is. It is simply the product ΘM. The individual states would be $\theta_1 m_1$, meaning electron spin up with meter reading "up," and $\theta_2 m_2$ for an electron with spin down and meter reading "down." Quantum mechanics lets us conclude this without having to know anything more about the details of the apparatus. You see how easy the most complicated mathematics can sometimes be![8] Obviously, if our experimental apparatus is any good, then $\theta_1 m_2$ and $\theta_2 m_1$ should both be zero. A good measuring device does not say the electron spin is up if in fact it is down, or vice versa.

Well, now, maybe you can see it coming. If we can only find a way in which a system interacts with itself, then we can get interactions that have $\psi^*\psi$ terms, which will signal automatic state vector collapse. This will give us something physical to pin the actualization of potential states to. If we are lucky—if we have been right so far—we will find that this process is somehow connected to the things that go on in the brain of conscious observers. This will then tell us why we are the observers and not simply a part of the interaction that creates unresolved potentialities.

To see how the physical system can look at itself, to see how $\psi^*\psi$ could occur naturally—let us consider again how the Einstein-Podolsky-Rosen paradox works, remembering that this is also the experimental arrangement for tests of Bell's theorem. Look at Figure 16.2, which has two parts. Part (A) shows how Einstein, Podolsky, and Rosen envisioned it, but they took something for granted. They talk about the result of comparing the measurement on one particle with a measurement on its twin, but they do not include any analysis of that part of the experiment. One thing twentieth-century physics has taught us is not to take any measurement for granted. It has to be included in our analysis. Part (B) of Figure 16.2 completes the EPR arrangement. It includes the all-important final comparison of the outcomes of measurements.

Figure 16.3 distills out the essential points. If we follow the series of interactions around the bottom part of the diagram, we see that the interaction has produced an object described by a state vector Φ_A—the set of possible pictures, one for the atom with spin up, and one for the atom if its spin is down. Next comes the interaction with the measuring device described by a state vector M_1, with its own set of pictures showing what it may display as the result of measurement. Along the top path, we have a state vector Φ_B for the other particle that interacts with the measuring device M_2, which then interacts with the comparator. The comparator is simply a device that checks whether the two measurements agree with one another. It could even be ourselves looking at the two measurement reports. This is the step that Einstein took for granted—and omitted. We know the two systems are tied together so that Φ_A will be in its first state when Φ_B is in its first state; this is the theoretical result. It is not real, however, until it is real—that is, until it is observed to be so. That is why the "comparisons step" in the EPR experiment, the step Einstein forgot to discuss, is so important. When that comparison is finished, then the state of the overall system is given by $CM_1M_2\Phi_A\Phi_B$. This says that the whole system's outcome depends on the state of the comparator, the two measuring devices, and the two parts that came out of the original interaction. You can see it coming, can't you?

Now we already know that for the EPR-type experiment, both Φ_A and Φ_B must be basically the same. If Φ_A is in its first state ϕ_{a1}, showing an atom with spin up, for example, then Φ_B has to be in its first state ϕ_{b1}, showing an atom with spin

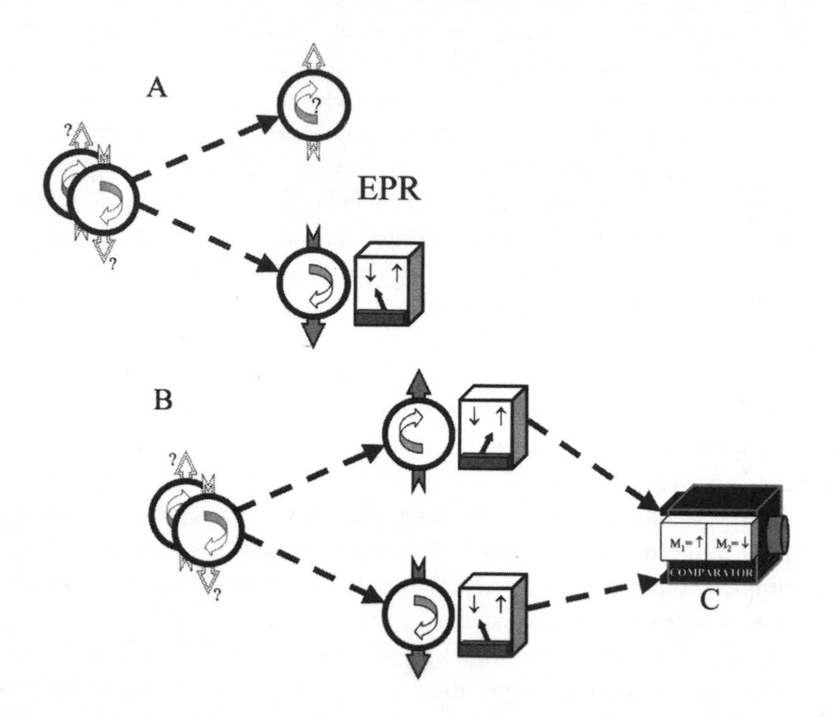

FIGURE 16.2 (A) The EPR paradox as envisioned by Einstein, Podolsky, and Rosen. The original molecule is known to have a spin-up atom and a spin-down atom, as indicated by the arrows. But which is up and which is down is in quantum uncertainty (dashed arrows). After separation, if we measure the spin of one atom and discover its spin is "down," (that is, 0 on the meter), then without measuring the spin of the other atom, we know its spin must be up (that is, 1). Supposedly, it was always in spin-up condition even before being measured. The tests of Bell's theorem, however, show that we cannot assume anything about a system until it is actually measured. In part (B) we see that this means a measurement should be made on each atom and then both results compared—for example, by being examined by a person or a computer *C*.

down. But the Φ_A and Φ_B are complements of each other. They just view the world from their perspective. The vectors Φ_A and Φ_B are really the same pictures but viewed from their respective parts of the dual space. Moreover, the probabilities they contain must be the same, because for whatever happens in the one, the corresponding must happen in the other's picture.[9]

What we do now is to interject a postulate! We say that whenever two objects interact and then separate, one of the objects (either one) goes off in a state Φ, while the other goes off into the dual space as $\Phi^\dagger$. There is no reason not to postulate this. There is no counterindicative experiment in all physics. What it does for us here is

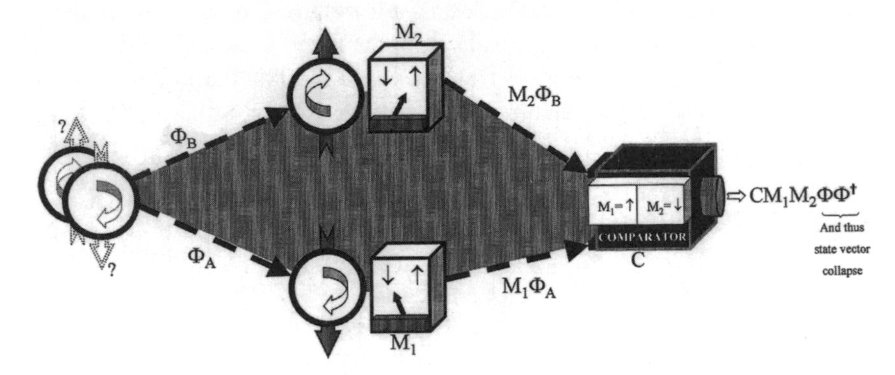

Figure 16.3 A complete measurement loop. Upon separation of the molecule into two atoms, atom *A* represented by state vector Φ_A and atom *B* represented by state vector Φ_B, the atoms are each measured for their spin orientation in the interactions with the measuring devices M_1 and M_2. This leads to states $M_1\Phi_A$ and $M_2\Phi_B$. The measurement loop is completed when the results of these measurements are compared in the comparator computer *C* to give the state $CM_2\Phi_B M_1\Phi_A$. Upon rearrangement and substitution of $\Phi^\dagger_A$ for $_B$, this gives us $CM_2M_1\Phi_A\Phi^\dagger_A$ or, more simply, just $\Phi\Phi^\dagger$ for the last part. This shows how a measurement loop serves as a physical interaction that creates this $\Phi\Phi^\dagger$ combination that causes state vector collapse automatically. Such interaction loops, loops in which the information from quantum mechanically firing synapses are compared in loops formed by our brain's neural circuitry, causes our observation of events in the outside world to produce state vector collapse.

to give us the form $\Phi^\dagger$, which will contain the $\phi^*\phi$ components that cause state vector collapse when they are put into that information-dispersion term we added to the Schrödinger equation.[10] When a loop measurement interaction is completed, the form $\phi^*\phi$ appears automatically. This, of course, is the $\psi^*\psi$ we spoke of before.

THE COURTSHIP OF Ψ AND $\Psi^\dagger$

We have to pause just a moment to explain how this is going to fit together. There are generally two parts to an equation used to solve problems or describe nature. There is an unknown quantity, that *x* that represents the answer we are looking for, which is a part of nature that we must know in order to quantify what happens. Then there is the "other side" of the equation that enables us to solve for *x*. In Chapter 2, we talked about Newton's equation $F = dP/dt$, which expresses the fact that the force applied equals the rate of change of the momentum. That is the equation, and *P*, the momentum of an object, is the value that tells us how the object moves at any time (it is that *x*). Having found *P*, the momentum, we can use

P to find just where the object will be, how fast it will move, or anything else there is to know about its behavior that results from the force, *F*, applied to it.

Schrödinger's equation is a more complicated expression than we have had to deal with in the classical mechanics of Isaac Newton. Now we have an equation that looks like this: $H\Psi = i\hbar\partial\Psi/\partial t$, where *H* tells us how the energy of the object depends on the location of the object, loosely speaking. Thus *H* takes the place of Newton's force term, *F*, but it is Ψ that gives us the answers we seek (see Appendix II). The factor $\partial\Psi/\partial t$ simply represents the rate of change of Ψ with time.[11]

Now then, let us say we are solving an ordinary problem in the motion of two billiard balls. We know that if nothing interferes, the total momentum of these two balls, *P*, has to be the sum of the momentum of the first ball, P_1, and that of the second ball, P_2, and that this sum has to remain constant. Now we can put that into Newton's equation, and this bit of extra knowledge can help us sort out what happens to each of these balls.

The same sort of thing happens to the molecule in Figures 16.2 and 16.3 that, as we noted earlier, has a solution Θ all by itself but that, when acted on by a measuring apparatus, has to be described by the combination $M\Theta$. $M\Theta$ is put into the Schrödinger equation in place of Ψ. For the system shown in Figure 16.3, the whole expression $CM_1M_2\Phi\Phi^\dagger$ must be put in the Schrödinger equation in place of Ψ. When we do all this using the regular Schrödinger equation, we get an equation in terms of $\Phi\Phi^\dagger$, which is what we were looking for.[12]

HOW GOD LOGS ON

But what if we change Schrödinger's equation ever so slightly? What if we put in a term that expresses the transfer of information that takes place during state vector collapse? Remember how we could measure information in terms of the logarithm of the probability? Well, maybe you don't; it was only mentioned in the notations to Chapter 14. Still, when we were finding how much information was needed to specify a single symbol out of, say, 64 symbols (upper- and lower-case letters and 12 other punctuation symbols), we came up with the answer of 6 bits. What we were really doing was calculating the value $\log_2 64 = 6$ to get the number 6. We did that so we could calculate the information capacity of a page of print in a book. We were taking something of a liberty when we did that, because when it becomes important to be very accurate, it is necessary to correct for the fact that symbols do not all occur with the same probability in print. As a result, one has to calculate the information contribution of each symbol. That contribution is given by $\log_2 p$. Then we have to add all the contributions up, the contribution of each possible symbol being weighted in proportion to its chance of occurrence, p. That is, we sum up all the $p \log_2 p$ contributions.

What we do now is simply put a term that looks a little like that information term into the Schrödinger equation. We add a term $\Psi \log \Psi$ to the Schrödinger equation (don't worry, that $\Psi \log \Psi$ is going to turn into $p \log p$ in a minute). This addition would ordinarily make the Schrödinger equation do two things: nearly always act wild and almost never be real. Only when we carry out a *complete measurement loop*, as in Figure 16.3, does the equation change. Then this $\Psi \log \Psi$ term that we added turns into a $\Psi\Psi^\dagger \log \Psi\Psi^\dagger$ term.[13] This is the result we had hoped for when we added the extra term to the Schrödinger equation. It gives us one of these $p \log p$ terms for each of the possible outcomes of any measurement loop.[14]

Remember, $p \log p$ is that measure of information that is also is a dispersion measure. If, say, state "1" happens, then there will be no dispersion,[15] and, moreover, we will have acquired information—so many bits of information—about what did happen.

In other words, when we carry out a complete measurement loop to make Ψ take the form that contains $\Phi\Phi^\dagger$, the math will force the whole thing to have nice, steady, real solutions only if one of the states happens and all the other states vanish.[16] This modified Schrödinger equation forces state vector collapse when there is a measurement loop, and then with its job done, the extra term that we added vanishes.

When the loop closes, state vector collapse is forced to happen. This, of course, means that consciousness and will have to become a part of the loop. This is the physical side of what observation is all about. It is this measurement comparison, this coming together of the two sides of reality that causes state vector collapse and that brings about consciousness as a part of the physical world.

One more crucial piece in this puzzle tells us why these loops are so very important to us. These loops are the link between measurement loops as the cause of state vector collapse and the observer as the cause of state vector collapse. The measurement comparison process in the EPR-type loop is exactly the kind of operation that goes on by the billions every second in the brain. Billions and billions of times each second, the neural network compares the inputs of sensory data along channels that involve quantum mechanical synapses. These loops require state vector collapse to occur, and at the same time, they carry with them the state vector collapse of the events that we observe.

You will note that the change in the Schrödinger equation did not involve putting in the information term $p \log p$; instead, we used an expression that contains the state vector, Ψ. This is because Ψ consists of quantities that are imaginary[17] and therefore represent potentialities rather well. The quantity p is a probability and would represent something that is already real. We want the Schrödinger equation to give us imaginary answers—state vectors that represent only potential states—until the measurement loop is closed and the state becomes a reality represented by the real quantities we get when we write $\Psi\Psi^\dagger$.

Now let us look at what we have in these measurement loops. What we have shown is that before the measurement is completed by a comparison with some other part of the system, the system exists as a state vector. But we know that if we are there, and if we make one of the two measurements on the particle at *A* or at *B* in Figure 16.3, we will get a number. It seems foolish to suggest that state vector collapse does not happen until we compare what we saw at *A* with what happened at *B*. To understand what we have here, we must be very careful to limit ourselves to what is going on in the state vector collapse process. Remember the business about Lorentz invariance, the principle in physics that space and time have to be treated on the same footing? That principle demands that the time at which state vector collapse happens be independent of observations, just as it is independent of the spatial separation of observers in the EPR experiment. The point is that for the measurement loop, there is no such thing as a spatial separation. This is what we learned from the tests of Bell's theorem.

Likewise, we must think of all the observations at the various parts of the *loop for that particular piece of the event that has to do with this measurement* as existing at the same moment. The loop happens as one event, and associated with that loop there is a single "piece" of consciousness-will experience that all the different observers at various points in the loop share. It is this nonlocal, time-independent characteristic of state vector collapse that makes will-channel effects possible as a result of the observation of events having macroscopic quantum mechanically disperse potentialities.

Time does not flow as a stream but passes by as chunks. The pieces of time that we consciously experience are ordinarily brief, a few hundredths of a second long. Perhaps in the nuclei of atoms, there are things that happen that also involve such loop interactions that are consciously experienced in pieces of time that are only 10^{-23} second long. I do not know. But at the other extreme, in phenomena in which one mind transfers information atemporally and nonlocally, there are events, consciousness-will events, that exist as single moments of time. These are pieces of time—involving perhaps only one or two bits of information submerged in the whole background of ongoing conscious events—that span hours, days, even years as measured by other events going on as a part of these loops. Although we ordinarily do not notice their presence, these grand moments of time for loop interactions overlap the billions of loops that occur in our brains as we observe, and they are also a part of our brain's functions (as data input) and a part of our stream of consciousness. It is not everything that happens during the loop that has this spread-out time, only the piece of information that causes the state selection determining the result of the comparison at the end of the loop.

Maybe here you can see something of the fabric of time and of reality. At least perhaps you can see some of the cords, fibers, and threads. Perhaps you can see many of the colors that dance across reality. Perhaps you can see patterns, lines, surfaces—the space and sweep, the depth and motions. Perhaps you can see certain of the forms, forms that have life. Perhaps you can see, among them, those forms that look back to cast a glance your way from the fabric of reality and the quantum mind.

Maybe now you can see something of what time is about as well. The idea that time is but an infinitesimal moment that only separates future and past will not work. You already know that this cannot be all there is to time. There must be more. There must be something that drives that boundary away from the past and into the future. There must be some structure, some granularity, that would keep the present from vanishing into an infinitesimal moment, just as any searched for infinitesimal point on a rule vanishes between its atoms. There must be some complexity to time's make-up that makes it sweep our *now* away from the past and into the future.

ONE MORE TIME

The Greeks saw time as a man who, riding backward on a horse, could see his past but not his future. Classical physics gave that rider the means to see, like Janus, in both directions. The equations of Newton's mechanics, perfectly symmetrical in time, represented time in a new way, as part of an idealized gridwork of directions in space and in time. With Einstein and Minkowski, these have turned out to look exactly the same—time being space, space being time—but with that one change we mentioned before: Only for $t = 0$, only for the present, are there any real distances. There, in that picture of nature, the infinitesimal now has no reason to go backward or forward. It does not look like our time. It does not look like the time in which we live. We know intuitively that there is more to time than physics has pictured for us. Time has structure. It does not flow as an infinitely divisible smooth stream. It has grains made of loops of interactions; each one that is finished jumps reality forward for everything touched by that loop. And these loops interlace with others by the billions each second, tying things together in the brain's interaction with itself and with the world outside so as to form patterns of colors created by the conscious loop events scattered throughout its intricate moss of neurons lacing over the cortex. Each loop of interactions, small and quick or large and lasting even through the ages, has consciousness and will. It is this weaving pattern of loops that marks the jumps of time—the individual pieces and bits of time.

Odd that from the time of the ancient Greek atomists, people have suspected that matter must be made of pieces, and yet even today, most physicists picture time as an infinitely divisible stream—albeit restricted by the energy required to actually measure brief divisions of it. But of course, the real reason is that time is just so elusive. We do not seem to be able to grasp its pieces with our fingers. There is no way to cut off a chunk and pass it around to be examined.

And we mean so many different things by the word *time*: physical time, biological time, psychological time. These are all common uses of the word, and each has its place. Both biological and psychological time reflect ways in which the brain or the body measures physical time in order to keep the organism—our bodies—in step with the world around us. We can tell these apart by the kinds of things that cause them to run faster or slower than accurate physical clocks. Things that physically alter or modify the bodily functions in a way that affects our ability to keep track of time are things related to biological time. Drugs and physical injury are obvious disturbances that impair our biological clocks. But physical time marches on, unchanged by what happens to the body, and actually marking off the rates at which the chemical and physical processes of the body proceed.

But the brain can also be affected by the information that feeds into its vast network of neural circuits. And in many cases, that information can alter the way the brain keeps track of time; thus pleasant activities often pass quickly, while boring tasks seem to never end. But the biological processes of the brain are only modestly affected by these things, while the chemical and physical processes proceed according to the laws of physics, quite irrespective of what the brain may think.

In addition to physical, biological, and psychological time, there is also consciousness time, the time we experience. This time is of course tied to the biological time and the psychological time, but, as we saw in earlier chapters, it has its own role to play in reality and is as much related to basic physical processes as, say, the rates of chemical reactions are. It is a kind of time that goes beyond these other kinds. It is the kind of time we used when we calculated the expressions for *C* and *W*.

◆

Saturday, May 3, 1952: I finally had enough materials to put together a primitive version of my color organ. It only had a 3-inch screen, but the colors it produced were good enough to create moods, as though one were viewing sunsets. I went over to Merilyn's to see how she was doing and to tell her about my color organ. On Sunday my parents went to Selma again. I went to church with Merilyn and spent the rest of the day at home alone until my parents returned.

School was beginning to slow down. I was a senior about to graduate. There were still tests and the final exams to be taken, but so much was behind us that all we could think about was getting out. Still we went through the motions. On Monday, I took Merilyn a rose again. There was a half-hour presentation in eighth period on the construction of a space station. Back then, despite my own enthusiasm for the idea of space travel, everyone really knew this was just science fiction. Who would ever spend all that money to put things into space? Why?

About the only thing of note the next day was the Hendrix Hi-Y meeting during Activity period. We made plans for a party on Saturday, May 17, at Double Oak Mountain. I didn't know it then, but once again Merilyn and I would not be able to go. As much as hayrides and picnics in the mountains appealed to Merilyn and me, they must have been anathema to our parents. So we would not go. On Wednesday, Economics test results, piano duets in the auditorium during Activity, a Latin test, results of an English test. I wrote a paragraph about what we had learned in English about poetry. Thursday, I got sick during the afternoon. I went up to Merilyn's house, but I did not feel well enough to stay. I went home and called her. Friday: "Called Merilyn when she got home from town. We had an argument . . . . Went to the new Homewood library and found out there are sixteen million Baptists and eleven million Methodists." There are always so many foolish things to argue about. They seem so important at the time, important enough to make you hurt those you love so much. But years later—what difference did it make?

Monday: . . . Got a letter from Maitland . . . . Called Merilyn at 3:00 P.M. She got mad at me for rubbing things in. Talked till 4:00 P.M. . . .

> Tuesday: Beautiful day . . . but Merilyn acted no good today. [She told me at school that] she dreamed that I had fallen down and cut my cheek—and that I had to have plastic surgery. She was afraid it might actually happen since some of her dreams come true . . . . Walked Merilyn home. Mailed a letter to Maitland and went to town to find something for the Clavilux . . . . Went to Merilyn's house to get my books, but there was no one there. Her parents had gone riding in a new car they bought today. I had to walk home.
>
> Wednesday: Auditorium period with a Fashion Show. Played tennis with Merilyn: 4–0, my set.

The next day I got together with Merilyn for a while at Activity period. We talked about our future. The conversation got around to our pets. I had had several dogs, mostly Heinz breed, mongrels, identified as German shepherd or spitz or chow for their strongest characteristic, rather than any true breeding. They were our dog "people" that were just part of the group anytime we went anywhere—animals with names like Chow, Pal, Scarlet. Scarlet was the spitz. She was named for the red spot on her side. At least that is what I had always believed, but maybe it was as much a "tribute" to her way with the males in her life. They were constantly coming over to our house and behaving in peculiar ways. Whatever, they were our friends when no one else was. Anyway, Merilyn said that she wanted to get a dog for her birthday. I asked her what kind she wanted. She said she wanted a cocker spaniel. It was not one of the dogs I thought necessary to round out the perfect family.

"A big dog; a man needs a big dog; a family needs a big dog." That is what I said, or words to that effect. That, of course, was supposed to settle the issue. It didn't, of course, but at the time it seemed that it should have settled everything.

Friday, May 16: "Mother drove Merilyn and me to the Senior Prom." I didn't write down anything about the Prom itself. There is a note that "The teachers kept people from coming back after they left—stinky, isn't it?" So I probably felt the whole thing was too much of an exercise organized to suit the teachers and honor custom than those graduating. Hence one of the major milestones in my life passed with no more note than that Merilyn had to wake up her mother to get in when we got back to her house. There is no note of (2) or (3), certainly no (4), only "Had to wake her mother to get in house." The next morning, my parents and I left for Atlanta. Merilyn and her parents left for Jackson, Mississippi. Merilyn and I did not get to go to Double Oak Mountain on the Hendrix Hi-Y picnic. We never made it to any hayride, nor did we picnic out in the woods. Only the oversupervised proms.

> Monday, May 19, 1952: We had our final exam in English today. Merilyn was absent from school. Bad weather.
>
> Tuesday: At Auditorium period two men from the American Legion spoke and awarded a copy of the Declaration of Independence to us. After that, there were tryouts for next year's cheerleaders. Then the new alma mater song was sung by the choir for us to vote on. I voted "No" but it passed anyway. I went to Merilyn's house at 5:00. Stayed till 7:00.

On Wednesday, I got into trouble. One of the teachers took me to Mr. Walker, the vice principal, for "crimes of high treason and gross impudence—whistling in the hall!" That wasn't the end of my troubles. Somehow at "seventh period I took a plug out of the third digit of my right hand. I had to go to the clinic. Shirley Love fixed it for me."

On Thursday, "we went directly to the Auditorium after roll call to practice for the graduation exercise. At 9:45 the class day program started." This was the occasion for various skits and takeoffs on the peculiarities of the people in class.

> Elliott the Lawyer said I would leave my Goldberg invention to the Smithsonian. Alas! Then we practiced the entrance for the sermon and the Annuals were passed out. Afterward, I got a ride with Elliott and Allan Leland to the class party at Cosby Lake, somewhere out toward Tarrant City. There wasn't too much to do. Merilyn wasn't with me. I got a boat and rowed around until it was time for the barbecue supper. I made a nickel bet with Buzzy Sattlewhite about what we would be doing in twenty years. We are supposed to meet then to settle up.

We never settled up. Whatever the bet was, I now owe him five cents.

> Merilyn stayed home [on Friday] to put up the scenery at the Pickwick she had been working on. [That evening] I took Merilyn to the leadout at the Pickwick. I gave her a beautiful white orchid—it cost fifteen dollars! Merilyn and I doubled with Mary Kinman and her date, Scot. Her scenery was a knockout—an arbor archway, with a fence and gate painted in blue and white covered with sparkling sequins and glitter, all in a garden setting . . . and the music, 'Dancing in the Dark.'

The music played into the night. There was an intermission party. Later we scampered off to somebody's house on Red Mountain for a breakfast! We got home at about 2:30 A.M. as it began to rain.

> Sunday, May 25, 1952: The Baccalaureate sermon was held today in the school auditorium. Merilyn was with the choir. Service started at 11:00. Everyone got mixed up in the processional . . . . Merilyn and I had lunch at the Molton Hotel. Went to Merilyn's house afterwards and stayed until 8:00. I had supper at her house. Monday the library called me about a book that was two hundred eleven days overdue.

I was out of school except for a couple of tests I still had to take, but I went over to the school so I could sit with Merilyn during her classes. After school, we played tennis.

On Tuesday, I had "finals in Latin and American History . . . . Passed them both—Mr. Ogle said I did quite well . . . . Merilyn was supposed to meet me at 12:00 in front of room 217 but she did not. I called her when I got home and we went out to play tennis. Pat Turner was at the tennis court," and my diary says that she told me "she and Henry Beatty would get married and have curly haired babies!" Later Merilyn went to her sorority meeting and picked up our leadout pictures.

"I walked around school Wednesday with Tommy Whitson." He was feeling pretty low. "He said that he and Betty had broken up for good and permanently!" The diary comments, "Ha, ha!" They were always breaking up, but never for long. [They are together and happily married to this day.] Merilyn had a French final exam, and she went home afterward. I went to town. "She met me at the library at 12:30. Later we went to the art gallery at the city hall. I came home after that while she went to get me a graduation present. That evening we went to Violet Bailey's house," a stucco house in the "Spanish" section of Hollywood, for another graduation party.

Thursday, May 29, 1952: "This was the day I graduated from high school. No more Mr. Shelton. No more Mrs. Steinbridge." We had to be at the school for graduation exercise at 7:00 P.M. "We picked up Mrs. Wilmont," our next-door neighbor from years earlier, and Merilyn, and then we went over to the school. At 8:00 it started. Mr. Buchanon, the Baptist minister from my church in Five Points, gave an address for half an hour. I was on the back row with Joan Wall cutting up and talking the whole time. My diploma had the supreme insult: "Even Harris Walker" penned in large script letters across the middle of the genuine parchment scroll. After the ceremony, Merilyn gave me two presents, a tennis racket and a book. Mrs. Wilmont, who was like a second mother to me, gave me a shirt. My graduation present from my parents was an Elgin watch. Mama gave me the expansion band to go with it. And that was it. The last days, each day filled with trivia. That was graduation from high school. Over, dead, buried—into the past. Gone and gone. Still, after all this time, I open the school annual, and the faces jump out as though it were yesterday.

On Friday, May 30, 1952, the day after my graduation exercise, Merilyn came home with me. I don't remember that she had ever been to my house prior to that. I had to mow the grass, so we went outside. I showed her the very tall pine tree in the back yard that I had always considered *my* tree. Wisteria had climbed the tree, covering the lower branches, branches 30 feet from the ground, with the purple blossoms of that fragrant vine. The back yard also had a small flowerbed, a birdbath, a garage—half of which served as my mad scientist laboratory as well as storage for odds and ends from the house. Merilyn had become an incredible new luxury. It seemed that suddenly she was free to go with me and be with me in a way that had always been restricted before. She was a beautiful luxury who could now wander about my world and pause and rest with me beneath the mimosa tree as if she too were one of the flowers in the garden.

I went over to Merilyn's the next day, Saturday. She was down at the creek with Joanne Ellard. I had to go looking for her. They were walking around in the water under the wooden bridge not far from the tennis courts. They were barefooted and chasing minnows and crawfish around in the shallow water. "I thought you were going to wait for me," I hollered. She got out of the creek. "Let's go back to your house." I sort of whispered, hoping that the two of us could go alone. Joanne took the hint. We went back to her house, back up to her garage where we could be alone and talk. I started working on some plans to make a small "ruling engine," a device for cutting optical diffraction gratings. She sat beside me quietly for a while, watching what I was attempting to do. I told her how the ruling engine is used to diamond-cut very accurately spaced parallel lines on glass or metal surfaces, 50,000 to the inch, each perfectly accurate to a millionth of an inch. Diffraction gratings are now used even in costume jewelry, but then they were an exotic scientific tool used by astronomers. I worked a while longer. She scurried off down the hill to her house to get us some iced tea. When she got back, we talked about when we could be married. We both knew I had to go to college. I don't believe there was ever any question about that, but I knew I could finish college in three years, just as I had high school. That would not be too long. I had already enrolled in Birmingham Southern for the summer quarter. That meant I would get a head start on college and also that I would be close by at least for the summer. I wanted to be a scientist—a physicist—too much to consider not going to college. I don't think she wanted me to do anything else, either, but we both felt it would be such a long time. So much could happen, even in three years. Later that day, we walked up the hill behind her house to Kite Hill. We stood there on Red Mountain in the summer breeze and looked off into our future, full of love.

THE ARROW OF TIME

Physical time itself, of course, is more than just the ticking of a mechanical or atomic clock. In classical physics—Newton's physics—time is that infinitely divisible stream that everywhere marks the changes in orbits and motions, days and nights, passing smoothly with no trace of any structure, moving exactly as Newton's equations prescribe—moving always in one forward direction without visible structure or motive action to make it go. In a way, that has been the problem. Why does time go one way and not the other? We are forced to move according to the arrow of time that points one way as forward, despite the fact that Newton's equations show no asymmetry in time. Going back should be as simple as going forward, like walking uptown or downtown. More than this, our mind keeps asking us, "If time is infinitely divisible, and the present spans an infinitely narrow *now*, how can we have any real existence, limited as we are to that infinitely short thickness of time?" Such questions that at first seem silly often turn out to be the ones that show us the clearest view of what is wrong with a picture of reality we have long held true.

If we make a movie of two or three billiard balls colliding on a billiard table and then run the picture backward, what we will see is just as physically correct, according to Newton's equations, as if it were viewed running forward. But take a film of traffic in the street, of birds flying, or of fire burning wood and run *it* backward, and things hardly seem right at all. How can this be? Where does nature's arrow of time come from?

Richard Morris, in his book *Time's Arrows*,[18] argues that it is changes in entropy (disorder) that determine the arrow of time. What does Morris mean by this? Let us go back to the billiard ball movie, but this time we will put more balls on the table. Let us say we start out with the usual 15 balls gathered in a triangular array. Now let us shoot the cue ball hard into this mass of billiard balls and film what happens. Of course, we know what happens. The balls scatter in all directions; 2 or 3 may fall into the pockets. All of this is just as it should be.

Now let us look at this movie backward. What we will see is 2 or 3 balls jumping out of the pockets and 15 racing to one end of the table to collect all in a neat triangular pattern of balls, while the cue ball suddenly races over to attach itself to the end of the cue stick. Everything we see is perfectly consistent with the classical laws of physics. If we were to measure the impact angles and velocities of the balls "before" and "after" any of the collisions, a simple calculation would show that everything satisfies the laws of conservation of momentum and energy quite well. The trouble is that there is virtually no chance that the balls would have started out with exactly the right speeds and directions to wind up so neatly collected. The way things actually happen is that any random scattering of balls makes their pattern become ever more random. Disorder—entropy—increases. What we see as time moving forward when we run the film forward is the natural tendency of things to go toward disorder. Reverse the film, and everything looks wrong because disorder just doesn't lead to order. Moreover, physicists in the last century figured out the laws by which all this works (the laws of thermodynamics, of entropy, and of how work is extracted from the energy of fuels) and showed that entropy—disorder—must always increase in any complete system. We can have some order happen, but it always occurs at the expense of a greater disorder somewhere else. The order we create when we set up the billiard table and neatly line up the balls is paid for many-fold by the increase in entropy brought about as our bodies burn the food we eat as fuel. This process entails more than just our bodies extracting energy from the food. The stored energy in the food must also have enough order for it to be used. The food must be collected. And the energy must be concentrated so that our bodies can use it. That is how our bodies, just like any other engine, use their fuel to do work.

Thermodynamics always controls the way things work. Entropy always increases overall. Time always goes along with that increase in disorder in the world. The proof lies in the fact that if we reversed everything perfectly, the way it was reversed in the film, then all the action would run backward. But we can't reverse the real world, so its time always goes in the direction of time's arrow.

But it isn't so! Morris's idea about how time's arrow works is not so. Entropy is not time's arrow. Yes, thermodynamics does govern everything. Yes, entropy and disorder do always go in one direction—for the overall system. But we don't live in a classical world. Not at the atomic level. Not at the macroscopic level.

Let us imagine that by some magic, by some suspension of all the laws of physics momentarily, or even by an act of God, in the street scene we talked about filming a moment ago, we could somehow reverse perfectly everything: the vehicles, the wheels, and pistons; the chemical reactions and the heat; and every molecule, atom, electron, photon, quark, and lepton. Imagine that we could suspend the Heisenberg uncertainty relations *momentarily*, just long enough so that we could exactly reverse the motions of everything except ourselves, standing on the street corner watching the world go by—backward. Reverse everything and then let 'er rip. What would we see?

We would not see what Morris pictured. We would not be able to match what we would see there frame by frame with the movie film running backward. The first few frames would match up all right. Some objects would keep going backward for quite a way, matching up with quite a few frames. But very quickly, Heisenberg's uncertainty relations—once again at work—would get in the way of this backward-running world down at the atomic level. Quantum mechanics, with time put in reverse, would take over, and all those perfect trajectories that were needed to make the billiard balls neatly assemble with the one ball racing over to stick to the cue stick—well, that just would not keep going at the atomic and molecular level very long. The uncertainty relations would mean that the backward-running atoms that might at first turn a few hot molecules of carbon dioxide and water into a few molecules of cold fuel and fresh air just would not continue to do that. The backward trajectories just could not stay perfect enough for the process to keep going. The backward-running engines would cough and quit. The engines would drag the backward-rolling cars to a halt. People walking backward would stumble and fall as neural messages quickly failed to reverse their paths. A few things would crash, and then thermodynamics would take over. The perfect backward order necessary for thermodynamics to run backward would not be there, and so thermodynamics in this negative-time world would—in just a few minus microseconds for molecules, a few negative milliseconds for neurons, a few backward seconds for people and cars, and maybe a few minutes in reverse for ships and trains—but sooner or later, everything would pick up the pieces and begin to work in just the way we see the world work.

Entropy, then, is not time's arrow. Thermodynamics sets the pattern, but it is quantum mechanics that actually points the way. Quantum mechanics points the way, but what that way is, that is another piece of the puzzle of the nature of reality.

The real arrow of time is our old friend state vector collapse. That is the thing that truly works only one way. Interactions prepare states, and observation causes state vector collapse, and that cannot go backward. State vector collapse is the real engine that drives time the way it goes, and, as we will see in a moment, it is that state vector collapse and those measurement loops we spoke of before that give time structure, texture, granularity, and "thickness."

The physical interactions in nature are all reversible by themselves. Even the Schrödinger equation works this way, even though it continually develops more and more potential states. But the observation causes state vector collapse, and this process goes only one way. There is no way to "back up" the observation. Observation and state vector collapse drive the clock forward and point the way. Entropy and disorder are only a part of the necessary pattern. Because of quantum mechanics and state vector collapse, that pattern of increasing entropy we see as time ticks by, would have been the same regardless of time's direction. Entropy (disorder) gives us macroscopic signposts that tell us things are different as we use up our energy supply, but it does not give the push or point the way. Strangely, we do that ourselves.

Figure 16.4 summarizes the way quantum mechanics, measurement loops, and observation all go together to make time run. This figure gives us a feel for the elaborate structure of time; it is more than a past and a future divided by some vanishingly thin present. It has a structure that is laced through with consciousness and will. And, oddly, it works backward from the way we would have expected. Our consciousness experiences the springing into being of possibilities like a myriad bubbles shining across a pond's surface. Then the will selects and the bubbles burst, driven by their own bursting—and the past is gone. We, the conscious entities, are like time's zippers: Our minds pulling together the future's infinite possibilities into yesterday's secured past.

But sometimes one of those loops spans through our lives, collecting distant moments into our experience. Sometimes these pieces of time are large. Sometimes the loops that thread some future enable us to see our tomorrows. Sometimes there is a magical glimpse into the future.

LOOPS OF THE ALL-SEEING EYE

Imagine that you are an executive. Decisions are your life. You have an important decision to make that will determine your future, and there are no facts to help you decide which way to go. Intuition is your only guide. You make your call. You make your choice for some reason that wells up from quiet subtleties inside your brain and around the edges of your consciousness. You make that call for some reason that wells up from your will. It is the right call. The decision works out right. How does that happen?

Days later, you learn what the correct decision had to be, and what it would have meant to have chosen any of the alternatives. At that moment, you experience certain aspects of the moment of decision you made days earlier—not every aspect of that earlier moment, but the few bits of information you needed to make your choice. These moments are the same *moment* in time. At least for those few bits of your total stream of consciousness that made you make the choice and feel the consequences—at least for that little part of your being—the two are the same moment. That is the miracle that makes us different from any machine we might have imagined people to be.

Exactly the same thing can happen for others who are a part of this decision loop that causes state vector collapse. A piece of time can reach into the future and touch anyone, anywhere, anytime.

These pieces of time, the causal conscious connections between the past and the future, span lives, and structure history as though guided by an unseen hand. They even span the whole course of evolution to make us our own progenitors.

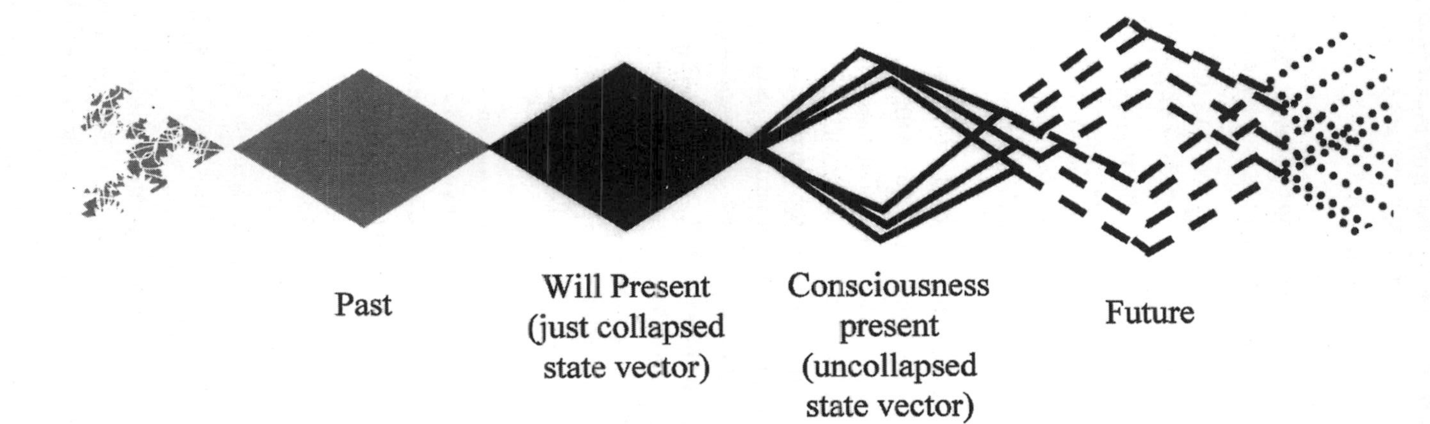

FIGURE 16.4 This diagram of a single sequence of measurement loops shows how state vector collapse on completion of measurement loops gives rise to the passage of time. To the far left, we see gray loops that are already in the past. Next, cross-hatched, is a state vector that has just collapsed into a single state. To the right of that is a loop that we should imagine as in the process of being completed. It has not yet collapsed and so consists of a collection of states represented by a series of overlapping parallelograms. These represent the will present and the consciousness present. To the far right in our diagram lie the potentialities of the next selected state, the unrealized future. Their separation out of the present will and present consciousness is done for the sake of clarity in the graphical representation of the flow of time.

And we, in our present moment, now complete that first moment of creation when all this reality began.

> Sunday, June 1, 1952: Went to church with Merilyn. Bought a ruler at the drug store. Briggs called about my debt at the Frat. Maitland called. She got home this morning. She said that they had had a raid at her school.

I didn't know what she was talking about at first.

"A panty raid," she said.

"Oh." As I said before, even the dance scenes in *Singin' in the Rain* seemed racy to me. Panty raids were, well, civilization in the final throes of decadence. Maitland's *joie de vivre* was sometimes shocking to me, and I could see no good coming of it. But, alas, she was tempting!

Monday morning: "Watched ants until about 1:00 P.M." I had graduated from high school. This was the life. Not a care in the world. No school. Free and in love. I could take the whole morning just to watch ants! They are interesting, you know. One single, tiny ant has a whole world—and today that ant had her "summer" off as well. I watched her. She took her time as though time were free forever just to groom every stubby stick of hair on her bone-hard skin, preening until her body armor glistened black. She preened; she paused. She tasted the softness of the summer air with long-waving antennae, looked myopically at me, and then suddenly, seemingly, she remembered something. Abruptly, she had some thought that we have no way of peering into, and then she scurried away as fast as her six little legs could carry her. I had spent the whole morning watching tiny dramas and manikin interludes. School was over. I was free.

> Went to play tennis at 1:30 P.M. We couldn't get a court, so we went across the street to Hamburger Heaven. She bought me a hamburger and a coke. I bought her a fried pie. We went to Homewood and mailed a package. We walked to my house to get something for my ruling engine. We took a moment to kiss—just a moment to kiss—then we walked back to the tennis courts.

"We walked back to the tennis courts, played a little, and then went to her house. I stayed until 6:30. I got into a watergun battle with the kids next door." I wanted to stay a while longer, I remember, but I had to go home. "I called Merilyn at 7:30. She called me back a little later. Bed at 12:15 A.M., 6/3/52."

That's all. There are no more entries on through the summer. I enrolled at Birmingham Southern. Merilyn and I spent the days and some evenings together. Time passed, and the diary lay idle—past Independence Day, through July, and on into August. Nothing until August 7.

The summer passed. Then this entry: "August 7, 1952: Broke up for good with Merilyn at 4:00 P.M. Went to her house to talk. Ralph gave me an envelope with my ring in it and her note." There was nothing that presaged the breakup. Her note referred to things that had never been a problem before. She mentioned something about being put on a pedestal, but I remember too little of the note now to reconstruct any hidden messages. There certainly was nothing that I could understand then, and nothing of it remains now. Looking further in my diary, I see that she had an operation to remove a tumor in September. Could she have already known that something was seriously wrong? Did her disease have anything to do with what she did then? Questions.

THE SUBSTANCE OF SPACE

Let us return to a subject we talked about several chapters back. Let us go back to the ideas of Einstein and of Minkowski about the nature of space and time. Largely as a result of Einstein's work, Minkowski showed how time could be treated as though it were another dimension. The fact that space has three dimensions—length, width, and height—affects the form of the equations of physics. This was known long before Einstein, of course. But Minkowski showed that the equations of physics could be rewritten so that time appears as a fourth dimension. It is, however, a very peculiar dimension.

To understand a little better how this works, it will be useful to remember how we calculate distances in space. This goes back to the Pythagorean theorem. A town may be reached by going 3 miles east and then turning north for 4 more miles. But if there is a direct road, the distance we travel will be 5 miles instead of 7. Pythagoras expressed this fact in a general form. He stated that the sum of the squares of the sides of a right triangle equals the square of the hypotenuse—the same Pythagorean theorem we used when we talked about Bell's theorem. More simply, if the distance east is x, the distance north is y, and the actual distance (along the direct path) is s, we know that $s^2 = x^2 + y^2$ (that is, $5^2 = 3^2 + 4^2$). In three-dimensional space, we can add the distance up, z, to this expression so that the distance s is given by solving $s^2 = x^2 + y^2 + z^2$.

Now what Minkowski found was that all the fundamental equations of physics look as though time can be included as a distance as well. The equations of physics had always been written so that the three directions in space were on the same footing, but time was always something different, something separate. Minkowski found that these equations look as though the events of space and time always involve a distance s given by an expression $s^2 = x^2 + y^2 + z^2 - c^2t^2$, where c is the speed of light and t is the time. Note that this equation is quite similar to the Pythagorean expression, except that we have a new negative term that includes time. This expression can be used to obtain a kind of *signature* for the structure of space-time. It is sometimes written as (1, 1, 1, –1). However, if we are to be more exact, we should note also that there are no terms that look like *xy* or *yz*. To do that we need a kind of map that has 1 for the term we add in, –1 for the time we subtract out, and 0 where there is no term.

The result is called a matrix, and for the space-time of our physical reality, it looks like this:[19]

$$(g_{\mu\upsilon}) = \begin{pmatrix} 1 & 0 & 0 & 0 \\ 0 & 1 & 0 & 0 \\ 0 & 0 & 1 & 0 \\ 0 & 0 & 0 & -1 \end{pmatrix}$$

where $(g_{\mu\upsilon})$ is the usual name given to this mathematical object.[20]

Now, what does this have to do with how nature, with the help of an observer, brings about the occurrence of quantum events? Earlier, we pointed out that by adding an information term to the Schrödinger equation, we could get an expression that would force state vector collapse. What happens is that all the probabilities become either 0 or 1, which means that each state either does not occur or does occur. Obviously, only one of the states can have a probability of 1, and all the rest must be 0. The way this works is that this added information term acts as though it came with a guiding matrix that selects which state happens. To make this aspect of quantum event selection explicit—to show how the quantum mind selects what happens in detail—requires that the information term we added to the Schrödinger equation have a matrix in front of it that is +1 for all the states that vanish and is –1 for the one state that is selected, the state that does occur. For an event that has just four possibilities, that matrix will look exactly like that $(g_{\mu\upsilon})$. The –1 being in the fourth position down would mean that the fourth possible picture would have been the one that would come into being.[21]

Once again, maybe you can see it coming. To select the third possibility, for example, would require the –1 to be in the third row and column. But the order of the terms in the state vector is arbitrary. We can reorder this matrix just by moving the names around. If we always write the state that actually occurs as the last term, then the matrix for an event with four possibilities will look just like the $(g_{\mu\upsilon})$ matrix.[22]

There is one further limit on this expression, this matrix that selects the state that actually happens. Remember that this expression must be capable of describing a dual space. The dual space comes from the fact that we have to have a comparison operation to have a state actually occur. The state vector collapse, including this matrix of 1s and 0s, comes out of "contributions" made by both sides in that comparison loop we dealt with earlier. This means that just before the completion of the comparison loop, this matrix of 1s and 0s that designates which state actually happens must be two separate matrices. When put together, the two make this 1s and 0s matrix, with the –1 down in the bottom corner. Before the state vector collapse, we must be able to divide this matrix into two separate matrices that, when multiplied together (that is, when $\Psi\Psi^\dagger$ come together in a comparison), will give us this 1, 1, 1, . . . , –1 form.

Now, it can be shown that this form for the matrix cannot be created out of two separate matrices[23] unless we have, as a minimum, exactly the form we have for the ($g_{\mu\upsilon}$) matrix.[24] The minimum matrix for observer state selection is the same as the signature matrix for the four-dimensional space-time matrix. Of course, there can always be more complexity to the state described by quantum mechanics (there almost always is), but the process of observation and state vector collapse alone is sufficient to limit the lowest level of our universe structure to a space-time structure that has the three extensions of width, length, and height—the three dimensions of space—and one of time.

But you may have noticed that the order is imposed by placing the event that occurred at the end of the list. That is to say, we chose to write the –1 in the lower right corner and rearrange terms accordingly. What this means is that the way we interpret the world is such that we have imposed this space-time order on it. The reality is simply that events occur, and in the lowest level of their occurrence, there are four coordinates that exist and that arise out of the underlying time-like substrate. But whichever time direction occurs, we see this as being the one time coordinate, and the others, where the event did not occur, are seen as space. Conceptually, the idea is perhaps beyond visualization. We see things in space that change in time. But the reality is much simpler than that space-time concept itself. *There is no space.* There are only observational events. Events can occur or not occur. For events to merely satisfy that condition and satisfy the dual-space requirement that must be met if observation is to exist requires a four-dimensional matrix, space-time, with one time coordinate for the events that occur. Thus space is merely the ordering of those things that could have happened but did not happen.

So you see, there really is no such thing as space. Time exists, but space is just the ordering of events that had probabilities—potentialities—but that did not become events. The structure of space actually is a product of Hilbert space, that space that describes probabilities in physics. This is why the Hilbert space of probabilities and our everyday real space look alike. They have the same beginnings. And this is the basis on which quantum mechanics and general relativity can be unified.

MATTER

Let us now go one small step further and then hopefully we will see that everything then comes together. Remember that earlier in this chapter we talked about the fact that the information term, $p \log p$, could not be used in the Schrödinger equation. I said then that what we need is an expression that has imaginary terms so that ordinarily state vector collapse could not occur. State vector collapse would occur only when measurement loops occurred—when things that had interacted with each other in the past came back together to interact again and produce the real quantities $\psi^*\psi$. This gives us quantities that do not have imaginary numbers in them. As you will remember, that is why we used the form $\Psi \log \Psi$ for the new term in the Schrödinger equation. But it turns out that adding this term can introduce some real quantities as well. They turn out to be just sort of add-on quantities that ordinarily do not interfere with the rest of the mathematics, a kind of grit that is left behind.[25] When I first found that these real quantities could appear, I was bothered. Then I realized what they were. Although they are a bit of a nuisance in the ordinary nonrelativistic Schrödinger equation, that equation is not the most fundamental equation.

Schrödinger's equation does not satisfy Einstein's relativity requirements. The relativistic form of the basic equation, the equation that replaces Schrödinger equation, is Dirac's equation. Dirac's equation is the cherry on top of the chocolate sundae of physics! It is more elegant and more nearly perfect than anything else in all of physics. It is the equation that combines Schrödinger's quantum theory and Einstein's special theory of relativity; it is that remarkable equation that predicted the existence of positrons. What I realized was that we could introduce this information term into the Dirac equation. We could do it by letting it replace the mass term—that is, the term for the electron or the positron in Dirac's equation. Doing this achieves something quite remarkable. This information term now becomes the source of the masses of particles in physics. By writing an equation with only two terms—the space-time term, which we have seen is tied to the dual space of probabilities involved in observation, and an information term tied to the process of consciousness that causes state vector collapse—we get the Dirac equation back. If we assume only that conscious observation exists, that alone is enough to let us understand where space-time and matter come from. There is no space as such, no matter as such. There exists only the observer, consciously experiencing his or her complement. And in doing so, the observer weaves the illusion of space-time, and matter falls like snow from the conscious loops of the mind.

I have not written the foregoing as a technical exhibition of the mathematics of this theory. (That can be found elsewhere.[26] I went to Vienna to talk about the technical details of the mathematics at a symposium marking the 100th birthday of Schrödinger.) What I have tried to do here is show something about the nature of reality that we can discover when we extend the prior work on consciousness and the measurement problem to round everything out—to explain how consciousness causes state vector collapse and illuminate the physical basis of the interaction that we have been calling observation.

A lot remains to be done. String theories, superstring theories, and supergravity hypotheses exist. I doubt that the answers lie in those directions, except somewhat incidentally. I think the direct path to an understanding of all the forces of nature, all the particles and interactions, is to be discovered by understanding what observation and consciousness are all about and understanding the measurement problem in quantum mechanics. In particular, I think the answer arises out of a very elementary equation that has two terms: a space-time operator and an information term operating on the state vector. This is the form we get if we merely say that "we observe events," or, even more simply, "conscious observation exists." This requires that something be observed that we can most simply and symmetrically assume to be of the same form as the observer. There is an observer-observed duality that is also exactly the same thing if viewed from the other side, the observed side. Then the original observer becomes the observed, and the observed the observer. The information input selects the states, which gives us the change in the states. And that process is the origin of space-time. That space-time has an indefinite number of pieces of what has already been seen, particles, pieces of information—the pieces of matter that are you and me. This dualistic idealism lies at the very core of existence. It is the loom that weaves the fabric of reality. And behind the loom, the quantum mind.

So we at last find that reality is the observer observing. It is the two parts of our great separation coming together. There is a separation. There is a dreadful and vast separation. But there is no space and really no matter to die but that our own minds did not first come together to create it. Our observation—our coming together—created matter. Observation itself is the stuff of the space that reaches out past the vast clusters of galaxies. Reality is the fruit of love's embrace.

17

The Causal Mind

If I had been, in the beginning,
God brooding upon the womb of
absence, I might well have
pondered: matter, plucked
from emptiness.

—Robert Pack, "Big Bang,"
Poetry, January 1988

In the beginning God created the
universe. . . . "God said, Let
there be light;" and there
was light.

—*Genesis* 1:1, 3

We have always looked in one of three directions in order to understand what our existence is all about. We have looked at the pieces of matter to find the make up of the world. We have looked outward, seeking to see infinity and believing that if we could, we would understand the universe and the grand design. And we have looked inward, into our own mind, searching there for origins and meaning. The answer will come when we can look in any one of these directions and understand the same thing.

The night sky seems filled with stars. Viewing it from a completely dark location on a moonless night, far from city lights, a person with good vision can see about two thousand stars. Most of these stars are much like the sun. They consist mostly of hydrogen gas, which fuels the sun's great nuclear furnace that gives us light and life here on earth. A few of the stars are the still-glowing ash of this hydrogen fuel converted into helium, which in turn burned into heavier atoms until the fire went out.

We look out into the night sky of stars like Spica, in the constellation Virgo; it is a star blue from its 20,000K temperature, radiating 20 times as brightly as the sun. Giant stars such as Antares, 10 times the mass of the sun and nearly 400 times the sun's diameter, burn red in the night sky. There are white dwarf stars such as Sirius B, as massive as the sun in a ball the size of earth. Stars such as Altair, Procyon, and Alpha Centauri A, like our sun burning brightly in the night for billions of years, line the "main sequence." Our telescopes enable us to see dimly glowing, wispy clouds of gas where new stars are being born, perhaps to harbor life like ours on planets that will move about some new sun billions of years from now. They let us see the remains of supernova stars that may have brought an entire world of life to an end in one sudden second on some planet that once was there. Out there among these twinkling points of light, there are stories that we cannot imagine.

Twenty-five thousand light-years away from us lies NGC2808, a cluster of thousands of stars glowing a warm red from the hues of its many old red-giant stars. Star cluster M3 contains half a million stars, many of which are burned out, remnant dwarves. The cluster M13 contains more than one million intensely hot stars that give it its bluish hue.

Yet all of these stars and the hundreds of clusters of stars are just parts of a much greater collection that make up our own galaxy. It contains 100 billion stars! It stretches 100,000 light-years across and has a thickness of 10,000 to 20,000 light-years.

Beyond our own galaxy there are other galaxies, galaxies of all sizes and types. All together there are about 50 billion galaxies scattered throughout the whole space of our universe, which extends perhaps 14 billion light-years in distance from us and 14 billion years back in time.

Scattered among the more distant galaxies are the mysterious and powerful quasars, so called because of their quasi-stellar appearance. Believed to be the most intense sources of light in the universe, quasars may consist of a very dense collection of stars having a total mass 10,000 to 100,000 times the mass of the Milky Way galaxy.

From the tiniest pieces of matter that we have studied, we have seen, at each level, the coming together of clusters that form the next larger objects—quarks forming nucleons, nucleons and electrons forming atoms, atoms forming molecules that cluster into all the forms of life and minerals on earth. The planets clustering into orbits about the sun, the sun among stars in an outer area of our Milky Way galaxy, the galaxies clustering into groups, and these into clusters of galaxies, and then these into superclusters of millions of galaxies. Just when we might think that this must go on forever in both larger and larger clusterings and into smaller and smaller constituents, we discover that all this ends. And just as it ends, we find suddenly a rather different shape; we discover the largest structures in the universe. We discover that the galaxies have formed into membrane-like structures enclosing enormous voids, as if the galaxies were the molecules in the foam on some giant mug of beer.

But that foam of galaxies cannot extend too far. We are already looking out into space and back in time almost to the edge and to the beginning of the universe. We are looking at a universe that has a structure that is itself a clue to its origins—a clue to the basic structure of its most elementary particles.

GENERAL RELATIVITY

Just as electromagnetic forces govern the structure of atoms and molecules, the force of gravity dominates the structure of the large objects in the universe. More than this, though, just as we had to have quantum mechanics in order to understand how atoms and molecules are put together, there is a special kind of mechanics that we need to know in order to understand how the universe as a whole works. That special mechanics is known as the general theory of relativity.

As you will remember, we talked about Einstein's special theory of relativity in Chapter 3. The special theory of relativity is the theory that shows us how the speed of light can always be the same no matter how fast we travel with respect to the source of the light, or, as it is usually expressed, the speed of light is independent of the velocity of our "frame of reference." Frame of reference is simply a means of discussing the way we go about measuring distances in any particular experiment, in which, of course, there must be some common reference point for all the measurements. Einstein's theory of relativity is all about how frames of reference are distorted by their motion relative to our own frame of reference, so that, as we said before, the speed of light remains a constant, the mass of objects increases, and forces that act on objects in those frames of reference appear to be strangely altered.

Despite the success of the special theory of relativity, Einstein perceived it to be an incomplete theory because it said nothing about what happens when a frame of reference is accelerated. In the special theory, frames of reference have different velocities, but those velocities are always assumed to be constant.

But what if the frame of reference is accelerated? How could the laws of physics be rewritten so that they would be valid regardless of what the frame of reference did, even if the frame of reference were being accelerated?

The problem Einstein faced, however, was the fact that if our frame of reference gets accelerated, you know about it! Remember when we talked about the way a pen behaves if, while flying at 600 miles per hour on an airliner, you let it drop to the floor? It falls just the way it would if you had dropped it while the plane was sitting on the ground at the airport. The laws of physics look the same in both cases. But what happens if you drop the pencil while the plane is taking off? You certainly can see that what happens to your local frame of reference has an effect on the laws of physics that you see then. The fact that the airplane is accelerating makes it look like there is a force acting on the pen that throws the pen toward the rear of the airplane. Einstein was faced with the problem of finding a way to write down the laws of physics so that they would look the same, somehow, even when one is in an accelerated frame of reference.

The problem, of course, is that it seems as though things that happen in an accelerated frame of reference are entirely different from what happens when one is moving uniformly. But Einstein noticed something that tied these two kinds of experience together. He noticed that an object dropped inside a uniformly accelerating room or frame of reference would move as though it had a particular kind of force acting on it—a gravitational force. He noticed that any two objects released at the same moment would seem to fall exactly the same way, just as though in a gravitational field, exactly the way Galileo had shown that gravity acts on objects when he did his Tower of Pisa experiment. If these effects are so much the same, they must be related in a deeper way. One must actually be a form of the other. Somehow, the gravitational field affects an inertial frame in exactly the same way that acceleration affects an inertial frame. At some deeper level, gravity and acceleration must be the same thing.

This surely seems like a good idea, but how do you take this equivalence between accelerated frames of reference and gravitational forces and turn it into a new general theory, a new set of laws governing the mechanics of bodies?

Einstein noticed, of course, that accelerated frames of reference and gravitational fields are not identical everywhere and for all time. A man in a small elevator that is uniformly accelerated up cannot tell whether the force is due to gravity, to his acceleration, or to some combination of both. But if the elevator were large enough, then he could detect that "down" is in a different direction at the sides of his elevator from its direction at the center. What this means is that the effect of gravity and of acceleration on a frame of reference will be equivalent only if we consider a very small region of space for our frame of reference.[1] Einstein's general theory, therefore, has to be formulated in terms of some quantity that characterizes any kind of frame of references in tiny regions of space, point by point throughout space.

Space and time—the stuff of frames of reference—are about distance, the distance that separates any two objects, and the time that separates events. Remember the Pythagorean theorem, the one about the square of the hypotenuse of a right triangle being equal to the sum of the squares of the other two sides, which we have talked about several times already? Well, that comes into all of this also. If you want to know the distance between two points, and if you can make measurements along two reference measuring rods, x and y, that lie in the plane of these two points, then you know that s^2 equals $x^2 + y^2$ because Pythagoras said so. Also remember that in special relativity, which deals with space-time events, the distance between two events is given by adding the squares of the x, y, and y distances between the two events and also subtracting out the square of the difference in time between their occurrences, measured in terms of the speed of light. The expression we had there was

$$s^2 = x^2 + y^2 + z^2 - c^2 t^2$$

which, as you can see, is a generalization of the Pythagorean theorem expressed for a four-dimensional space-time frame of reference. In general relativity, however, space can be curved because of the presence of matter or because we are being accelerated. In a curved space, the Pythagorean theorem no longer works. In such a space, we also need other possible combinations of x, y, z, and t in order to calculate the distance between points or events. We may need terms like $2xy$ or zct, and terms like x^2 and y^2 may have to have coefficients like 2 or $\frac{1}{2}$ or other quantities. In general relativity, then, in order to be able to talk about the shape of space, we use a shorthand way of expressing how we calculate the distance s. Remember the map for special relativity? The map for $s^2 = x^2 + y^2 + z^2-(ct)^2$ looks like this:

$$(g_{\mu\upsilon}) = \begin{pmatrix} 1 & 0 & 0 & 0 \\ 0 & 1 & 0 & 0 \\ 0 & 0 & 1 & 0 \\ 0 & 0 & 0 & -1 \end{pmatrix}$$

where $(g_{\mu\upsilon})$ is this matrix's name, and the $g_{\mu\upsilon}$ are the individual values in the matrix.

Now, Einstein used this same kind of map, this $(g_{\mu\upsilon})$ matrix, to represent what space-time would look like in the presence of gravitational fields or when we are being accelerated. In that case, many of those 0s and 1s change, though usually by extremely tiny amounts.

Thus Einstein could use this map in which the individual elements are the quantities $g_{\mu\upsilon}$ as a way of characterizing the shape of space-time. All he needed to complete his theory was an equation that could be used to relate changes in the local shape of the $g_{\mu\upsilon}$—that is, changes in the space-time metric—to the presence of matter, momentum, energy, and electromagnetic radiation, the stuff that causes space-time to warp. With such an equation, it would then be possible to figure out the shape of space-time everywhere.

He found the equation he was looking for in something called the Bianchi identities. These expressions are identities—equations that are always true because they are really just tautologies. They are mathematical expressions that talk about the $g_{\mu\upsilon}$ and that are constructed so that they always sum up to zero, regardless of the way we represent our frame of reference in space. However, Einstein felt the Bianchi identities were "too high a derivative" for what he needed. That is to say, they give the rate of change of the rate of change of the rate of change of the $g_{\mu\upsilon}$—a bit too much. Isaac Newton's equation is a second-derivative equation. Maxwell's equations are second-order, as are the equations governing the flow of fluids.

Einstein clearly saw that what he needed was a "second-order equation," an equation that has "second-order derivatives" in it—much, but not too much! That is to say, he wanted an equation that would give the rate of change of the rate of change of the radius of the curvature of space.

Einstein took Bianchi's identities and "contracted" them. That is, he massaged them a little to get them into the second-order form he was looking for to describe the changes in the curvature of space and time. He took the resulting expression and instead of saying that it always equaled zero, he set it equal to the sum total of all the energy (or matter) in the electromagnetic fields and that in all other forms of matter (ponderable matter) as well.

Figure 17.1 shows an illustration that is frequently used to explain the ideas behind Einstein's theory. It shows a heavy mass that has been placed on an elastic membrane so as to cause this otherwise flat surface to be distorted. This is, in fact, very much like the way the space-time curvature happens, according to general relativity. Of course, in Figure 17.1 we are talking about the distortion of a two-dimensional rubber surface, whereas in general relativity, we have to deal with the four dimensions of space and time that are made out of nothing.

Let us look for a moment at what happens as the mass of an object gets larger. In that case, the distortion of space and time gets more and more severe, corresponding to the successively greater and greater "necking down" of the elastic membrane in Figure 17.1.

Ultimately, we should expect this necking down to go all the way. The entire space-time metric would fold up around the object, and it would "go over the event horizon," which means that nothing that goes in can ever come out again.

Einstein recognized that the matter in the stars throughout space might also cause a distortion of the entire space of the universe. The equations of general relativity that he derived in order to generalize special relativity so that it would work even in accelerated frames of reference, and in order to explain gravitation, now permitted him to study the cosmology of the whole universe. He could find out from astronomers how densely stars are distributed in space and use this information to determine the shape of the entire universe. But when he made the calculations, he discovered something that he personally found entirely unacceptable. He found that his equation said that the universe is not some stable eternal

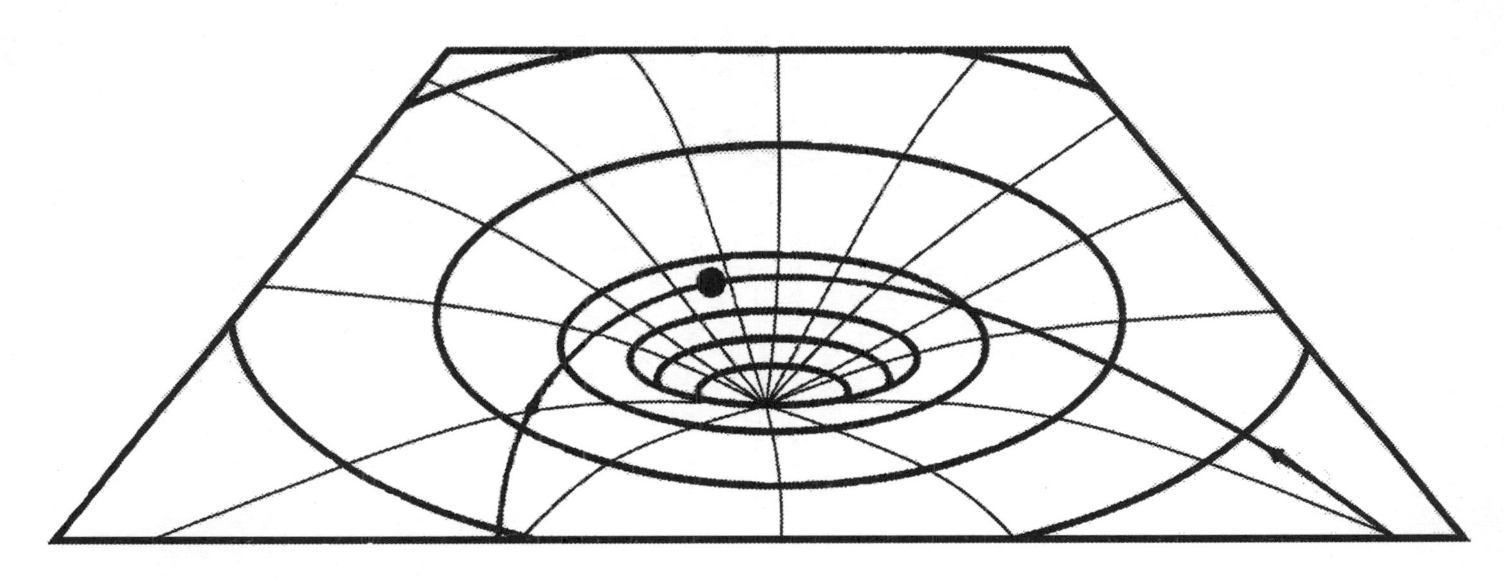

FIGURE 17.1 Imagine a sheet of some elastic material stretched out and supported along its edges. If a heavy ball is placed at the center, the elastic material will be deflected downward as indicated by this drawing. This distortion is much like the distortion that matter causes in space according to the general theory of relativity. For example, the sun distorts space itself like this, so that a comet hurtling in from the depths of space is deflected as it simply moves along the most direct path past the sun, just as a little ball rolling along the elastic sheet will follow a curved course, as shown.

configuration but, rather, that it must have had a beginning. It must have been created some 10 or 15 billion years ago and is now rapidly expanding outward as though there had been some long-ago explosive creation event.

Einstein did not like that idea. If you look out at the night sky, it all looks so permanent and eternal. More than this, and despite Einstein's frequent references to how God would do physics, Einstein's scientific instincts were those of a materialist, and the materialist ideas at that time tended toward the idea that matter was forever and had never been "created." Creation ideas, then as now, were considered to be notions espoused by religious fundamentalists. Science had no place for such ideas. Certainly the last thing any physicist or astronomer would have expected to find was evidence of a creation event! And yet that is what was discovered. Einstein's equations revealed that the universe should be expanding out from a creation event. But he felt so strongly that the universe had to be in a steady state that he made, as he himself called it, the biggest blunder of his career. In 1917, Einstein changed his equation to force it to give him the steady-state solution he wanted. It was just a few short years later that he discovered how badly he had erred.

In 1922, Edwin Hubble, at Mount Wilson Observatory in California, began to find evidence that nearly all the galaxies seemed to be rushing away from one another as if the entire universe had been created in one grand explosion billions of years ago. This was the beginning of the "Big Bang" theory of the universe's origin. Of course, the Big Bang theory was not accepted quickly. Steady-state theories of the universe continued to be proposed and studied. But in 1965, these steady-state theories were shown to have been wrong. Arno Penzias and Robert Wilson detected a microwave radiation coming in from all directions in the sky—a radiation due to the leftover heat from this Big Bang explosion. In a sense, they were looking directly into the original flash of creation at the beginning of time. Calculations confirmed that the temperature and the distribution of this radiation were just what the Big Bang theory predicted.

◆

"Thursday, November 6, 1952: I went to school, Shades Valley, to see Merilyn on her birthday. I know now how close she was to death even then." The entry in my diary was written on Thursday, December 19, 1952.

> She acted very nice toward me, almost as if she still liked me. She spoke condescendingly about Dick, the boy she had been going with. I saw her one more time after this, about a week later. Mary Kinman greeted me with, 'Oh no, not you again.' Merilyn said, 'Hello,' in an unenthusiastic voice. I promptly left.

On December 26, 1952, I made this entry for "Tuesday, November 11, 1952. This is the last time I saw Merilyn alive." I would have a glimpse of her at the hospital later, but this was the last that I saw her as she was. On that same day, I tried to put my feelings into a poem, but it just would not work.

The ending is at the end of the diary. The last entries fill the few remaining pages like last breaths. These last pages are difficult to read even now, filled with the anguish of the moment, filled with impotence, filled with the fact that the world goes on even when it seems it should end.

> Monday, December 15, 1952: Dear Diary, My love, stupid as is the term, is dying. The time is 10:07 P.M. I learned ten minutes ago that she is dying, and I hate the world for it. She has cancer, leukemia, which developed from the tumor she had removed in September from under her right arm. I have been sick today, my stomach. I should be sick. I was talking to Whitson on the phone before I called Ceil Keeley and I found out. I took it so lightly when I was talking to him. She is dying and you, this book, know what she meant to me.

The next day I had a German test at the college. I got sick during the test, and I had to leave.

> Went to South Highland Hospital to see Merilyn but didn't get to. I saw Mr. Neal instead. I went with him to their house. I saw the dog, Merilyn's puppy that she got for her birthday. It's a cute little animal . . . . I cannot force myself to believe her dying is a reality. Went to Shades Valley to see Ceil Keeley. The world is ridiculous.

I wrote a poem that day. It is not much of a poem, but it recorded comments made to me that day, mainly by Ralph—title: Death of Merilyn.

'Makes you think twice, doesn't it?'
But you don't know what you are thinking—
'She's always been a little less active, hasn't she?'
I feel empty. Maybe something I ate; maybe nothing.
'People dying all the time.'
But she's one in a million, she's one in a million.
'Her throat is sore, very sore.
Sometimes she can't speak,
and she's afraid it will affect her voice.'
A girl sixteen is dying.
She doesn't know . . .
She won't live, and she won't live . . .
And I hear myself say,
'She'll live in me, she'll live in me.'
'Makes you think twice, doesn't it.'

Wednesday, December 17:

> I had a chemistry test today. I thought about Merilyn the whole time. I went to the hospital at noon. The nurse said that Merilyn couldn't see anyone. Her mother didn't feel like seeing me either. Ralph wasn't there. He was at lunch. I came home and spent the rest of the day in tears and praying for her. I don't know why. The nurse said that Merilyn was talking, but she didn't know what she was saying. Merilyn probably won't be able to see anyone again. She will die by Christmas.
>
> Thursday, December 18, 1952: At noon I went to Shades Valley to see Ceil at lunch. I wanted to talk to her about how bad I feel, but she was too happy and I left.

Ceil was not happy about Merilyn, of course, it was just that I was so hurt by everything that I could not stand to see others behaving normally.

> I went to town from there and left my blue suit to be altered. I went to see about getting a spring to fix my 32 cal. pistol. I went to the Post Office, went to the library to get a book by Oscar Williams. Then I went to see Merilyn. She was no better. Instead, she is getting worse slowly. I went to her room. The door was cracked open and I saw her—only from her waist to her shoulders. She didn't have anything on. At first I didn't realize I was seeing her. She looked so small, and so thin.

There was a large discolored area of skin on her side.

> She was bare from her waist to her shoulders. The doctors had been giving her oxygen but it didn't do any good. She has been given a narcotic. At least I saw her one last time.
>
> At eleven that evening, while I was writing a poem to Merilyn, I felt as though for a moment she was behind me. It seemed as if she were looking over my shoulder. Friday . . . Elaine Ellard woke me at 7:35 this morning, calling to tell me Merilyn had died. She died at 2:55 A.M. of lymphoreticuloma, Hodgkin's disease. I went to town to get a pair of black shoes to wear with my blue suit. I went to the Zehnder's . . . , that is to the Neals' house about one. I met Marty, the girl Merilyn was always telling me about. Ralph talked to me; we stood outside. He said that before she died she had said to him, 'I feel like I'm two people . . . one of them is lousy . . . the other drinks water.'

The reference to feeling like two people is reminiscent of the near-death experiences that have come to light in recent years with the works of Kübler-Ross, Moody, Ring, and Greyson. The reference to one who feels lousy is of course obvious, referring to her sick body. The other, who "drinks water"—perhaps that is symbolic of spiritual things; perhaps it was all part of a delirium. Ralph told me that "this was the only thing she said that made him believe she knew she was going to die. She never said anything else to make them feel she knew. They had never told her. But Merilyn was very intelligent and quite mature. She knew. At that age, most of us worry about the slightest sign of sickness—that it might be the beginning of the end. We may not say anything to anyone, but the worry is there. Merilyn knew from the time they told her she should have this "simple procedure" to remove the swelling under her arm.

> I left the Neals' house about two in the afternoon. I walked to Valley Chapel. She, Merilyn's body that is, wasn't ready. Jeanne Connally came to see her. I went in. She didn't look like Merilyn. She was too thin. She had on a blue dress with her wide gold belt. Her L/A pin was on wrong. Her hair was pulled back flat. Her lips were thin. Her lipstick was pale. She looked strange. I stayed with her about a half-hour, alone. Then the morticians came in to fix her. They put on a darker lipstick . . . a thin oily lipstick quickly painted on the lips. They had everyone leave while they fixed her arms. Her hands were white, chalk white. The body is not that same Merilyn. Merilyn has left it. They let us back in and I stayed some more. A mortician came in and again put on a still darker lipstick. She never had so much lipstick when I was with her. They fixed her hair better. I put her L/A pin on straight. Her dress was blue, the belt was gold, her bracelet was gold, her hands were white, not olive. The club pin that I put on her was gold—and black: 'How do you like me in black?' I remember her asking me, 'How do you like me in black?' Later, my mother came. I left with Tommy Whitson for a while. They moved her into the chapel and I went in there and stayed with her a lot longer. Mrs. Waldrop came and she held my arm. Miss Gaston came (what mean things we had said about her and she turned out to be so much better than so many). Mrs. Hightower was there. Pat Turner, Violet, Mary Kinman, Jo Ann Bowden, Hildegarde Spears. Dick, the guy she had dated just before getting sick, came in, sat in the back a couple minutes, and left. She had two preachers for her service that began at four. One was Reverend Calvin Pinkard, the Methodist minister at Trinity Methodist Church in Edgewood. He read from the Twenty-Third Psalm: 'Yea, though I walk through the valley of the shadow of death . . . .' The second was a Catholic priest who conducted the rest of the service. Merilyn's body was placed in the hearse at six that evening and she left for Jackson Mississippi where she will be buried beside her father. I will go see her, and go to the cemetery. I must see you Merilyn. I will go to the cemetery and write it down. I love you, Merilyn.

The last entry in my little black diary for 1952 is dated January 1, 1953, entered in the diary's Cash Accounts section: "Total, Nothing. Why did God want it?"

On February first I wrote this in my notebook collection of poems:

Time, distance, death, Merilyn;
Memory, thought, question.

Question one be one—nothing.
No answer, no time, no Merilyn—
No death.
Personal God?
Mechanical or Puppet God?
Or will the ball bounce and fall by chance
On my geometry
Just because it just so happens?
That Merilyn is dead forms some tensor
And that is that.
But even dead she is only one infinitely
Minute instant
Less possessed by consciousness of the world than I.
But it is this instant
That memory says my past looks to
And Life says
My future worships.
Oh me,
It seems rather terrible—but,
Be out with the light
And the moth will leave.

And so, nearly half a century ago, Merilyn was dead.

THE INFLATIONARY THEORY OF THE UNIVERSE

Despite the successes of the Big Bang theory in explaining such things as Hubble's data and the background microwave radiation that Penzias and Wilson discovered, the theory also has some problems. For example, there is the *flatness* problem. Unless the original density of material in the universe at the time of the Big Bang had exactly the right starting value, the universe should not now seem so very nearly flat. Space should instead be curled up in a tiny ball, or it should have such a negative curvature that the Big Bang would have long ago blown everything away from everything else. Either the universe should have blown away—that is, expanded so fast that matter would have all been swept away—or the expansion should have been so slow that it would have stopped long ago and gravity would have swept it all back up, curling everything into a tiny knot. In the one case, there would not have been sufficient matter around to form the stars and galaxies; we could never have been here. In the other case, there would not have been enough time for these things to have formed before it was all over.

Alan Guth, of M.I.T. has pointed out that for the universe to have made it this long, some 14 billion years, requires a fine tuning of the density of matter to 1 part in about 10^{55}! This requires knife-edge precision in the "adjustment" of the initial conditions of the universe.

Another problem with the Big Bang theory is something called the *horizon* problem. Regardless of what kind of starting point one chooses for the initial universe, no matter how small, the explosion behaves as though the various parts of the universe all decided to explode at the same time. When an ordinary ball of explosive (say, a ball that is 1 meter in diameter) is set off by placing a detonator on the surface of the ball and firing it, it may look like the whole thing goes off at one instant. But if we take a high-speed movie of the event, we see that before the fireball expands, the detonation has to propagate away from the detonator. As a result, the blast is in reality decidedly asymmetrical. Our universe, however, is so homogeneous that we must assume the whole thing was "detonated" all at once. In fact, to give the result we see now, there would have had to have been about 10^{83} "detonators" all going off at the same instant. That detonator problem is something that the Big Bang theory by itself just cannot explain.

And apart from all these issues, there is a much bigger one: the *singularity* problem. "Where did the universe come from? How could everything appear from nothing?"

In the last few years, physicists have developed a new theory of the Big Bang's first moments, a history of the universe as it existed between the age of 10^{-43} second and 10^{-35} second. The theory, first developed by Guth, is called the inflationary theory of the universe. It describes how random quantum fluctuations in an infinitesimal vacuum would have led to an incredible expansion, an expansion in size of the universe by 24 orders of magnitude in only 10^{-35} seconds. At the end of this time, matter would have precipitated out of that vacuum fluctuation state, superheated as the particles collided, and then undergone the Big Bang explosion of the universe that is still going on.

This inflationary theory of the universe combines Einstein's equations, which describe the way space-time changes its shape depending on how much matter is there to distort it, with an equation from quantum mechanics that describes how matter behaves in a vacuum. This equation from quantum mechanics tells us there are continuous fluctuations in the vacuum state. Remember that quantum mechanics does not permit us to make absolute statements such as "There is exactly zero energy in this place at this exact time." As a result, we have to view empty space as filled with virtual particles darting into and out of existence. According to inflationary theory the shape of space-time would have been very rapidly altered by the presence of these virtual particles—these quantum mechanical potentialities—in such a way as to begin the universe.

Moreover, there is a peculiar twist to this story. The conservation laws of matter require that something called the *continuity* equation holds true everywhere. This law tells us that any of the matter content in a region of space cannot change without matter flowing into or out of that region of space—very reasonable. But during the inflationary era, the amount of *space* changed. The amount of space in the universe itself rapidly expanded. Einstein's equations require that during this expansion, the continuity equation still had to hold everywhere. That is to say, the density of matter everywhere in space had to stay the same, while the total amount of *space* itself increased explosively. This is what created the matter in the universe, all the matter, all except for a tiny seed that was in the beginning.

From that seed of perhaps only a few kilograms of initial matter—matter that then, indeed, existed only as a potentiality—from that, all the matter that now forms the billions of galaxies and billions of people came into being.

What existed before that first moment when the universe was only 10^{-43} second old? Well, physics is still silent on that question. At a time of 10^{-43} second (the Planck era, as it is known), both gravitational and quantum mechanical processes would have been at work, and we do not know the physics of that era. But there are some possibilities . . . .

With the inflationary theory as a starting point, together with the subsequent Big Bang, we are now able to trace the events in the formation of our universe from the Planck era to the present time. We have discovered an incredible interrelationship between elementary-particle physics and our understanding of the structure of the universe as a whole. Tracing the creation of the universe back to well within the first second of its existence, physicists have discovered that the observed structure of the universe has a lot to say about particle physics and about the unification of forces that is now believed to have been a characteristic of the universe during its first moments. In these moments, one would not have been able to tell that there were any differences among electromagnetic forces, weak interactions, strong interactions, and perhaps even gravitation. Rather, it is thought that all these existed simply as a mass of look-alike particles that had not yet had time to cool into the different kinds of particles that now make up our world.

It is also possible now to use these theories to calculate, on the basis of our knowledge of the elementary constituents of the universe, just how much hydrogen, helium, and lithium should be present in the universe and to show that our list of how many kinds of quarks and leptons that should exist is already complete. That is, it seems that the search for the kinds of particles that now make up our universe is complete. And our knowledge now lets us determine the distribution of galaxies throughout the universe, showing us that the frothy distribution is just what we should expect to find. All these theories, then, are finally beginning to let us see the entire pattern, the entire cloth of the fabric of reality.

But there have been some more recent problems with the Big Bang theory and with the inflation scenario. The difficulties involve the fact that inflation requires a very specific density of matter to be present in the universe today. Observation, however, shows that there is not that much matter in the universe. The result has been that theoreticians, in trying to solve this problem, have been making too many trips to the "tooth fairy." They have been proposing that there exist in the universe fantastic new kinds of unseeable matter called "nonbaryonic matter." But even with these assumptions, things still do not quite work out. The predicted structure of things in the universe does not correspond to observations. At the largest scale, the universe contains immense structures—great cavities and walls of galaxies—and galaxies that came into being when the universe was far too young. Further, and perhaps worst of all, it has been found that there are stars that are older than the universe itself is supposed to be, according to the Big Bang theory. To fix things up, astrophysicists have thrown the cosmological constant, "Einstein's worst mistake," back into the mix—with speed and alacrity—to make things fit the latest facts.

All of these problems, however, arise from a failure to take into account the fact that just as inflation can create matter, deflation during the end phase of stellar collapse and black hole formation can lead to a loss of matter from the universe. That basic equation Einstein gave us that tells how the Big Bang works shows that the universe is very sensitive to having just the right amount of matter in it, and Guth's inflation scenario, created to solve the flatness problem, shows us that at the end of the inflation era, the universe would have been poised on the knife edge of that matter content. Matter loss as a result of deflation, if it occurs early in the life of the universe, can therefore be significant. A loss of only 1% when the universe was just 100,000 years old (at the end of what is called the radiation-dominated era) would cause the universe to go suddenly into superexpansion, leaving the universe to seem now to be 90% deficient in matter (compared with what the inflation theory leads us to expect). This seems to have happened. The cause was the fact that the first generation of stars formed were supermassive, superfast-burning million-solar-mass stars clustered into incredibly large collections (the first generation of quasars, each containing some 10 billion of these million-solar-mass stars). The collapse of these stars at the end of their short lives did two things. First, the cores of the stars turned into black holes that in the end underwent deflation and mass loss[2]—the 1% mass loss we mentioned before. Second, this mass loss caused the rest of the star, suddenly released from full gravitational confinement, to explode; this *super*nova explosion released the heavy elements that now make up 3% of the mass of the universe and form such things as the planets and our bodies.

As mass was lost from these collapsing superstars, the universe began a superexpansion process, which resulted in these regions becoming the centers of the giant voids that form the "beer froth" structure of the universe that we see as its largest structural features. We can still see the flashes from these explosions. They are observed as "gamma ray bursters" that satellites detect about once a day at random locations in the sky—incredible bursts of energy reaching us from the edge of the universe and the beginning of time.

The Big Bang theory does in fact work, but it has to be tweaked.

The inflationary theory, together with the Big Bang theory, lets us trace everything back to the very beginning. That beginning is called the Planck era. And what do we find there? We find that the universe began as a primitive, pure, intense—a very intense—quantum state that existed in an infinitesimal point of space and time—a moment when all matter potentiality, all space, and all time existed as one thing, one quantum state. But what is it that we have learned that such quantum states in fact are? What have we found, as we have searched point by point to understand the nature of reality? We have found, above all else, that quantum states and mind are one in the same thing. We have found that in their essential nature, quantum fluctuations are the stuff of consciousness and will. And now, here, we find that this mind stuff was the beginning point of the universe—the stuff that out of a formless void created everything that was created. We find that in the beginning, there was this quantum potentiality. We discover that in the beginning, there was the Quantum Mind, a first cause, itself time-independent and nonlocal, that created space-time and matter/energy.

18

A God for Tomorrow

I am He who exists from the Undivided . . . . It is I who am the light . . . . I am the All. Everything came from Me and Everything extends unto Me. Split a piece of wood, and I am there. Part the stone, and you will find Me.

—The Gospel of Thomas

It used to be that I would read a book such as Harvey Cox's *Secular City* and feel the emptiness, the vacant eyes of the hollow men, the stuffed men. I used to believe that only a few held onto some religion and worshiped what they believed in. I used to feel alone—alone or lost—because I feared that perhaps I was just one of a dwindling few, leaning together—one of a last few who still felt a need to look in wonder and in awe at the world and, in doing so, make his own kind of supplication. But I do not make that error anymore. Everyone worships—everyone.

Everyone worships reality. Each person looks about him, listens a moment—listens as long as life will let him pause to listen—and then he falls down and worships whatever it is that looks like this is what it is all about. He does anything, and to this final meaning, he falls down and worships. In his own twilight kingdom of hoped-for dreams and voices speaking promises in his head, he falls and worships. And this is something wonderful, too.

A few years ago, I attended a conference in California on "New Visions of Reality." The conference was sponsored by the Department of Physics of the University of California at Berkeley and the *Journal of Time, Space, and Knowledge*. Most of the conference took place at the Claremont Hotel, a huge gleaming-white Victorian resort in the hills of Berkeley overlooking the Bay Area. Among the people at the conference were Fritjof Capra, author of *The Tao of Physics*, Gary Zukav, author of *The Dancing Wu Li Masters*, Geoffrey Chew, who proposed the "Bootstrap theory of elementary particles," David Finkelstein, editor of the *Journal of Theoretical Physics*, Rupert Sheldrake, of "morphogenic field" fame, and Durk Pearson, author of *Life Extension*. For several days we talked, debated, and mulled over ideas ranging from particle physics, Buddhist thought, consciousness, evolution, and inner visions to physical fitness and vitamins. On the last day, the organizers held a news conference. About 20 reporters came to hear what had been happening. All the participants gave a synopsis of the talk each had made, a kind of last chance to publicize his or her piece of the puzzle. As I remember, toward the end, Professor Chew was once again saying things about his Bootstrap theory, about Feynman diagrams, and something about an infinite number of lines representing an infrared disaster. We had really explored some wonderful ideas at that meeting, but little of any meaning seemed to come across to the reporters. The reporters seemed to be searching for a way to yoke all these divergent ideas into something coherent. But I felt I knew what the speakers were wanting to express. It is on everybody's mind in some form, though we do not usually express it any more clearly than to say we are searching. In one way or another, we are looking for our God.

When the palaver finally worked its way round to me, I said quite directly, "What we have been doing here is laying the foundations for a religion of the twenty-first century." I was surprised at how quickly the other speakers agreed with this assessment of a meeting in which neither God nor religion had figured as the primary topic. But the reason for their agreement is clear. The fabric of all our conventional beliefs has been severely strained in recent years. For many, little more than tatters remains. We were there to put some of the pieces back together, knowing that if we could understand those pieces of reality, we might then better know the mind of God and the future of religion.

Mankind has traveled a long way in the search to understand reality. We have looked deeply at ourselves, at our religions, and at our science. We have looked at what we are made of and looked at our origins. Beginning with the myths and superstitions of antiquity, we have followed a path of discovery into the modern age of scientific understanding, and we have helped to open the door to new realms of science. Ancient peoples looked into the sky. Someone pointed at the eclipsing sun and said, "That is the eye of Baal closing" . . . and the people were afraid and believed. Others at other times saw an image in a gnarled branch and cut a god free with a stone knife, or chiseled gods out of whole rock . . . and the people were afraid and believed. They fashioned myths and gods, philosophies and religions as tools for their minds to understand a beautiful and fearsome world . . . and the people were afraid and believed.

But over the generations, people have tested these ideas little by little. They developed pictures of their world that could give them a better understanding of reality. New ideas moved them step by step closer to answers about their origins—our origins—and about the meaning of life. These pictures, these ideas are still a part of our lives, and the questions are still with us as well. We still ask about the meaning of life. We still believe and we are still afraid.

We still search to find who we are. We still hunger to find some god that we can know. We still ask, "Is there a God?" We still look to see where. Our Hubble telescope peers to the beginning of time, and we ask, "Where could He be in this vast, empty universe?" And in a universe with 100 billion galaxies of 100 billion stars, if He does exist, somehow entwined in these laws that turn the wheels of time, can there be anything there in Him that cares about us, about me? Do we have some meaning that transcends this moment in time? Do we have souls? Do those we have loved, who have now died, still exist somewhere? Is there a God to watch over us, some God of Abraham who would intervene in His great clockwork to spare even the whole world even for a moment? Is God a personal God? Was there a resurrection of Christ? And where is her soul? Where is Merilyn?

How many of us fear we already know the answer? How many live tied and baled by the skeptical judgment we fear true? The words were spoken by Lucretius. The words are as ancient as humanity, and as modern as today's science news: The world, the universe—all of it is made of matter. When the matter is lifeless, we are dead, and there is nothing else. We are no more. Death is the end. There is no God. There are no miracles. And she no longer exists.

But it isn't so. Now we know that the world is something entirely different from what we imagined before. Now we know from our study of Bell's theorem that the simple picture of a box-like space containing pieces of matter is not adequate to explain how the universe works or what we are. We have discovered that there is much more. We have examined the world, the physics of particles, the nature of mind and will, and the things that tie it all together.

We have seen how consciousness is something involving the creation of potentialities as described by quantum theory. We have seen how will is a part of the conscious mind that selects what will happen to those possibilities. We have seen how all of us—our minds—are connected. We have seen how this collective consciousness is the source of psychic knowledge, is the power that creates miracles, and is infinite in its span of space and time. We have seen that our universe came into being in the first moment of time—in an explosive expansion of potential states described by that same quantum theory. In an instant, the mind of the cosmos created our universe in a thought, in a conscious act of will.

We have searched back to the beginning of time and to the origin of the universe to find the first thought, the first word of God springing into existence as consciousness and physical matter. Consciousness, will, mind—these were the first moment, the potentialities that continue to this day. Our consciousness, our mind, and the will of God are the same mind.

Over and over, scientists have searched for and found the most fundamental truths of reality amid those ideas that resemble the most beautiful forms of nature—creations that are the lilies of the valley. Even as science would remove one imperfect god created in our image, we have caught, in the same new threads of this greater fabric, a glimpse of a more nearly perfect pattern woven by a greater weaver. That the laws of motion set down by Isaac Newton could span so much of nature seemed to Newton more like a sermon than like the agnostic retort others have turned it into.

Einstein, who did not believe in any personal God, nevertheless had an odd faith in the orderliness of nature, a faith that induced him to seek the truths he found for us in an inner sense of an almost mystical conception of the material world. But it is in his efforts to understand matter—to unify all our experience of reality into a single conception that would unify all the forces and all the laws of nature into a single edifice, reaching, he hoped, from the atomic to the cosmos—that we first see that objective reality is a flawed and incomplete conception of reality. We find that what we had thought was an independent and objective physical world is in fact contingent on the observer, on our combined consciousness, in a most basic way. And it follows that the God we find must be a personal God. It is here that we first see that the observer—that we ourselves—must feature in this picture of reality. It is in Einstein's relativity and in his space-time that matter begins to lose its sharp edge. It no longer cuts antithetical views quite so deeply nor quite so well. Now we see that the independent existence of matter and the absoluteness of space were false dogma.

But it has only been with the advent of quantum theory that we have discovered proof that we exist as something more than pieces of matter. In the development of quantum theory, the observer emerges as a co-equal in the foundry of creation. Just as the clock tender is implied by the clock's turning hands, so the quantum view of matter requires the presence of an observer. The tests of Bell's theorem have shown us that objective reality as it has been conceived is not the true fabric of reality. The observer interacts with matter. Consciousness, the substance of this new-found reality that defines the observer, has fundamental existence. It is the quantum mind that is the basic reality.

We have explored the nature of consciousness and of mind. We have found the measures of consciousness, and we have found out something about the machinery of the brain that is the source of our human consciousness. We have found ways to take the measure of consciousness, to study its contact with the functions of the brain, and to discover how the mind, through the agency of the will, affects matter and transcends the limits of space and time. We have seen how our individual wills, each of which can intervene in the material events of the world, can and must be constrained at some common source. We have gone just a bit further to see, perhaps, where matter comes from and what space itself is, and we have even seen something of the nature and structure of time. We have seen matter and space as the natural consequence of nothing more than the fact that conscious observers exist.

I do not know how much of all this will remain and stand the test of time. I do know that the philosophy of science that ignored the question of consciousness is standing there dead. I know too that science can no longer ignore the fact that our conscious observation affects the quantum potentialities of matter, and we can no longer ignore the significance of what our will does in bridging minds and affecting phenomena in the physical world. The links that exist among the measurement problem in quantum mechanics, the concept of the observer in physics, information processing by the synapses of the brain, consciousness, and mind—these are going to remain.

I believe that in all of this we can see a few answers to the many questions about our place in the universe. We can see the "secular," material aspect as only a part of a larger reality. We have a mind that is indeed our own to govern a brain and a body. We have a quantifiable will that controls the brain and reaches out to influence the world we observe. And more than this, there is a collective will that has some of the characteristics of what we call God. What happens to our consciousness on the death of our body? The answer is not yet clear. Our will is transcendent. It has existence that is not dependent on any limits of space and time, an existence that is shared with others. Thus, to a certain extent, it seems something of us must survive. There is, of course, a great deal to be learned before we will know the answers to the question of survival, but at least we see that there are valid approaches to such questions.

We also now know that miracles occur, and we know the mechanism of such intervention in the workings of the world. The quantum mind is what lies behind the selection of quantum events and determines the outcome of quantum mechanically entangled states.[1] It is at once a part of a more fully understood natural world and at the same time the direct intervention of the mind of a greater Consciousness.

We do not know whether God would intervene to speak to Moses or lead a people. Of course, we do not know from what we have studied here whether Jesus healed the lame and gave sight to the blind. We do not know whether Jesus rose from the dead. But we do understand that things that would seem miraculous can be brought about by the intervention of mind. Low-probability states from the state vector allowed by quantum mechanics do occur, and their occurrence happens because we are part of the selection process. More than this, we know there is a Consciousness that at least would have had the power to do these things—a consciousness so great that it was able to create all the potentialities of the universe in one thought, in that first 10^{-43} second . . . in the beginning.

Certainly I want to know the true answers. But it is only through a careful and faithful study of science that we can find the answers and the basis for accepting as fact any of the stories of our religions. Contrary to what many may conclude, I have not sought answers that would satisfy any prior commitment of my own. Far from it! I began as the most ardent student of objective science. I did want to search for answers to the fundamental questions of philosophy. But just as it should be in science, the facts have pointed the direction. Consciousness is a fact that had been overlooked in conventional science. Understanding what consciousness is led me into the intricacies of the measurement problem on the one hand and into an understanding of how the mind selects quantum (and real-world) events on the other. At every turn, the requirements of physics have dictated the conclusions. Different observers cannot see different realities. The hidden variables of conscious observation must be collectively constrained. These are facts that shape the reality we live in. Whatever future and more nearly perfect description of physics may be discovered, it must still embrace these same facts. And if it says this, then there must exist something that with time we can only come to know better, something that stands in the role of God.

Many scientists will deplore my saying this and my saying it this way. Most scientists think we cannot speak of God in science without losing objectivity. The abuses of dogmatists in religion are still fresh in the collective mind of science. Undoubtedly, these abuses are just cause for concern, but it will not do for us to have thrown off one group of dogmatists only to deliver science into the clutches of another such group. The tools of science permit us to question, test, and dispute atheistic doctrines posing as scientific principle just as much as they permit us to question, test, and dispute theistic doctrines.

Scientists have long believed that science had to embrace the doctrine of materialism. Well, science has had a long and glorious history discovering the secrets of matter. It has studied, classified, theorized, and measured matter in all its manifestations and has given us great understanding. Along the way, many false visions, including many scriptural teachings, have fallen victim to the scrutiny of science. As T. H. Huxley put it, "Extinguished theologians lie about the cradle of every science as the strangled snakes beside [the cradle of] Hercules." Materialism in science has served us well. It has enlightened us. It has brought us closer to many truths. But science's investment in materialism has itself turned into a creed, with its own high priests ready to torment the unorthodox. Many phenomena have been ignored in the name of this materialism. The obvious—such as consciousness—has been shut out, exactly as if such ideas were the teachings of a heretic. Phenomena that would not fit materialistic concepts have been made anathema and estranged.

Science has rolled its war wagons over the crushed myths of so many religious beliefs. It has marshaled its mechanics to explain the motions of the sun, moon, and stars. It has mapped the heavens, leaving no place there for gods to live. It has rolled out its Darwin, its chronologies and atomic dating methods, and its Freud to destroy the ideas of our creation in the image of God, of an earth believed to have been made in 4004 B.C., of religious visions, and of divine inspiration.

The consequences have been devastating to religion. Religion—all religion—has for some time been dying a slow death. Many will greet such a statement with elation that the earth may be freed of the self-righteous, the fearmongers, those who raise the specter of eternal condemnation, those who instill guilt in children and the frail of mind, and those who have inflicted pain, inflamed hatred, and fostered wars in the name of this or that religion. Sadly, these have indeed been legion. Surely what I have to say here, what I see for tomorrow, will never tolerate the return of such perverted minds. We cannot understand those like the conquistadores, who crossed the oceans to enslave the aborigines, steal their gold, and save their souls. We cannot understand those who went to war to capture Jerusalem for Jesus. We cannot understand those who could even imagine a "holy war" as anything but sacrilege. We cannot understand those atrocities any more than we can comprehend the twentieth-century world with its 100-million-killed militarism inflicting fascism or dialectical materialism. But at least we have risen high enough on our hind legs to know that those were truly perversions—perversions that had nothing to do with religion.

Religion to me is not merely one particular creed, though perhaps it is best experienced through one's own creed. It is not the condemning of the sins of others. It is none of the vile acts of hate and war and death. It is not the harbinger of misery. All that is the treachery of usurpers who use whatever device they can find to turn people to their own purposes.

But for all this terror, there is one thing that is worse: the thought that all the suffering and all the pleasure of life have no meaning. And that is the sad corollary of our vanishing religious life. Science has the capacity to show us the path to truth. We must go down that path and face whatever is there. But I think that just as some high priests of past religions have sought to impose their personal wills by distorting the teaching of their own prophets, so too scientists have often guided our steps down equally false trails. We have often presumed the direction of science first and cut the path later, before checking to see whether we were going the right way. Science has ignored all issues that might have suggested some middle ground or that might have compromised its secular bias. It has now brought us to the very edge of a world stripped of all innate moral values, without giving us anything to take its place.

While humanism and existential philosophies may be formulated in the universities, the ignorant thug in the street has already reached the conclusion that awaits the ponderous thinker: "You have nothing else but what you get. When you're dead, you're dead." Bankrupt of values, civilization is on a course to disaster—a disaster to kindle nostalgia for the relative tranquility that has been the twentieth century. Science has warred with religion, and for many of the world, it has won. It has pushed back superstition and ignorance, but it has left no principles on which to build lives or any new system of values. Science has won not by showing the concerns of religion to be false but by ignoring those areas it could not deal with. It has ignored questions of God, mind, and morals and replaced them with material pursuits. This has made it appear that there is nothing else to reality but a material world. I would like to believe that what we have done here at least indicates that there are other worlds for science to find and explore. I hope we have shown that questions of religion can be pursued and answered by science and that somewhere in the future, perhaps, we will have a full understanding of our reality.

I think one thing at least is certain: It is clear that consciousness can be broached scientifically. Consciousness exists. And for the first time, we have used the instruments of scientific investigation to fit its existence into the overall tapestry of reality. For the first time, we understand what consciousness is. We can understand mind as including conscious experience and will. We can see how these fit into the physical processes of the brain that are involved in thinking, the data-processing operations at synaptic junctions in the brain.

Our knowledge of consciousness enables us to understand how things must work in the brain. For once, we can extend our knowledge of physical processes and quantum mechanics into realms we never dreamed could be fathomed. We see how paradoxes in physics, such as the Wigner's Friend paradox and the puzzles of the test of Bell's theorem, can all be put together to show that we as observers of reality play a basic role in the structure of the universe. But more than all of this, surprisingly, we have discovered that every path we have taken to learn something of the structure of the universe finally comes around to the same result. Whether to understand the interconnections of will, to understand the most basic facts in quantum theory, or to discover the beginnings of the Big Bang universe, each path leads to the fact that there must exist a supreme Consciousness out of which everything else springs. It is Consciousness that began everything, that grows matter into a universe of existence; it is Consciousness that unifies and constrains all of us as individual beings; it is Consciousness that orders space and time out of a chaos of random events.

Perhaps, too, we have seen something more than all of this. Perhaps knowing that consciousness is something real and ponderable, perhaps we see that the Kingdom is here and it is now. We may be attached to bodies, and maybe we can see that is well and good, but our existence, this infinite space of consciousness, this Hilbert space of many places, is the house of many mansions—not figuratively, but quite literally.

All of us have known the materialist's vision of reality. All of us have spent most of our lives boxed up in it. Maybe now, some of us at least can see beyond this. Maybe now, some of us can see the justification of faith. Maybe some of us can see deep in all of this the God of Abraham, see the Trinity, see miracles as more than myth in shaping who we are. Perhaps in what we have learned, we can see something of the beliefs that inspired the great Buddha, see something of the Taoist's way, something of the ancient ways of the Hindu. And if we have learned well, we will see all of this as one vision that does not lead to competitive struggles or conflict.

There are many questions to which we do not have answers. We have the beginnings of many answers, but many questions remain about the structure of reality, about other realities, about life after our bodily death. We have many questions about just how we should sort the wheat from the chaff of all the religious literature that has come down to us through the ages. We should know that if we can understand the message clearly, it has something to tell us. But we need more tools to find the comfort of truth that faith alone has been able to give us in the past. Faith was never meant to be blind faith. Faith was always meant to be a faith guided by revealed truth—revealed through the experience of something beyond our own physical self; revealed through the lives that many have lived as examples; revealed in histories, in prophesies, and in the poetry of scripture.

But the demands are so much greater now. Now we can see better how easily we err and how easily we stray. We need a better way to seek out truth, to assimilate the jewels of all our religious teaching into one universal faith founded in knowledge that we can verify as we do the facts of science. I hope that the discoveries recorded in this book are the beginning of such a mission. No one who believes in the truth of any of the world's great religions should fear losing any essential part of that faith by testing its truth against what we can learn with this new science. Those willing to discover an even greater truth in their religion will find untold wonders hidden in what they already believe.

◇

A god waiting to be chiseled free from the stone or exposed from within the fiber of the tree . . . the mechanics of pendulums, clocks, planets and stars . . . images of objective reality made of rocks, atoms, and quarks—these have marked our passage. But there is a greater reality. We have seen that the discoveries of the twentieth century have altered the view of reality as something that can be understood apart from the observer who is a part of that reality. Space, time, matter, and energy—the very stuff of objective reality, as it turns out—depend on the perceptual participation of the observer. His or her motion, place, and frame of reference alter that structure in an oddly democratic way, with no special preferred observer. Even the atoms and particles of matter have structures that depend on what can and what cannot in principle be observed by us. But it is with the tests of the EPR paradox that we have been forced to abandon the concept of objective reality. After so many hints, we see at last that we must restructure our concepts of nature and put ourselves back into the picture. Finally, we see that we must understand consciousness, so obviously a part of reality, in order to understand fully any part of the design.

Our knowledge of how quantum mechanics works with state vector collapse on observation ties in with a quantum mechanical picture of consciousness, consciousness arising out of the very observer-dependent processes that go on in the brain as they do in the laboratories of physicists, in the hearts of atoms, and in the cores of stars. And with an observer in the brain, this consciousness selects the things that happen in the external world.

Out of this arises a special picture of what the fabric of reality is. Because of that signal-to-noise disparity between the will and the consciousness of our mind, only about a thousandth of what could happen happens out of an ostensive correlation with what we would wish for. All the rest is stochastic or moves inexorably forward under the laws of physics for microscopic bodies. Matter, objects—a physical domain exists that is governed by immutable laws. But these laws leave open a range of happenings that are left to the selection of the mind. Behind this selection is the will. The will works much like a communications channel that links all of us to a common control center. Within the power of our combined will lies the power to do essentially anything. Within the power of this will lies any knowledge—of anything known, or knowable—of the past or of the future. And it is a power that is always present. This is the collective will of all sentient beings with the power of determining the state of all events, with the knowledge of a seer. This is the Omnipresence, the Omniscience, the Omnipotence of Abraham's God, at once personal and supreme.

This is a God of our collective will and of the collective will of the universe. This is a God that has the potential of any knowledge that we know. A God that has the power to make any event occur and yet is restrained by the limits of our own minds. A God that pervades all things and yet acts through our vision.

We have seen the connection between consciousness and the workings of the brain. We have studied the role of consciousness in state vector collapse. We have fathomed the relationship between mind phenomena and quantum mechanics—between miracle and physics. In all this, we can see that the miracles of our faiths, which seemed to defy all reason, now lie within the grasp of our understanding. These things do not violate the laws of the universe but rather form a part of a reality that bonds man, mind, and God together in a lawful and knowing universe. We can now understand that these miracles are the inevitable consequence of our own being.

But this trinity of man, mind, and God—of brain consciousness and collective will—is not the full realization of the fabric of reality. There is a structure that I as a Christian find a justification of the central features of the Judeo-Christian conception of nature. But I also see, as one and the same, a Buddhist conception, even in its most solemn expression—a unity in nature, all things being aspects of mind. But now we see more of the underlying structure, of the engine that drives our struggling souls. We see the separation that lies between any of us and the rest of reality. We see space and matter as springing from the brow of God.

A universe that has only matter cannot have consciousness and cannot have will. The picture painted to explain the material world, orderly but without God, has failed to work. The closer conception of a Spinozan universe, matter that is ordered into a divine-like creation, is also a flawed philosophy. It does not have the machinery to account for its order, and it does not represent the real universe in which conscious beings dwell. Einstein could see the print of God's hand on creation extending to the edges of the cosmos, but he failed to see us there, he failed to see the implications of mind for physics, and he failed to see anything but the shadow of God.

Some have failed to see any place—any space—where God could reside, and others have failed to see where any consciousness could hide within the atoms of matter. But we have found that reality. We have found that hiding place. We have seen that the universe springs from every thought of God and matter from the very existence of mind. We have looked to find reality. We have seen beyond the open door.

Monday, October 17, 1988:

Often I had walked up to Clermont, past the house at 1414, past that steep lawn with brown pine needles mingled with ivy leaves spilling down the bank and over the wall by the sidewalk. Often I had walked there, each time hoping to find some feeling of her presence still, some tangible fragment of those memories, hoping for something that would come to me as though a ghost of her being were still there, somewhere along the walk to her house. But it was always just old memories—old memories of all those same houses. Memories of those same yards, grown full with vegetation just as it was then, all the same, but with the children now changed to old people, bent, turning slightly at their door to see me walk past.

But this time I stopped at the park where we had played tennis and strolled down the same paths in the park, past the same tennis courts, past the same chain-link fence around the courts where people were playing as we had played years ago. There were the same half-buried flagstone walkways there then that now zig-zag as if for no reason across a culvert that has since been covered over. Some of the cannas that once lined this culvert are still there, but they now are separated into two rows, seemingly with no reason for being where they stand. I walked past the stone walls and past the same picnic table pavilion where we dashed to wait out one of those heavy Alabama downpours. I walked across the wooden bridge we had crossed so often, stepping up the same oversized concrete steps to cross the creek there, and, once again, I walked around the tennis court.

Suddenly, I felt as if half a dozen souls had rushed through me, causing time to shudder as if I were only a jittering image on a TV screen or some Kirk starship caught in a rift in space. I was physically shaken. Her soul had passed through me. Shaking, I walked through the same park, past the same fence, the same flagstone wall, and I felt her presence as though we were still playing there . . . she waiting for me to die . . . waiting for some future when my soul would be free . . . waiting for me . . . .

I left, driving away still tingling in my fingers and on my face, my legs still shaking and nervous from this strange meeting that we had just had across the decades. Something of her was still there waiting. And where else would she have gone to play and to wait?

◈

As I walked down the steps from her apartment, my mind went back over the years. My mind went back to things that have been and that I have done, the things of my life and the things of that day. My mind went back to the things I have told you about Merilyn, back to an image, to a wonderful image, to a vision forever of my future and of my destiny. My mind went back there.

As I reached the landing, my mind went back to that time long ago, back to her memory within me, and I knew, at last, that she is still with us.

appendix I

Consciousness Equations

In Chapter 12, we listed four items that when multiplied together give us the coupling factor Q, the number of electrons that, tunneling from an initial active synapse, are able reach a suitable receptor site on other active synapses during the time that the first synapse is still active. These four factors are

1. The number of synapses that are active (about to fire) at the same time our donor synapse is active.
2. The chance that any particular donor electron will be on one of the 200,000 donor molecules in another active synapse.
3. The number of donor electrons traveling out from the donor synapse.
4. The number of hops each electron can make during the time the donor synapse is active.

The product of these four values gives the coupling factor Q for the donor synapse.

The first of these (item 1 in the foregoing list) is the number of synapses that are about to fire at the same time as the donor synapse. It is the product of the total number, N, of synapses in the brain, about 2.35×10^{13}; the frequency, f, with which they become active (essentially, this is the firing frequency); and the length of time an average synapse stays in this state before firing or reverting to the rest state—that is, the synaptic delay time, t, which has a value of 0.3 millisecond. Thus item 1 is the product Nft.

The second item in our list is the chance that any particular donor electron will be located on one of the 200,000 acceptor molecules or gate molecules (the same number as donor molecules) in any synapse. Now because there are 7.45×10^{20} soluble RNA "stepping stone" molecules, there is only a small chance that the electron will be on one of these 200,000 sites at any given moment. The chance is 200,000 divided by 7.45×10^{20}, but let us write it symbolically as n/M (where $n = 200{,}000$ and $M = 7.45 \times 10^{20}$). The third item is just n: 200,000 donor electrons start out from the first synapse "looking" for a second synapse to fire. Finally, the fourth item: Because the electrons can make a jump in a time, τ, that is only 8.4 trillionths of a second, each electron can make a very large number of hops in the 0.3 millisecond available. That is, each electron can make t/τ hops.

Now, putting these four factors together yields the product $Nft \times n/M \times n \times t/\tau$ as the coupling factor, which we will call Q:

$$Q = n^2t^2Nf/M\tau \qquad (1)$$

Given that the electron finds a suitable site on another synapse—that is, given that the electron gets to a donor molecule across from a gate molecule on another synapse—the chance for it to fire that synapse is just the same as it would have been in a normal firing of a synapse. Because that electron is just one of some 200,000 electrons that it usually takes to fire a synapse, we must multiply the coupling factor Q by 1/200,000. That is to say, we must divide it by n.

We have found that the chance that a neighboring synapse can be fired by the original synapse's electrons is the coupling factor Q divided by n (that is, Q/n). The process becomes self-sustaining when this value, Q/n, equals or exceeds 1. This means there is essentially a 100% chance that a connection will occur. We can now express all these factors in an equation:

$$nt^2Nf/M\tau \geq 1 \qquad (2)$$

This equation determines whether or not the basic conditions for the quantum mind hypothesis will be satisfied. It sets a severe limit on the occurrence of consciousness in a human or similar brain.

We know already that the value of n is 200,000, the number of available electrons on the dark matter of Gray; that t is the synaptic delay time, 0.3 millisecond; that N is about 23.5 trillion synapses; that M is 7.45×10^{20} RNA molecules; and that τ is 8.4 trillionths of a second. If the normal firing rate, f, of the average synapse is high enough—that is, if the level of brain activity is high enough—then this quantum mechanical process will be self-sustaining.

At this point, the theory takes shape almost by its own momentum. We know that mental activity—that is, mental stimulation—wakes us up and helps keep us conscious if the synaptic activity, f, is high enough. That is just what this formula says. The only question that remains is whether or not the rate at which neurons and synapses fire is high enough to agree with this equation.

We can use the value of Q/n to calculate the minimum value of the synaptic firing frequency that will sustain consciousness. If we rearrange the terms in equation (2) to solve for f, calling the value of f that makes Q/n equal unity f_{min}, we have

$$f_{min} = M\tau/nNt^2 \qquad (3)$$

When we put in the values of M, τ, n, t, and N, we get

$$f_{min} = 0.0148/\text{second} \qquad (4)$$

This, then, is the minimum synaptic firing rate that gives rise to consciousness. Thus, this theory of consciousness not only gives us the reason why both a state of consciousness and a state of sleep (unconsciousness) exist; it also tells us the firing rate needed to cause consciousness. Now let us see what the experimental data give for this quantity. Let us see whether the theory predicts the right value for the synaptic firing frequency.

Experiments using electric probes in various unfortunate creatures such as mice, rats, cats, and an occasional human victim have provided data on the activity of neurons during sleep and wakefulness. A typical value for the firing rate of an average neuron during the waking state, as given in experiments by Crentzfeldt and Jung,[1] was 12 per second. There are 10 billion neurons in the human brain. From the value for the number of synapses—23.5 trillion—we see that the overall average number of synapses per neuron is 2350. Now if every time a synapse fired, it fired a neuron, and if every neuron fired all its synapses, then within a few milliseconds, all 10 billion neurons in the brain would be firing. The synaptic firing would have cascaded into an avalanche of neural convulsions. In epileptic seizures, something like this does occur. But ordinarily, only some synapses fire. In addition, there are inhibitory synapses that exercise further control over such cascade firing. It takes about 7 synapses to fire another neuron (this, again, is an average number derived from experiments). The result is that only about 3/1000 of the synapses fire, giving an average firing rate for synapses in the conscious brain of about 0.036 per second (12 × 3/1000). Thus, during the wake state, the average synapse is firing at a rate that is about twice that required for the onset of consciousness. For the onset of consciousness, Crentzfeldt and Jung[2] obtained neuronal firing rates in the range of 3 to 9 per second. For the synapse, this would give a range of 0.009 to 0.027 per second, which neatly overlaps the number the theory requires. For example, the average of 0.009 per second and 0.027 per second is 0.018 per second, which is very close to the 0.0148 per second given in equation (4) computed from equation(3).

The coupling factor Q that gives us the number of synapses interconnected at any instant allows us to derive an expression for the consciousness channel capacity. All we need to do is divide that by the time, t (the time it takes for that coupling to occur), and we get the number of synapses that contribute information to our conscious experience each second. If we then multiply that number by how many bits of information one synapse can transmit at a time, we get the consciousness data rate—the channel capacity that carries our stream of consciousness! This is all present in the theory already; there is no need for additional assumptions. If we use the letter i to represent the information that one synapse can handle each time an impulse arrives, then the consciousness channel capacity, C, is iQ/t. Writing the factors out, we have

$$C = in^2tNf/M\tau \qquad (5)$$

The values of all these numbers have already been given except for the information per synapse, i. If we put these values in, we get for C the value i times 667 million each second. Here we used the computed value of 0.0148 per second for the minimum frequency, f, that initiates consciousness. The nominal value for consciousness, using 0.036 per second as the full conscious state firing frequency for synapses, would lead to $C = i \times 1.62$ billion bits each second. Now the fact that the firing probability for a synapse must be low means that it really cannot handle much information. Its firing (or not firing) does not even convey 1 bit of information. Using the firing probability of 7/2,350, there is a formula that gives us the information conveyed. Using that formula, we find that $i = 0.0293$ bit of information. This gives, for the consciousness channel capacity, $C = 47.5$ million bits per second. Experimental values for this number were given in Chapter 11. The values obtained ranged from 45 million to 200 million bits per second. The nominal value was about 50 million bits per second. The calculated value of just under 50 million bits per second fits the theoretical equation almost perfectly. The result falls out of the very simple assumptions made at the beginning of Chapter 12 about how consciousness must work.

Now we are ready to discover where the individual consciousness images come from. In Chapter 13, we saw that the presence of melanin in the brain gives us a minimum consciousness time interval in agreement with what we had obtained experimentally, about 0.04 second. Now because data flow through the consciousness at a rate of nearly 50 million bits per second, the data received in an interval of time equal to the length of the minimum consciousness time interval is perceived as part of one image. We referred to this in Chapter 11 as the consciousness field information capacity. In Chapter 13, we represented it symbolically by the letter F. Multiplying the data rate, C, by the time interval gives us the value of F. The result is 1.9 million bits in a single consciousness image. This theoretical value for the consciousness field information capacity is almost exactly the same value as the 2 million bits we obtained in Chapter 11.

appendix II

Quantum Equations

A road map to the various ways in which symbols are used in quantum theory may be helpful. For a given collection of physical objects that are interacting, the Schrödinger equation yields solutions that are collections of "complex" (that is, made up of real and imaginary parts) mathematical pictures, each complex picture being a possible observable configuration of the physical objects. These collections or sets of pictures are called state vectors, and the individual pictures are known as states. These pictures are part real and part imaginary because they are rotated out of the "real" space where we exist.

Complementary to the Schrödinger equation is a mirror image equation. This mirror image equation has mirror image solutions called the Hermitial adjoints. They also consist of individual pictures that are mirror images of the first set of pictures. There is no explicit reason for this; the math just works out that way.

When these two sets of pictures are joined (multiplied together), so that the individual pictures are joined, the imaginary parts of the pictures become real, and the pictures are the real-world pictures that we can see and measure.

When we talk about several physical systems that interact, as when a measurement device makes a measurement on some system of objects, we must use different symbols to represent the different parts that are interacting. Capital letters are used for the state vectors (collections of pictures) that represent each part separately, and another capital letter is used for the combined result. The combined system then consists of a set or collection of pictures or states, each of which is a product of one each of the individual pictures from each of the interacting systems. A hydrogen atom with a state vector Θ, with individual pictures or states θ_j, collides with a probe with its own state vector P, and set of states p_i. The result is an overall set of pictures, Ψ, that is represented by $P\Theta$ and has individual states $p_i\theta_j$ representing every possible combination of the individual component states. If the interaction is with a measuring device with state vector M, the result will still have the same form, $\Psi = M\Theta$, but the individual pictures will consist only of corresponding pictures, $m_i\theta_i$, because if it did not work that way, it would not give us reliable measurements on the system Θ. Note also that there may be a succession of interacting systems resulting in a series of terms, such as $\Psi = CM\Phi N\Theta$ with individual states $c_i m_k \phi_j n_l \theta_k$, for example.

The following table helps sort these facts out:

$H\Psi = i\hbar\partial\Psi/\partial t$	the Schrödinger equation
Ψ	solution to the Schrödinger equation; a collection of "complex" pictures
ψ_1, ψ_2, ψ_3, and in general ψ_i	individual pictures that make up Ψ
$H\Psi^\dagger = -i\hbar\partial\Psi/\partial t$	the conjugate Schrödinger equation
$\Psi^\dagger$	Hermitian adjoint solution to the conjugate Schrödinger equation consisting of complex conjugate pictures
ψ_1^*, ψ_2^*, ψ_3^*, and in general ψ_i^*	individual complex conjugate pictures that make up $\Psi^\dagger$
$\Psi\Psi^\dagger$	the product of real and conjugate picture sets; a collection of real pictures
$\psi_i^*\psi_i$	the product of real and complex conjugate pictures; a real picture
Θ, P, M, C	alternative symbols for solutions to the Schrödinger equation
θ_i, p_i, m_i, c_i	corresponding symbols for individual pictures or states

notes

INTRODUCTION

1. Carl Sagan, *Broca's Brain: Reflections on the Romance of Science* (New York: Random House, 1979), p. 9.

CHAPTER 1

1. Harvey Cox, *The Seduction of the Spirit* (New York: Simon & Schuster, 1973), pp. 32, 34.

2. After J. Gresham Madren. See Harvey Cox, *Religion in the Secular City* (New York: Simon & Schuster, 1984), p. 73.

3. Harvey Cox, *The Secular City* (New York: Macmillan, 1966).

CHAPTER 3

1. Max Planck, "Scientific Ideas, Their Origins and Effects," in *The New Science*, trans. James Murphy (New York: Meridian, 1959), pp. 298-299.

2. Gary Zukav, in his book *The Dancing Wu Li Masters* (New York: Bantam, 1979), gives a nice description of the Michelson-Morley experiment and addresses some of its implications in more detail.

3. J. C. Hafele and R. E. Keating, "Around-the-World-Atomic Clocks: Observed Relativistic Time Gains," *Science* 177 (1972): 168-170.

CHAPTER 4

1. Interestingly, Albert Einstein's future wife, Mileva Marić, was at the time of these experiments a student of Lenard in Heidelberg; the extent of her contribution to the work on the photoelectric effect is not known. In their divorce agreement, signed in 1917, Albert agreed to give Mileva *all* the money from any future Nobel Prize. In 1922, Albert Einstein was awarded the Nobel Prize in physics for this work on the photoelectric effect. He honored the terms of the divorce agreement.

2. R. L. Pfleegor and L. Mandel, "Interference of Independent Photon Beams," *Physical Review* 159 (1967): 1084-1088.

CHAPTER 5

1. Note that I have used $2\pi x$ instead of x. The reason is that these quantities refer to positions of electrons as though they were on circular orbits. In essence, $2\pi x$ is the circumference of the electron's orbit indicated by the radius x.

2. The value can be obtained in another way. We multiply Ψ by its complex conjugate. This mathematical concept of the complex conjugate will become important to us. It is the quantity $R - iS$. *Complex conjugate* is a long term that simply means we change the sign of the imaginary term: we change $+iS$ to $-iS$. The complex conjugate is represented by the symbol *.

3. Because $|\Psi|^2$ is given by the product of Ψ and its complex conjugate Ψ^*—that is, $\Psi^*\Psi$—we can say that the probability of finding the particle in the volume dv is $\Psi^*\Psi dv$. Because Ψ is $R + iS$, which is the way we set it up, and Ψ^* is $R - iS$, this gives us, for the probability in the small space dv, $p = (R + iS)(R - iS)dv$.

4. This is very much like what one does with the electromagnetic wave that represents the density of photons in a light beam. There, the probability of finding a particular photon at a given location is obtained by squaring the electric field, E, and the magnetic field, H, and adding to get $(E^2 + H^2)dv$.

5. That is to say, each of these quantities (ψ_1, ψ_2, ψ_3, etc.) consists of two factors: ψ_1 equals $a_1\phi_1$, $\psi_2 = a_2\phi_2$, and so on. This ϕ_1 is a mathematical function describing the object (an electron, say) as it looks if it is in fact in the first allowed configuration or orbit; ϕ_2 is that description for the second; ϕ_3 is for the third, etc. Therefore, we can write the Schrödinger solution as $\Psi = a_1\phi_1 + a_2\phi_2 + a_3\phi_3 + \ldots$ where the numbers a_1, a_2, a_3, etc. tell us about the probability that the atom will be in the first, second, third, etc. possible state, if we make a measurement on the electron. Each one of the functions, ϕ_1, ϕ_2, ϕ_3, etc. is itself a mathematical picture of the electron in its own particular "orbit," or special configuration in the atom, each with its own particular energy for that configuration. The probability of being in state 1, 2, 3, etc. moreover, is given by the squares of the numbers, a_1, a_2, a_2, etc.

6. If we add up the chances that the object will be found in any given state in any volume of space throughout all space—that is, if we sum up $\psi_1{}^*\psi_1\, dv$ for all space—we get ${}_1{}^*\int\psi_1\, dv$, where the symbol $\int$ means that we are to carry out this addition process for every element of volume in the space. The result will be $a_1{}^*a_1{}^*\phi_1{}^*\phi_1\, dv$, because $a_1{}^*a_1$ is a constant. ϕ_1 is defined so as to make $\int\phi_1{}^*\phi_1\, dv$ equal one. Thus the probability that the atom is in state 1 is just $a_1{}^*a_1$. However, the possibility that the object is in both state 1 and state 2 at the same time—that is, $\int\psi_1{}^*\psi_2\, dv$—is exactly zero. Although these states exist together before measurement, they are still orthogonal pictures of the object. The object can never be found in both state 1 and state 2 at the same time.

CHAPTER 6

1. Richard Feynman, *QED, The Strange Theory of Light and Matter* (Princeton, NJ: Princeton University Press, 1985), p. 7.

2. It might be thought that there would be nine gluons: red-yellow, red-cyan, red-magenta, blue-yellow, blue-cyan, blue-magenta, green- yellow, green-cyan, and green-magenta. However, there are only eight. The reason for this is that these "colors" really only represent the way the gluons behave. These colors are a conceptual shorthand for the way certain ma-

trices—arrays of numbers—behave when they are multiplied by other matrices that represent the color of a quark. There is a constraint on the way the colors can be transformed that means there will be one fewer entirely distinct gluon than we would otherwise expect. It is as though the color of any quark depended on the lighting used to see the quark. Each observer might call the quark a different color depending on his or her lighting, but he or she would always know there were three different quark colors. The gluons transform the quark from one color to another, but with the added constraint that absolute color, like absolute position and absolute time, has no meaning. This constraint, when expressed mathematically, reduces the number of different gluons that exist, regardless of lighting, to eight.

3. Chris Quigg, "Bound Bosons," *Scientific American*, 251 (July, 1984): 66.

CHAPTER 7

1. Ronald William Clark, *Einstein, the Life and Times* (New York: World Publishing Company, 1971), p. 252.

2. Banesh Hoffmann, *Albert Einstein, Creator and Rebel* (New York: Viking Press, 1973), p. 187.

3. Ibid.

4. A. Einstein, B. Podolsky, and N. Rosen. *Physical Review* 47 (1935): 777.

5. Ibid.

6. Planck's constant equals 6.626×10^{-27} erg-seconds. For an atom of iron with an uncertainty in velocity corresponding to its thermal motion at room temperature, this leads to a position uncertainty of about a tenth of its diameter.

7. If we want to know what an initial Ψ is like at a later time, where

$$\Psi(0) = \psi_1(0) + \psi_2(0) + \psi_3(0) + \cdots$$

at the time $t = 0$, quantum mechanics gives us tools—operators—to make such conversions. A change of the state vector from time zero to a later time, t, is represented symbolically using an operator T by writing simply

$$\Psi(t) = T(t)\,\Psi(0)$$

The tool, and the mathematical machinery that goes with it, does all the work for us. The result is expressed simply by means of a new set of pictures:

$$\Psi(t) = T(t)\psi_1(0) + T(t)\psi_2(0) + T(t)\psi_3(0) + \cdots$$

Even though the thing that is changing is a probability description for each of the possible states, the change with time is entirely deterministic.

8. Von Neumann expresses this symbolically as

$$\Psi(t) \Rightarrow \psi_i(t')$$

where $\psi_i(t')$ gives us the *i*th picture, picture ψ_i at the time t', the next moment of time after t.

9. Johann von Neumann, *Mathematische Grundlagen der Quantenmechanick* (New York: Dover, 1943).

10. John von Neumann, *Mathematical Foundations of Quantum Mechanics*, trans. Robert T. Beyer (Princeton, NJ: Princeton University Press, 1955), p. 419.

11. E. P. Wigner, *Symmetries and Reflections: Scientific Essays of Eugene P. Wigner* (Bloomington: Indiana University Press, 1967), p. 172.

12. E. P. Wigner, Private communication.

13. E. P. Wigner, *Symmetries and Reflections*, p. 172.

CHAPTER 8

1. D. T. Suzuki, *An Introduction to Zen Buddhism* (New York: Grove Press, 1964), p. 32.

2. Bernard d'Espagnat, "The Quantum Theory and Reality," *Scientific American* 241 (November 1979), pp. 158-181.

3. Douglas Adams, *A Hitchhiker's Guide to the Universe* (New York: Ballantine Books, 1995).

4. For simplicity, we have not discussed what happens to the magnetic field in the light wave, but basically it goes hand in hand with the electric field.

5. The number of phrases in the book that say, "Dog something blue"—that is, *N*(*Dog something blue*)—must be exactly equal to the number of photon codes that said "Dog is blue" plus the number that said "Dog was blue." Thus

$$N(\textit{Dog is blue}) + N(\textit{Dog was blue}) = N(\textit{Dog something blue}) \qquad (1)$$

Also, it is obvious that

$$N(\textit{Dog was something}) = N(\textit{Dog was blue}) + N(\textit{Dog was red}) \qquad (2)$$

If we leave out some of the sentences that say, "Dog was red," then the right side of this equation must either stay the same or get smaller. The count of phrases that say "Dog was something" will always give a number that is at least as large as the number of photon codes saying, "Dog was blue." Thus

$$N(\textit{Dog was something}) \geq N(\textit{Dog was blue}) \qquad (3)$$

Because there are at least as many sentences that say, "Dog was something" as there are that say, "Dog was blue," if we put *N*(*Dog was something*) in place of *N*(*Dog was blue*) in equation (1), that side of the equation may get larger—it cannot get smaller. Therefore, by substituting the left side of equation (2) into equation (1), we can write

$$N(\textit{Dog is blue}) + N(\textit{Dog was something}) \geq N(\textit{Dog something blue}) \qquad (4)$$

Additionally, the number of statements "Dog is blue" is no larger than the number of statements that say, *something* "is blue," Therefore, we know that

$$N(\textit{Something is blue}) \geq N(\textit{Dog is blue}) \tag{5}$$

so we can replace *N*(*Dog is blue*) in equation (3) with *N*(*Something is blue*), because that will make the left side of equation (3) still larger. This gives us

$$N(\textit{Something is blue}) + N(\textit{Dog was something}) \geq N(\textit{Dog something blue}) \tag{6}$$

6. d'Espagnat, "The Quantum Theory."

7. David Mermin, "Is the Moon There When Nobody Looks?" *Physics Today* 38 (April 1985): 38-47.

8. d'Espagnat, "The Quantum Theory."

9. Quoted by Max Jammer in *The Philosophy of Quantum Mechanics* (New York: Wiley, 1974), p. 151. See also David Mermin, "Is the Moon There When Nobody Looks?" p. 38.

CHAPTER 9

1. D. T. Suzuki, *An Introduction to Zen Buddhism* (New York: Grove Press, 1964), p. 10. At the time of publication, Suzuki was professor of Buddhist Philosophy, Otani University, Kyoto, Japan.

2. Ibid., p. 13.

3. Douglas R. Hofstadter, *Godel, Escher, Bach: An Eternal Braid* (New York: Vintage Books, 1980), p. 246.

4. Ibid., p. 250.

5. Ibid., p. 251.

6. Ibid., p. 251.

7. Alan Watts, *Three* (New York: Pantheon Books, 1961), p. 106.

8. Paul Reps, *Zen Flesh, Zen Bones* (New York: Anchor Books, 1955), p. 154.

9. Hofstadter, *Godel, Escher, Bach*, p. 254.

10. Julian Jaynes, *The Origin of Consciousness in the Breakdown of the Bicameral Mind* (Boston: Houghton Mifflin, 1976), p. 1.

11. The joke in Hawaii goes, "When folks first come over here, they just go down on the beach, sit, and think. After six months, they just go down on the beach and sit."

12. Suzuki, *Zen Buddhism*, p. 49.

13. Suzuki, *Zen Buddhism*, p. 73.

14. Suzuki, *Zen Buddhism*, p. 59.

15. See Reps, *Zen Flesh*, p. 20.

CHAPTER 10

1. Lucretius, *On the Nature of Things*, trans. C. E. Bennett (New York: Walter J. Black, 1946).

2. Thomas Hobbes, *The Leviathan* (1651), ed. A. R. Walker (New York: Macmillan, 1904).

3. E. H. Walker, "The Nature of Consciousness," *Mathematical Biosciences* 7 (1970), pp. 131-178.

4. Paul M. Churchland, *Matter and Consciousness, A Contemporary Introduction to the Philosophy of Mind* (Cambridge, MA: M.I.T. Press, 1984), p. 15.

5. Gilbert Ryle, *The Concept of Mind* (London: Hutchinson, 1949), p. 20.

6. Churchland, *Matter and Consciousness*, p. 36.

7. Churchland, *Matter and Consciousness*, p. 37.

8. Paul Davies, *God and the New Physics* (New York: Simon & Schuster, 1983), p. 85.

9. Marvin Minsky as quoted in "Artificial Intelligence. Will Machines Ever Be Conscious?" by Andrew C. Revkin, *Science Digest*, October 1985, p. 42.

10. David Mermin, "Is the Moon There When Nobody Looks?" *Physics Today* 38 (April 1985): 38–47.

11. When we understand what consciousness is, we will be able to answer this question, but only after we have arrived at a new conception of measurement.

12. You can demonstrate this for yourself by looking at a large, very neutrally colored area—a gray or ivory wall, for instance—closing first one eye and then the other to notice the slight change. One view may look "cooler" (bluer) than the other, for example.

CHAPTER 11

1. Fritjof Capra, *The Tao of Physics* (New York: Bantam Books, 1984), p. 175.

2. The weak nuclear force and electromagnetic forces have now been successfully combined into one force. However, they play their roles in such different arenas that it is best, for our purposes, to continue to deal with these as two different kinds of forces.

3. Note that the forces can be equivalently represented by their particle fields: photons for the electromagnetic and gluons for the strong nuclear force.

4. The force of 1 dyne is about 3/1000 the weight of a U.S. penny.

5. Jack Sarfatti as quoted in *Space-Time and Beyond. Toward an Explanation of the Unexplainable*, by Bob Toben (E. P. Dutton: New York, 1975).

6. Julian Jaynes, *The Origin of Consciousness in the Breakdown of the Bicameral Mind* (Boston: Houghton Mifflin, 1977), p. 33.

7. Lawrence LeShan and Henry Margenau, *Einstein's Space and Van Gogh's Sky* (New York: Macmillan, 1982), p. 235.

8. The point is somewhat simplified. Along another coordinate perpendicular to the first, there are again two possible values that can be found by measurement, again with values of either $\frac{1}{2}\hbar$ or $-\frac{1}{2}\hbar$, that are independent of the first, but that is it!

9. This value is the base-2 logarithm of 100; that is, $\log_2 100 = 6.64$.

10. Bela Julesz, *Foundations of Cyclopean Perception* (Chicago: University of Chicago Press, 1971).

11. Strictly, this means in psychology the ability to discriminate via sensory input in such a fashion that it can affect one's response.

CHAPTER 12

1. E. H. Walker, "Quantum Mechanical Tunneling in Synaptic and Ephaptic Transmission," *International Journal of Quantum Chemistry* 11 (1977): 103-127.

2. B. Katz and R. Milhedi, "The Effect of Temperature on the Synaptic Delay at the Neuromuscular Junction," *Journal of Physiology* (*London*) 181 (1965): 656.

3. We will deal later with the quantum mechanical nonlocality property of Bell's theorem as it applies to this question. At this point, suffice it to say that the nonlocality property of quantum mechanics would not give the proper connectedness to account for consciousness that we have discussed here. What we seek to understand now is the source of a single individual mind that nevertheless contains all the varied displays of sensory input and thoughts in the brain. We seek to understand how all this constitutes one quantum process.

4. We could envision, for example, that the molecules of the dark areas of Gray actually are formed in the postsynaptic neuron and migrate to the particular synaptic site, that the formation of these molecules involves information storage, and that this information can only be read at corresponding sites in synapses. Assuming that such an encoding of information exists enables one to calculate the size requirements on these molecules. Such a calculation happens to give results in agreement with observations.

5. A dalton is a chemical "weight" unit defined to be 1⁄16 the mass of the average oxygen atom; it is slightly less than the mass of one hydrogen atom. Its value is x1.66024 × 10^{-24} gram.

6. This is the value obtained in Walker, "Quantum Mechanical Tunneling."

7. Stuart Hameroff at the University of Arizona has proposed that microtubules (the cell's cytoskeleton) play a role in consciousness. The possibility exists that these structures serve or augment the function of the soluble RNA. If so, the quantity M would refer to the number of the "tubulin" molecules in the brain. Taking the length of these molecules to be λ, the quantity λ/τ would be the electron conduction speed along the microtubules and n would be the number of microtubules per synapse.

8. One can easily see in this a connection to the psychological disorder of multiple personality as being the result of the breakdown of this control, much as epilepsy results from the breakdown of the brain's ability to control the level of neuronal firings throughout the brain.

CHAPTER 14

1. The formula is given by the "self-information" expression $I = \log_2 Q$.

2. $W = (\log_2 Q)/t = 58{,}700$ bits/second.

3. The law, stemming from relativity theory, that says that space and time act physically on a par—that is, they appear in all basic physical equations in the same way—except that time is multiplied by the imaginary i and the conversion factor c, the speed of light.

CHAPTER 16

1. In Chapter 5, each possible state, ψ_i, is added to form the "wave function"; that is, $\Psi = \psi_1 + \psi_2 + \psi_3 + \cdots$. Here Ψ represents the "state vector" notation in which each ψ_i is an element in a column matrix, or $\Psi = \text{col}(\psi_1\ \psi_2\ \psi_3 \cdots)$. The Hermitian adjoint is $\Psi^\dagger = \text{row}(\psi_1\ \psi_2\ \psi_3 \cdots)$. This alternative state vector representation leads to simplifications in the notation.

2. Each of the ψ_i and ψ_i^* actually consists of two parts: a real part R_i and an imaginary part iS_i where i is the imaginary square root of -1. In notes 2 and 3 of Chapter 5, we used

the notation $\Psi = R + iS$. Here we use R_i and S_i for the real and imaginary parts of ψ_i, respectively. That is, $\psi_i = R_i + iS_i$ and $\psi_i^* = R_i - iS_i$, the subscript i being just a number denoting which ψ picture potentiality is referred to.

3. Each ψ_i^* is made up of the real part R_i less the imaginary part—that is, $R_i - iS_i$. This is called the complex conjugate of ψ_i. Each ψ_i^* is also a picture of the possible world, but it is the "conjugate" picture. See notes 2 and 3 of Chapter 5.

4. That is, we get R_i^2-S_i^2, without imaginary quantities.

5. These dimensions in Hilbert space merely reflect the fact that the Ψ solutions to the Schrödinger equation consist of a set of individual pictures we call the ψ_i pictures. That index number i counts off the number of dimensions in this odd space.

6. That is, $p - p^2$. For example, assume there are two possible states, ψ_{up} and ψ_{dn}—the result of measuring the spin of an electron. There is a 50% chance that the spin will be up, which we can express as $\psi_{up}^*\psi_{up} = ½$. Call $\psi_{up}^*\psi_{up}$ the first probability value, or p_1. For finding the spin to be pointed down, the probability p_2 will be given by $\psi_{dn}^*\psi_{dn} = ½$. This means that $p_1 - p_1^2$ will be ¼. We also get a dispersion of ¼ for the down direction. Adding these together gives a dispersion measure of ½. We might say, therefore, that the quantum mechanical system, before observation when both states are possible, is halfway disperse. If we calculate the same quantity for a system with three equally likely states, we find the system to be ⅔ disperse. But the important thing is that when we observe the system, all the states go to a probability value of zero except one state, which then has a probability of 1 (certainty that the system is in that state). The dispersion measure is then zero for every state. For example, if we measure the spin in the example above and find the spin to be up, then we know that $p_2 = 0$, giving for the dispersion of the state $p_2 - p_2^2 = 0$—that is, no dispersion. Similarly, because the spin was up, the probability p_1 is now 1. Because $p_1 = 1$, $p_1 - p_1^2$ is also zero. There is no dispersion here either, because we know that the spin is up.

7. Past efforts at nonlinear modification of the Schrödinger equation have always had one of three faults: (1) They were too slow. (2) They produced results at variance with experiment. (3) They required an anthropomorphic switch to be triggered when the measurement happened.

8. Technically, for the modified Schrödinger equation with the added information term, higher-order terms in Ψ would also appear, giving $\Psi = \Theta M + \mathcal{O}(\theta_i, m_i)$. However, because the information term vanishes on the completion of measurement, all higher-order terms will vanish also. Thus we can describe how the modified Schrödinger equation behaves in the simplified manner shown here, using only "linear" expressions such as $\Psi = \Theta M$.

9. In Hilbert space language, they must "point" in the same direction.

10. We have taken some liberties with the mathematical representation here for the sake of simplicity. The quantity $\Phi^\dagger$ involves a special mathematical operation that has components $\phi_1^*\phi_1$, $\phi_2^*\phi_2$, ... In addition, some operations on a Hilbert space are valid because of restrictions that come into being only when complete measurement loops occur.

11. The symbol ∂ means "partial derivative." The rate of change, here the rate at which the quantity Ψ is changing as time passes, is to be determined while holding all other variables constant. This is simpler than a total derivative in which the rate of change with time could also involve other changes, such as location.

12. The equation separates out into one for the terms CM_1M_2 and a separate equation, $H\Phi^\dagger = i\hbar\partial(\Phi\Phi^\dagger)/\partial t$, simplifying to $H\phi^*\phi^* = i\hbar\partial(\phi^*\phi)/\partial t$.

13. Actually, it becomes a $(CM_1M_2\Phi_A\Phi_B)$ log $(CM_1M_2\Phi_A\Phi_B)$ term. It is easily simplified by separating out the CM_1M_2 factors. This gives $(\Phi_A\Phi_B)$ log $(\Phi_A\Phi_B)$, and with the dual space substitution, we obtain individual terms $(\phi_i^*\phi_i)$ log $(\phi_i^*\phi_i)$. $\phi_i^*\phi_i$ is the probability, p_1, for the first state to happen.

14. p log p represents the collection of the possible p_i log p_i.

15. The term p_1 log p_1 is the change in information that selection of the first state will bring about. If that first state ϕ_1 is selected, then p_1 will become one. That makes p_1 log p_1 become zero. For state 2, which then must have $p_2 = 0$ because it does not happen, p_2 log p_2 will become (0) log (0), which is also zero. Thus the dispersion becomes zero for all of the terms.

16. An additional mathematical detail concerns the manner in which our modified Schrödinger equation (MSE) generates the same probabilities as the usual quantum mechanics. The mechanism for this involves two facts. First, every ϕ_i used here represents a vector in Hilbert space that is itself given by an infinite sum of mutually perpendicular "event" vectors ξ_{ij} in Hilbert space, each of infinitesimal, but equal length *and probability*, that sum to equal the ϕ_i. The second fact is that the term added to the Schrödinger equation to give the MSE, Ψ log Ψ, is multiplied by a matrix G that operates on the individual ξ_{ij} event vectors. We will discuss this G matrix later in the chapter.

17. Actually, they are complex numbers that contain imaginary terms. They can be thought of as representing a reality that points out at an angle to the reality we see.

18. Richard Morris, *Time's Arrows* (New York: Simon & Schuster, 1984).

19. The terms are arranged in the order of x, y, z, ct across and x, y, z, ct down; these terms answer the following questions:

Do we need x^2? Do we need x × y? Do we need x × z? Do we need x × t?
Do we need y × x? Do we need y^2? Do we need y × z? Do we need y × t?
Do we need z × x? Do we need z × y? Do we need z^2? Do we need z × t?
Do we need t × x? Do we need t × y? Do we need t z? Do we need t^2?

with 1 for yes and 0 for no. If the coefficient for the term differs from 1 or 0, then that coefficient is given.

20. $g_{\mu\nu}$ is actually the value in the μth row and νth column. The parentheses indicate that the object is a matrix of values

21. In general, where there are many possible quantum states that can occur, the matrix has to look like this:

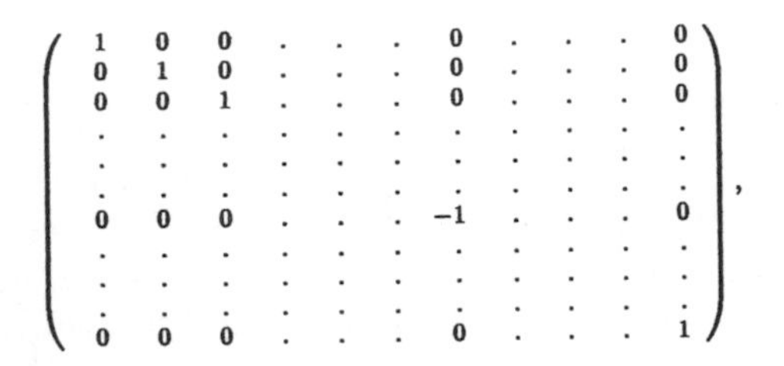

where all those dots represent some definite number of zeros or sometimes ones that we have chosen not to write out explicitly and where the position of the –1 term is determined by the chance or the quantum mind. This is the *G* matrix mentioned in note 16. In the complete treatment—that is, where we put in all the details of this state selection "machinery"—the position is entirely stochastic (occurring by chance), and it is this matrix that determines which state will actually happen and be observed.

22. In general, where there are more than four possibilities, this rearrangement will look like this:

$$\begin{pmatrix} 1 & 0 & 0 & . & . & . & 0 \\ 0 & 1 & 0 & . & . & . & 0 \\ 0 & 0 & 1 & . & . & . & 0 \\ . & . & . & . & . & . & . \\ . & . & . & . & . & . & . \\ . & . & . & . & . & . & . \\ 0 & 0 & 0 & . & . & . & -1 \end{pmatrix}.$$

23. That is, while also satisfying the requirements of observer symmetry.

24. J. M. Jauch and F. Rohrlich, *The Theory of Electrons and Photons*, 2nd ed. (Reading MA: Addison-Wesley, 1959), p. 427.

25. E. H. Walker, "Information Measures in Quantum Mechanics," *Physica B* 151 (1988): 332-338; see p. 333.

26. Ibid.

CHAPTER 17

1. More precisely, relativity holds differentially. It does not necessarily hold in the integral form. Thus, whereas in flat Minkowski space, motion is entirely relative, in our Big Bang universe described by the Robertson-Walker-Friedmann (RWF) metric, motion is absolute! One can measure the motion of the earth *relative* to the Big Bang background microwave radiation, but this motion cannot be arbitrarily transformed away—as would have to be the case for the motion to be relative—without imposing on the overall space a rotation that would be easily discernible as a distortion of the RWF metric. In addition, one cannot arbitrarily transform the time coordinate prior to the Big Bang event. Differentially, space, time, and translational motions are relative; in the whole-world integral form, they are not.

2. This process does not happen in the same way in the case of rotating stars; rotating stars can form semistable structures with both inflation and deflation coexisting.

CHAPTER 18

1. This is, in part, what the new technology of quantum computers is about. There, however, quantum "qubit" circuits are designed to suppress unwanted states by making them low-probability states, whereas in the case of the quantum mind, desired states that are of low probability occur because they are recognizable to the conscious brain.

APPENDIX I

1. O. Crentzfeldt and R. Jung, "Neuronal Discharge in the Cat's Motor Cortex During Sleep and Arousal," *The Nature of Sleep*, ed. G. E. W. Wolstenholme and M. O'Conner (Boston, MA: Little, Brown, 1961), pp. 131-170.

2. Ibid.

index